Nature, Aim and Methods of Microchemistry

Proceedings of the 8th International
Microchemical Symposium

Organized by the Austrian Society
for Microchemistry and Analytical Chemistry
Graz, Austria, August 25–30, 1980

Edited by H. Malissa, M. Grasserbauer, R. Belcher

Springer-Verlag Wien New York

With 134 Figures

Library of Congress Cataloging in Publication Data

International Microchemical Symposium (8th : 1980 : Graz,
Austria) Nature, aim and methods of microchemistry. Bibli-
ography: p. 1. Microchemistry-Congresses. I. Malissa, Hanns,
1920– . II. Grasserbauer, M. (Manfred), 1945– . III. Belcher,
Ronald. IV. Title.
QD79.M5I59   1980   540   81-5835   AACR2

ISBN-13:978-3-211-81653-0   e-ISBN-13:978-3-7091-8630-5
DOI: 10.1007/978-3-7091-8630-5

PREFACE

This proceedings volume of the 8th International Microchemical Symposium contains the plenary and keynote lectures delivered at the conference.

Besides basic and historic aspects the following major topics are covered:

"Microchemistry in Arts and Archeology"
"Microchemistry in Life Sciences"
"Microchemistry in Environmental Sciences"
"Microchemistry in Material Sciences"
"Instrumentation, Methods and Automation in Microchemistry".

The papers show the present state of microchemistry and the development of this field since the pioneer days of Fritz Pregl and Friedrich Emich. Today microchemistry is a different science as compared to the Pregl and Emich days, for it combines many disciplines like chemistry, physics, mathematics, informatics, biology and does not only mean microanalysis - even if it is still predominant and the best tool for elucidation of the microcosmos.

Due to this development modern microchemistry plays an important role in science and technology. It had been the intention of the Scientific Executive Committee to demonstrate this at the 8th International Microchemical Symposium with the goal to encourage interdisciplinary communication and stimulate discussion.

April 1981

H. Malissa, Vienna
M. Grasserbauer, Vienna
R. Belcher, Birmingham

CONTENTS

I. INTRODUCTORY PAPERS

NATURE AND AIM OF MICROCHEMISTRY -
AN INTRODUCTION TO THE 8TH INTERNATIONAL MICROCHEMICAL SYMPOSIUM

HANNS MALISSA
Institute for Analytical Chemistry,
Technical University,
Getreidemarkt 9, A-1060 Wien, Austria

Abstract. In order to stimulate the discussion for clarification of the
term "Microchemistry", and what are its nature and aim some thoughts and
examples of basic interest and practice are presented, without claiming
a complete depiction.

It gives me a feeling of exaltation that we are able to meet again, thirty
years after the first International Microchemical Symposium - organized by
the same society as today - and for the third time in this town of great
microchemical tradition. Some of us may feel the spirit of this place, where
60 years ago F. Pregl and his antagonist F. Emich laid down what is generally
meant by the term "Microchemical School of Graz". But we must not forget that
this is also the town where L. Boltzmann evaluated 100 years ago, the famous
thermodynamic principles which are ruling not only chemistry, but also physics
and information.

It is not my intention to go into the history of microchemistry; and I am not
claiming microchemistry as an Austrian "invention". We know only too well that
p.e. 300 years ago R. Hooke, Fellow of the Royal Society, published his famous
book on Microchemistry, entitled "Micrographia" and furthermore that French as
well as German, Russian and Dutch scientists were deeply involved in microchemistry some 150 years ago. A look at the exhibition will show you some more

details.

Up to now - it seems to me - microanalysis stands for microchemistry and has already become a synonym for it. So - apparently - , it is really high time to try to clarify the situation and to try to evaluate what is really meant by the term "microchemistry", and what is its nature and its aim.

It is interesting to note that there are no problems when one speaks about microphysics and of the microcosmos in general, but as soon as the word "microchemistry" appears, nearly everyone thinks of "microanalysis" only.

Today we feel a strong tendency to amalgamate again philosophy and science, as was usual in earlier times of human search for knowledge. We know very well that pure philosophy, pure science and pure technology without any reciprocity and without a high amount of reflection to ethics leads us towards a dangerous situation - either running out of goods or merging in a meaningless and, moreover, hopeless world. Prof. Frankl's lecture will give you some impressions.

This is certainly not the place, and time is too short for going into the aspects of "controlled research", but the temporary world-wide discussion, about p.e. gene technology or extensive atomic research, forces us to establish new ways for research patterns, whereby the framework will be the dialogue: scientific research ⟷ ethic reflection via proper assessment studies. That the best starting-point with the lowest probability of destruction of our environment would be certainly the work at the microscale level. This is quite clear and does not need further explanations.

As long as everything has been done and is done on an easily controlled "microscale", as long as things went well and new wisdom could be gained; but as soon as "larger" amounts become involved, beyond financial interest, political and even military aspects are often the dominating factors to further research - and discrimination begins.

First of all, let me make a statement:
"Microchemistry exists and plays an important role in our life and research work, regardless if we accept this or not. The major problem is to define its border lines".

It is just a question of our philosophy, what rank, which place and what dimensions we assign to microchemistry. Since Einstein gave us the fourth dimension, nature can have different pictures depending on the state of motion of the "observer". This principle makes every definition very difficult or even senseless. To define microchemistry is as senseless as the definition of inorganic and organic chemistry, because it is not the object which makes the problem, it is the subject. To define microchemistry is as difficult as it is to define "life". Very often it is more or less a personal problem to follow exact rules and dimensions, we are always attempted to use and to include some "private" fixation of limits. The thinking and working in thousands of light-years and mega-tons as well as in fractions of seconds and grams has forced us to scale units in over 24 decades and more and more it makes no problems for us to think and to work in a world of pico-grams.

Year by year rockets are sent into the space to "look" behind the "world" of our planet and in two years (1982) again a Spacelab will be sent into the orbit with apparatus of the Technical University of Graz on board. Minute chips are built in and the circuits and pieces of detectors are as small as 2 x 2 mm transmitting decades of signals. On the other hand transplantations and implantations under the microscope of very interesting biological minute bodies are going to lose their reputation of being impossible and worthless.

In order to understand the nature and aim of microchemistry we have to include in our considerations some philosophical momenta not only in respect to "macro" and "micro", but also to "classic" and "modern" as well as the meanings and definitions of terms like canonical systems and ensembles, causality and so on. And here starts what is meant by amalgamation of philosophy and sciences.

Both in thinking and in writing we can establish and model every system with every canonical ensemble, as large or small we like. In reality we are bound to nature - and the verification of an axiom or theory - we may better say the falsification of it - can only be done by observations and measurements. This means in first instance the careful elaboration of measurement of differences of thermodynamical and kinetics parameters of the initial and final state of our stepwise scaled down system until the mean free pathway is reached. These borderlines will designate the real nature of microchemistry. This means further the broad use of micrurgy and miniaturization of tools as well as volumes and a heavy fight against the ratio volume over surface will accompany the perceptibility of microchemistry in practice.

From the first formation of minute amounts of substances in small volumes to the coacervation of molecules to larger units, the whole evolution process is bound to thermodynamics, especially to entropy, and chemical kinetics.

The idea of entropy has been generalized and built into nearly every branch of sciences. In physics as well as in chemistry, biology, cosmology and also in information theory down to linguistics.

But the roots of this concept are in the physics of macroscopic thermodynamic phenomena - in the heat engines of the 18th century. The development has led us to an extended Boltzmann-Planck formulation in the generalized form of

$$\dot{S} = k \cdot \log W \, (\bar{E}, \bar{N}, \bar{V}, \bar{\beta})$$

where the thermodynamic weight W is now defined to the number of microscopically specified possible states, corresponding to a given macroscopic equilibrium state of the system under consideration ($\bar{E}$ = internal energy, $\bar{N}$ = average particle number, $\bar{V}$ = volume and $\bar{\beta}$ = average values for their internal state parameters). As M. Harrison pointed out in his consideration of the entropy concepts in physics there is one important difficulty in the application of the second law to our universe as a whole. It is clear from everday experience and observations that the universe as a whole is not in a state of thermo- dynamic equilibrium. And yet as a single closed system one might expect it to have reached equilibrium long ago. The temptation to regard the observable part of the universe as a very large fluctuation in a system in equilibrium as a whole can be resisted by realizing that comparable non-equilibrium conditions supporting biological development and human observers can occur on a smaller scale with even greater likelihood, because there is a greater probability for a suitable fluctuation in the structure of smaller systems, for it contains far fewer degrees of freedom than the galaxy.

Here the selection of microsystems of a proper canonical ensemble will produce a broad field for research and with its aim a lot of discrepancies between calculated and measured values of real systems can be cleared off and eventually new wisdom gained.

As far as chemical kinetics is concerned even more complexity may occur. Regard- less which theory is our favourite, collision - or activation theory, particle number and volume, this means particle and flux density, are the dominating factors.

In every case we have to consider the ratio N/V very carefully because here again is one of the keys to the nature of microchemistry.

It is of the greatest basic interest to learn more about the compatibility of calculated values in reference to measured data of fixed particle number by lowering the volumes to the smallest system possible at a given temperature. It is simply the problem to find for each chemical system the limit where the probability of the occurrence of a chemical reaction changes from "one" to a significant lower number.

From the practical point of view the borderlines of microchemistry are not only where the smallest number of reactants are present in the smallest volume to be handled for a giving equilibrium-condition, but even more where the physical properties change drastically with the volume. From this point is the smallest amount meant which can be handled in the vicinity of the equilibrium-condition. In 1950, Benedetti-Pichler, the well-known co-worker of Emich and his ambassador to U.S., had already mentioned that about 60.000 molecules are necessary to assure chemical reaction proceeding the same way as larger amounts. Taking a mean molecular weight of 100 p.e. leads us to the masses of 1,7 . 10^{-16}g. But it is a tremendous difference having this amount dissolved in 100 ml or 100 µl. How sensitive the ratio N/V is, can easily be demonstrated (for gases) by the formula for collision frequencies.

To learn more about the nature of microchemistry we must give attention to morphogenetic principles and we should do more research on conservative and dissipative reaction mechanisms. Whilst the former is (more or less) under consideration, also on the microscale is the latter not in the files of larger investigation programs.

We have only to think of homogeneous oscillating reactions of the Belousov-Zhabotinskii-type to be performed in the smallest volume, also, not only at low concentrations as has been done up to the present. The field of research on microchemical oscillators is still open and waiting for elucidations.

A practical example of the aims of microchemistry is the advanced research activity on the lifetime and chemical surface reactions of droplets of the order of 1 - 2.000 µm in diameter. T. Novakov and his school in Berkeley are doing a great deal of work in this respect and we will hear something about it in the session of environmental analysis of this symposium. But it is realy interesting to learn from relative simple experiments - microdroplet on a very thin quarz fibre under remote condition - that the lifetime of a droplet with about 10 µm diameter covered with soot and surface reactants can have at r.h. of 80 % a lifetime of more than 17 hours and is about 10.000 times larger as somewhat larger droplets. What this means for the environmental control activities can easily be imagined.

The aims of microchemistry are too many and cannot be demonstrated briefly; the plenary and keynote-papers will give you a very good impression. But in general, the aim of microchemistry is the elucidation of all the forces,

particles and arrangements at the starting point of all the compounds and
bodies which built up our macroscopic world. Microchemistry is not as "weak"
as we may believe, we are only too diffident to arrange the analytically
found data to a canonical system and to articulate our findings.

We will not make any progress in our efforts of so-called better life and
worthwhile world, if we do not give more attention to the elucidation of
microprocesses. The formation and conversion of pollutants p.e. cannot be
studied without considering micro-heterogenic catalysis. We are able today
to collect and to study particles as small as 0,1 µm at very low sampling
rates (that means with very good time resolving), and we are going to eluci-
date the reaction mechanism in spots of less than 0,1 mm; and it is one of
the other aims of microchemistry to give guidance in this broad domain of
research activity on life and material science.

We know that small amounts of masses have other physical properties p.e.
surface-tension of droplets - is quite different as flat surfaces. The
behaviour of whiskers as an example of what microchemistry and microtechniques
can do has influenced the solid-state technology to an enormous extent. Small
crystallites have always lower melting-points and higher heats of combustion,
etc. The study and the interpretation of these phenomena of the micro-range
are still a wide field of activity and need a vast amount of often very
sophisticated instrumentation. As a general rule we can say: the smaller
the amount or volume to be investigated the larger (and more expensive) the
machinery. And what we are interested in is not as much trace-chemistry
(chemistry at very low concentration), but microchemistry, where beyond
chemical parameters, physical forces are of great importance. This means,
the main aims of microchemistry lie in studying the problems:

a) Is there an overlapping and moreover an overruling of physical laws over
   chemical laws in reaction dynamics in building up compounds and phases,
   and

b) if yes, where are the borders as far as size of reaction volumina and
   masses are concerned.

This short introduction to our meeting is not to be considered as the explana-
tion of microchemistry, but rather a trial to open our mind in this respect.

We have invited outstanding scientists from various fields of science in order
to enlighten different fields of microchemistry and hopefully by the end of
this symposium you will be better informed about this domain. You must not
forget that the best teacher and the best scout in finding the proper way

to relatively unknown fields is always the analysis. Therefore microanalysis will be not only one of the most important branches in microchemistry, but serves also as a forerunner of new arrangements of very great simplicity as well as sophistication in handling small amounts and volumes. What has to be done in order to obtain a better elucidation of what microchemistry really is - is the conversion of data, not only to the composition of small amounts of material, but more to the physico-chemistry of reaction mechanisms especially in the range where the probability is much less than "one". The field of "uncertainty of chemical reaction", as F. Emich has named this area 70 years ago and where research has nearly stopped. Now we have to go in again to find out the borderlines mentioned before.

HISTORY OF MICROCHEMISTRY TILL 1945

FERENC SZABADVÁRY
Institute for General and Analytical Chemistry,
Technical University,
Gellért tér 4, H-1502 Budapest, Hungary

Abstract. Microanalysis and macroanalysis today. Beginnings of microscopic
identification of compounds. Boricky's, Haushofer's and Behrens' work.
Emich and the development of the Austrian school of microchemistry. Organic
qualitative and quantitative microanalysis. Microtitrimetry. Catalytic ultramicroanalysis. Development of colorimetric and photometric determinations.
Birth of activation analysis and chromatographic microanalytical methods.

The extensive spreading of microchemistry in recent times has created a
conceptual uncertainty both in methods and in applications.

Little by little, all chemical analysis turns into microchemistry, rightly
speaking into microanalysis. In fact, the confusion started with the term
itself, namely that the name microchemistry was adopted long ago, although
what is meant is microanalysis. Probably the term microchemistry was derived
from the abbreviation of the title of Raspail's book "Essai de chimie microscopique" which appeared in 1831. Microscopic chemistry: microchemistry.
Later, the classicists of microchemistry, among others Emich, clearly defined
microchemistry as the methods using micro amounts of sample, quite precisely
amounts of sample below 0.01 g. Thus the term does not include methods in
which micro amounts, that is, below 0.01%, of a component can be determined.
However, this distinction is not used subsequently in the literature. Methods
aiming at the determination of traces are frequently named microchemical
methods. Even the various microanalytical publication series like Hecht-

Zacherl and Benedetti-Pichler devote whole volumes to methods of trace analysis. In fact, it is frequently difficult to make a distinction. Practical science should always contemplate the aim, the usefulness and should not trouble about the coherence of some definition. I believe this to be valid for human society as a whole, in all its manifestations.

It appears that nowadays macroanalysis is applied only in industrial and agricultural routine analyses, in scientific analytical chemistry microanalysis and trace analysis gradually became practically the only methods. The subject to be discussed here is scientific analytical chemistry, the discipline which is the servant of other scientific disciplines: ancilla scientiarum.

In the circular announcing the publication of the first journal for analytical chemistry, the Zeitschrift für analytische Chemie, Remigius Fresenius wrote in 1861: "Ohne Mühe läßt sich nachweisen, daß alle großen Fortschritte der Chemie in mehr oder weniger direktem Zusammenhang stehen mit neuen oder verbesserten analytischen Methoden. Den ersten brauchbaren Verfahrungsweisen zur Analyse der Salze folgte die Erkenntnis der stöchiometrischen Gesetze, die Fortschritte in der Analyse der anorganischen Körper fanden ihren Ausdruck in den immer ge- naueren Äquivalenzzahlen, der genauen Methode zur Bestimmung der Elemente or- ganischer Körper folgte der ungeahnte Aufschwung der organischen Chemie, die Spektralanalyse führte sofort zur Entdeckung neuer Metalle etc. Die Ent- wicklung der analytischen Chemie geht daher der Entwicklung der gesamten chemischen Wissenschaft immer voraus, denn wie frisch gebahnte Wege zu neuen Zielen, so führen bessere analytische Mittel zu neuen chemischen Erfolgen. Eine ähnliche bedeutende Einwirkung verbesserter analytischer Methoden gibt sich auch bei den anderen Wissenschaften und Fächern, welche mit der Chemie verwandt sind, aufs Deutlichste zu erkennen. So wurden, um nur einige Bei- spiele zu erwähnen, die Löthrohrreaktionen bald wichtige Hilfsmittel zur Unterscheidung der Mineralien, so führte die Entwicklung der Analyse bald, infolge der hierdurch möglichen Prüfung der Arzneimittel, einen großen Auf- schwung in der Pharmazie herbei, so folgte der Entdeckung der Alkalimetrie, der Chlorometrie und anderer rasch ausführbarer, namentlich maßanalytischer Methoden bald der solide Gebrauch, chemische Waren von bestimmten, verbürgten Gehalten in den Handel zu bringen, so führten die vereinfachten Methoden der Stickstoffbestimmung und Aschenanalyse rasch zu einer Reihe der wichtigsten physiologischen und agrikulturchemischen Wahrheiten, so erhielt die Unter- suchung menschlicher Ausscheidungen für die medizinische Diagnose erst Wert, als die vereinfachten analytischen Methoden eine rasche Ausführung ermöglich- ten, so wurden die genauen Methoden zur Ermittlung von Giften, Blutflecken usw. bald die gefährlichsten Feinde der Verbrecher. Die analytischen Methoden

sind daher in Wahrheit eine große Errungenschaft, ein wichtiger wissen-
schaftlicher Schatz." [1]

/"It can easily be realized that all major developments in chemistry are more
or less directly connected with novel or improved analytical methods. The
first serviceable methods for the analysis of salts were followed by the
recognition of the stoichiometric laws, progress in inorganic analysis resulted in more and more accurate equivalent values, the exact method for
determining the elements in organic compounds led to the astounding developments in organic chemistry, spectral analysis immediately brought about the
discovery of new metals etc.   Hence, the development of analytical chemistry
always precedes the development of chemical science as a whole. Just as
pioneer routes lead to new ends, better analytical means lead to novel
chemical successes. A similar important effect of improved analytical methods
can also clearly be recognized in other sciences and disciplines related to
chemistry. To mention only a few examples: blow-pipe reactions soon became
significant aids for distinguishing minerals; progress in analysis soon resulted in the remarkable development of the pharmaceutical industry by enabling
drugs to be properly tested; the discovery of alkalimetry, chlorometry and
other fast (mostly volumetric) analytical methods was soon followed by the
trustworthy practice of commercializing chemicals with defined, warranted
compositions. The simplified methods of nitrogen determination and ash analysis
resulted in the recognition of a number of highly important physiological and
agricultural-chemical ideas; only by these means did the testing of human
secretions gain value for medical diagnosis; when accurate methods were developed to determine blood stains, poisons etc., they soon became the most
dangerous enemies of criminals. Thus, analytical methods are truly a major
scientific accomplishment, a real treasure for science and practice."/

The same is valid for microchemistry. Its development allowed rapid progress
in other scientific disciplines. The recent results in biochemistry, in
physiology, in electronics, in astronautics, in agriculture etc. were preceded by the development of microanalytical techniques, and the solution of
their present problems will again be largely promoted by the discovery of
suitable microanalytical methods.

One antecedent of microchemistry was the microscope, invented by the Dutch
bookkeeper Anthony van Leeuwenhoek (1632-1723). A new world unfolded itself
to his amazed eyes in his primitive instrument. He reported his observations
in seven volumes between 1685 and 1718. The micro-world included everything
that came within his reach: bacteria, blood corpuscles, plant structures etc.

The microscope was soon perfected and became an indispensable instrument of researchers, including among its users chemists who frequently observed crystals with it. The Prussian chemist Sigismund Andreas Marggraf (1709-1782) was the first to use the microscope for the identification of compounds. In 1747, he stated that beet sugar was identical with cane sugar on the basis of their crystal shapes observed with a microscope. [2] Tobias Lowitz, the tsar's court pharmacist in St. Petersburgh, (1757 - 1804) studied the shape of certain salt crystals, mainly chlorides, and the changes taking place under the effect of reagents, systematically under the microscope, and published his observations with the aim of identifying these salts. In certain cases he added the reagent, for instance sulphuric acid, on the slide proper, to note whether the crystal shape characterizing the sulphate was formed. [3]

After a longer pause, François Vincent Raspail (1794 - 1878) suggested the identification of compounds by observing the shape of their crystals, in his book mentioned previously, entitled "Essai de chimie microscopique appliqué à la physiologie". Later, Raspail boosted that he was the inventor of microscopic analysis.

Raspail's main field of interest, although he was no medical doctor, was in biology and drugs. The microscope, at that time, was already a much-utilized instrument in that discipline. After botanic sciences mineralogy took the lead, above all, in the use of the microscope, and the developers proper of micro-scopic analysis were mainly mineralogists. Let me mention three important personages: Emanuel Boricky from Prague (1840-1881), Karl Haushofer from Munich (1839-1895) and Theodor Behrens (1843-1905), professor of mineralogy at Delft in the Netherlands. Boricky attempted to establish an analytical system based on microscopic reactions, namely on the study of crystals formed by the action of hydrogen silicofluoride added dropwise on to thin mineral sections. The title of his book published in Prague in 1877 was "Elemente einer neuen chemisch-mikroskopischen Mineral- und Gesteinsanalyse"/Elements of novel chemical-microscopical mineral and rock analysis/, and only his early death prevented him from the development of the system. His method apparently showed exceptional promise. In a paper dated 1881, a certain Reinsch wrote the follow-ings: "Die Anwendung des Mikroskops zur chemischen Untersuchung der Verbindun-gen macht von Tag zu Tag größere Fortschritte und das Mikroskop nähert sich im Bezug auf die Erkennung der kleinsten Mengen dem Spektroskop, welches es aber darin übertrifft, daß es auch Anhaltspunkte für die vorhandene Menge eines Be-standteiles gibt."[4]/"The application of the microscope for the chemical in-vestigation of compounds progresses increasingly from day to day, and with regards to the identification of tiniest amounts, the microscope approaches

the spectroscope, surpassing it, however, in that it also yields a basis for
the estimation of the quantity present of a component"/:These expectations
were too optimistic, even if one considers that at that time quantitative
spectroscopy had not been realized.

For the moment, the triumphal career of microscopic analysis continued with
Haushofer's book "Mikroskopische Reaktionen" /Microscopic reactions/ published
in 1885. In this book he described the most characteristic reactions that
yield crystals for almost all elements, and presented the designs of the
crystals. In most cases he carried out the reactions in small test tubes
instead of performing them on the slides, for he found that the crystals
develop better in such conditions. However, the crystals had subsequently
to be transferred on to the slide, and therewith the problems of microchemical operations and devices came into existence. Haushofer himself developed some simple operations and devices, for instance a microfilter.
The technique was improved by Behrens. He applied a number of additional
reactions, systematically choosing the most reliable and selective ones.
He also made use of new operations: he was the first to separate the precipitates by centrifuging instead of filtering. His book "Anleitung zur
mikrochemischen Analyse"/Instructions for microchemical analysis/ appeared
in 1898, in two volumes.

The majority of reactions suitable for microscopic detection were selected
by Haushofer and Behrens. In Emich's book "Lehrbuch der Mikrochemie"/Manual
of Microchemistry/ 398 microscopic reactions of the elements are listed,
among which 69 originate from Haushofer, 173 from Behrens, 17 from Emich,
and the rest from various other researchers.

And now my story turns to Friedrich Emich (1860-1940), who was born in Graz,
studied there, was professor there for 42 years and during this time made
Graz the centre of the world in the field of microchemistry. [5]

Emich was a scientist of great distinction . Many professors of chemistry at
universities in Austria and in other countries, all over the world, were his
pupils. As a microchemist, his main interest was not so much in the field of
qualitative reactions, but rather in the development of operations and devices,
and his research activity provided many valuable contributions. He regarded
microchemistry truly as the discipline expressed by the term. When inventing
his devices and operations, he had in mind not only analytical, but also
preparative microchemical techniques, although undoubtedly, in truth, it was
analysis that dominated his work. In fact, classical qualitative microscopic

analysis, as the above-cited scientists expected, never became significant
in practice. Firstly, because qualitative inorganic analysis is hardly used
in our time, both at macro- and microscales. I personally, for instance,
carried out qualitative inorganic analysis last in my freshman year, in the
course of the obligatory laboratory exercises. Secondly, because if necessary,
there are many convenient instrumental methods available. Regarding inorganic
analysis, the same may be stated for the spot-test method developed by Fritz
Feigl (1891-1970). He attempted to create a system based on this technique,
working at it persistently for forty years, from his doctor's dissertation
onwards. [6] A largely similar method was also developed in another part of
Europe by Nikolay Tananaev (1879-1959). It goes without saying that this
method too, as practically everything, had its antecedents: when Robert Boyle
(1627-1691) first applied a drop of acid on paper impregnated with lithmus
extract, [7] he actually carried out a spot-test.

The microchemical operations and devices that were developed in the course
of research proved to be of greater significance than the microchemical
reactions themselves. They are being in use in many highly sensitive analytical
techniques developed since, for instance in radioanalysis, and gas chromatography.

In organic qualitative analysis, both the microscopic and the spot-test methods
attained higher importance. Quantitative data like thermal behaviour, melting
point, refraction, molecular mass contributed to qualitative identification.
For the microscopic, microscale implementation of these tests, the best-suited
devices and techniques indispensable in organic chemistry and biochemistry
were developed by Ludwig Kofler (1891-1951),[8] professor of Pharmacognosy at
the University of Innsbruck, and his wife Adelaide Kofler, between 1925 and
1940.

Emich's fundamental merit was the initiative to extend microchemistry to
quantitative analysis. He proclaimed its possibility and developed many
methods.

The major condition was the creation of the microbalance. It is known that
Emich also was preoccupied with the problem, which was finally solved
successfully by Pregl. Looking back, it appears that what he achieved was
only a small matter, but this small matter, attained with much patience and
work, is a date of great significance in the history of chemistry. Organic
quantitative microanalysis opened up a new chapter in many sciences, above
all in biochemistry, allowing the measuring and determination of very small
samples. Probably the determination of organically bound halogen by Emich

and Julius Donau (1877-1940) in 1909 was the first organic quantitative microdetermination. [9] The complete development of organic micro-elemental analysis was the accomplishment of Fritz Pregl (1869-1930).[10] He reported his work in 1911. [11] Alber, in his history of microchemistry, remarks, with slight indignation: "Emich... deserves recognition as the founder of quantitative microanalysis in both organic and inorganic chemistry. However, the Nobel prize was awarded to Pregl, who adopted Emich's techniques." [12] In the history of science, however, it is a common feature that everybody adopts something from somebody else, and others will adopt and improve all earlier achievements. Pregl adopted his techniques not only from Emich, but also from Liebig and from Dumas. On the other hand, the awarding of the Nobel prizes is subjective, as is the case with all prizes. I do not dispute that it could have been awarded to Emich, but it is a certain fact that Pregl greatly deserved it, for his method marked a new epoch in many scientific disciplines.

Let us also mention the titrimetric application of microchemistry. The first route attempted, conceivably with little success, was the dilution of the measuring solution. Franz Mylius (1854-1931) and Fritz Förster (1866-1932) were the first, in 1891, to try measuring solutions with normalities of 0.001. [13] The method developed in Graz, the miniaturization of the measuring apparatus, proved more successful. The first microburette was constructed by Fritz Pilch in 1911. [14] The first monograph on the subject was written by a pupil of Emich's, the Hungarian chemist József Mika (1897-1977) in 1939. [15]

Finally, I wish to mention the microanalytical or rather ultramicroanalytical methods that have no macroanalytical analogous at all: kinetic-catalytic methods

Chemical analysis is characterized by the feature that sooner or later it attempts to utilize all chemical and even physical phenomena for its own purposes. It utilizes catalysis so that the presence and amount of the catalyst can be concluded from the catalytic process taking place and from its extent, respectively. Guyard, as early as 1876, suggested utilizing the catalytic effect of vanadium on the aniline black reaction to detect vanadium.[16] Witz and Osmond, in 1885, proposed the same method for the quantitative determination of vanadium, by demonstrating that the strength of colour attained after a certain time is a function of the amount of vanadium present. [17] After a long pause, the method was revived in the 'thirties of our century, its main centre being in Minneapolis [18]; another important centre was in Budapest, in László Szebellédy's laboratory (1901-1944). [19] It was a great loss to science that this gifted analyst died young. [20]

The great age of microchemistry, which may be considered the "classical" age, was in the period between 1920 and 1940, and its emissive centre was Austria.

The old Habsburgian Austria, the imperial city Vienna was an important centre in culture and the arts, in music, in literature, in philosophy, around the turn of the century, that is, just at the time of its decline. The small Austrian Republic which, after the St. Germain Peace Treaty, started its new life under unfavourable conditions, first achieved international recognition and authority in the field of microchemistry. As early as 1923, the journal Mikrochemie was started at the initiative of Rudolf Strebinger. The Österreichische Mikrochemische Gesellschaft was established in 1936, only just preceded by the Microchemical Club in London and the Microchemical Society in the United States. The journal Microchimica Acta, the second periodical of this scientific body was started in Vienna in 1937. In the corresponding non-Austrian societies, in the institutes and microchemical laboratories established in turn all over the world, Austrian microchemists occupied important, frequently leading posts, and next to them, scientists from other countries who learned the art of microchemistry in Graz and later developed it further by their own results, like for instance Anton Benedetti-Pichler (1894-1964), Oskar Wintersteiner (1899-1971), Fritz Feigl (1891-1970), Joseph Unterzaucher (1901-1973), József Mika and many others. In this manner, the achievements of the Austrian school spread all over the world. This further development, however, differed widely from the classical pattern. Physico-chemical techniques, and more recently automated techniques replaced the earlier methods. The appearance of analytical laboratories, to be sure, has changed fundamentally. Instead of the laboratory tables full of flasks, racks and supports, blue gas burners, steaming water baths and sand baths, we now have tiny or huge electronic equipment, buttons are being pushed, instruments are being read, and computers print out the results. The laboratory characteristic for the first half of our century already belongs in a museum, and museums of technology would do well to collect what is still available, because after two or three decades there will be nobody to remember.

Some of the new methods were in existence, in a more or less developed form, or in embryo form, before World War II. The technique of colorimetry and photometry is very old and extensive. Carl Heine determined iron by colorimetry as early as 1845. [21] The first visual colorimeter was constructed in 1853 by Alexander Müller (1828-1906). He named his instrument "complementary colorimeter"; [22] it was subsequently improved by Dehm, [23] and later by Dubos cq in 1870. [24]

The official birth of emission spectroscopy was in 1859, at the lecture of Bunsen and Kirchhoff. Its prehistory, however, is much older and reaches far back into the past. William Talbot (1800-1877), one of the inventors of photography, found a relationship between spectral lines and chemical compounds as early as 1826: "Whenever the prism shows a homogeneous ray of any colour to exist in a flame, this ray indicates the formation or the presence of a definite chemical compounds .... a glance at the prismatic spectrum of a flame may show substances which it would otherwise necessitate a laborious chemical analysis to detect." [25]

The American, David Alter (1807-1881) wrote in 1854: "the light produced by some element differs from the light of all other elements regarding the number of lines, their intensity and location, so that is can be recognized by a mere glance." He also listed the lines of the individual elements in form of a table. [26]

However, only in the hands of Bunsen and Kirchhoff did the method develop into a truly microchemical technique suited for practical purposes. Even before publication, Bunsen - recognizing the importance of the moment - wrote an enthusiastic letter to Roscoe: "... If you have a mixture consisting of lithium, sodium, potassium, barium, strontium and calcium, just give me one milligramme of it, and I shall tell you, without touching the substance, what elements are present in it, by means of my instrument, just looking at it through the telescope." [27]

Quantitative spectroscopy was considered unpracticable for a long time. Its pioneers were Hartley, Pollok, Leonard and de Gramont towards the end of the past century, and subsequently Gerlach (1889-1980) in 1924.

Absorption photometry for analytical purposes started in practice in 1870, with the photometer of Carl Vierordt (1818-1884), [28] and analytical infrared spectroscopy with the activity of William Weber Coblentz (1873-1962), at the beginning of our century. [29] I cannot go into details on the numerous kinds of spectroscopy, nor on the many instrumental microanalytical methods created before World War II. Just two examples suffice: activation analysis, although its principles were established by Hevesy and Levi in 1936, [30] didn't attract much attention at the time, and the number of papers concerned with it did not exceed a dozen until 1940. The first analytical chromatographic techniques were developed during the years of World War II, and was followed by such a large number of microchemical methods that it would be difficult to enumerate them.

Obviously, smaller and smaller amounts must be determined both in the fields of science and industry, for knowledge on the significance of these tiny amounts in biology, agriculture, technology etc. is one of the current problems in these sciences.

I would gladly outline the history of these methods too, although in this respect much historical research is still required - sometimes it is the origin of the most recent things that we know least about - .

References

1. H. Fresenius: Z. anal. Chem. 36 X. (1897)

2. S.A. Marggraf: Opuscules chymiques, Paris, 1762 I. 216

3. T. Lowitz: Nova Acta Imp. Petropol. 8 (1794), 9 (1795), 11 (1797), Technologitscheskij Jurnal 1, 26 (1804)

4. E.H. Reinsch: Ber. Deutsch. Chem. Ges. 14,2325 (1881)

5. A.A. Benedetti-Pichler: Mikrochemie 36/37,17 (1951);id. Mikrochim. Acta 1960, 625

6. F. Feigl: Z. anal. Chem. 57,135 (1918), Spot Tests, Amsterdam, 1954

7. R. Boyle: Experimenta et considerationes de coloribus, Opera, Genf, 1680 p. 95

8. M. Brandstätter: Mikrochemie 38, 295 (1951)

9. F. Emich-Donau I.: Monatshefte 30,745 (1909)

10. H. Lieb: Mikrochemie 35,123 (1950), Benedetti-Pichler A.A.: Mikrochem.J. 6,5 (1962), Szabadváry F.: History of Analytical Chemistry, Oxford, 1966 p. 302

11. F. Pregl: Abderhaldens Handbuch biochemischer Arbeitsmethoden 5,1307 (1912)

12. H.K. Alber in A History of Analytical Chemistry, Amer. Chem. Soc. 1977, edited by H.A. Laitinen and G.W. Ewing p. 26.

13. F. Mylius, F. Förster: Ber. Deutsch. Chem. Ges. 24,1482 (1891)

14. F. Pilch: Monatshefte 32,21 (1811)

15. J. Mika: Die exakten Methoden der Mikromaßanalyse, Enke, Stuttgart, 1939.

16. A. Guyard: Zentralblatt, 1876, 120.

17. G. Witz, A. Osmond: Bull. Soc. Chim. France 45,309 (1885)

18. E.B. Sandell, M. Kolthoff: Mikrochim. Acta 1,9 (1937)

19. L. Szebellědy, M. Ajtai: Mikrochemie 26,87 (1939)

20. F. Szabadváry: Hung. Sci. Instruments 1979 no. 46.1.

21. C. Heine: J. prakt. Chem. 36,181 (1845)

22. A. Müller: J. prakt. Chem. 60,474 (1853)

23. F. Dehm: Z. anal. Chem. 2,143 (1863)

24. J. Duboscq: Chem. News 21,31 (1870)

25. W.H.F. Talbot: Phil. Mag. 4,112 (1834)

26. D. Alter: American Journal 18,55 (1854), 19,213 (1855)

27. H. Roscoe: Gesammelte Abhandlungen von Bunsen, Leipzig, 1904, I. XXXIV.

28. C. Vierordt: Pogg. Ann. 140,172 (1870)

29. W.W. Coblentz: Investigations of Infra-red Spectra, Washington, 1905

30. G. Hevesy, H. Levi: Kgl. Danske Vid. Selsk. Math. phys. Medd. 14,24 (1936)

II. MICROCHEMISTRY IN ARTS AND ARCHEOLOGY

MICROCHEMISTRY AND THE TECHNOLOGICAL ASPECTS OF PAINTINGS

FRANZ MAIRINGER
Institut für Farbenchemie,
Akademie der Bildenden Künste,
Schillerplatz 3, A-1010 Wien, Austria.

Abstract. The scope and the limitations of scientific methods in the field
of art history and the conservation of works of art are first discussed in
general. The important role of microchemistry in this field is illustrated
by a case study. The object considered is the altar piece by Michael Pacher
in St. Wolfgang, Abersee, Upper Austria. It is shown, that the determination
of the painting materials and techniques used, aids the art historian in the
difficult question of distinguishing different hands of artists.

Works of art are an intrinsic part of our cultural heritage and are considered
today as an essential component of the quality of life. But nowadays they are
jeopardised by a hostile,polluted sourrounding, by dramatic climatic changes
caused by hectic exhibition activities, which are nourished by irresistible
forces like educational and political intentions. Transportation hazards,
high illumination levels and proudly announced gigantic numbers of visitors
are likewise disastrous. The rate of natural ageing processes is speeded up
considerably by these factors. Conservators, who are in charge of the care
and protection of the cultural goods, early recogniced the importance and
usefulness of scientific examinations and accepted the assistence of scientists.
They began to investigate the causes and sources of chemical and physical decay
and since works of art are precious, the necessary samples were forbiddingly
small. So microchemistry and modern nondestructive physical methods of analysis

played an important and unique rôle. The chemists were confronted very often
with an unusual type of chemistry with the possibility of reaction times of
several hundreds up to a thousand years, which implied highly improbable
reaction products. They made also proposals for the application of new
materials in conservation, which were made available by modern chemical
industry and were disappointed, when the restorers were not enthusiastic
about it out of a esthetic reasoning. E.g. the gloss of a picture varnish
is a tricky thing and that of a modern acrylic resin is for a trained eye
inferior to that of a natural resin. An other problem was the durability
and ageing properties of this new materials and the reversibility of their
applications. Whereas the behaviour of natural products, like animal glue,
natural resins or bees wax, which were used by painters and restorers for
many centuries, were well known, for the new materials an adjustment of
the different time scales of art and industry was a necessary goal. In
industry durable or nearly eternal means 20 - 50 years, which provides
difficulties for the artist or the conservator, for the degradation and
decay starts within the span of his own lifetime. Tricky is also the
question of reversibility of a procedure, for it is a strict and sensible
rule in conservation that all manipulations should be reversible as far
as possible, because it is easy to preserve a work of art physically and
ruin it completely and irreversible from an esthetic point of view. Hot
sealing of pages of an illuminated manuscript with plastic foils or soaking of
a wheathered porous stone sculpture with an epoxy resin, which renders un-
natural refractive properties of the surface, povide an example.

But of course the scope of scientific methods is much broader than the
determination of painting materials and obnoxious agents. The goal of a
systematic investigation and description of the materials used e.g. in
painting (and their ageing properties) is a history of these and of
painting techniques for a certain period, landscape, school or even workshop; for the realisation of an artistic idea or intention affords a
certain technique. Investigations of this kind can help to secure the
authenticity, purity or alterations of works of art. Nevertheless, art
historians are still very reluctant to accept this kind of support for
their work. This is caused by several factors.

First of all they consider the positivistic approach of scientists as
inadequate for works of art [1]. Only the application of X-ray radiology
and IR methods, like IR photography or reflectography with modern image
converters found access, because they deliver images of invisible states,
which can be evaluated by the usual methods of style critics. Secondly,

only in rare cases the results of a point analysis speak for themselves. Mostly they require a broad reference material which must be assisted by a thorough comparison with the historic source books on art techniques of the different periods [2].

Here, of course, the main difficulties are encountered.

For the art historian and very often even for the restorer, the results of a careful and cumbersome chemical analysis mean nothing. E.g. he cannot understand why it is so difficult to distinguish between so very different materials like animal glue and yolk of egg from an elaborate quantitative amino acid analysis. On the other hand, the scientist is normally not informed about the historic use of these materials and is, of course, not familiar with the historic sources, not to speak of the inherent errors.

But even a derivation of a "terminus post" or "terminus ante" for dating purposes can be misleading. E.g. the pigment chromium yellow, which is a lead chromate, was first synthesized in 1797 by the French chemist Vauquelin and was used as a pigment from about 1819 onwards, which should state a reliable terminus post for authentifications of paintings [2].

Still there is in rare cases analytical and historic evidence for much earlier use, because the natural mineral chrocoite, called "plomb rouge de Sibérie", was available.

So the evaluation of chemical analyses of painting materials is full of pitfalls. For meaningful results a large number of analytical results obtained from a great number of paintings of different periods and landscapes [3] must be connected with the evaluation of historical literary sources of the still vast mass of unpublished documents. This affords a close cooperation of scientists with art historians, conservators, specialist historians and philologists. Such enterprises are rather difficult and cumbersome to start, because of the lack of a common language, but encouraging beginnings can be seen, like the new translation of Pliny's Historia naturalis under the patronage of the GDCh.

Now I would like to illustrate these introductory remarks by an actual example of such a multidisciplined cooperation.

In 1969 the Austrian Bundesdenkmalamt decided to undertake conservation work on a most famous treasure of late Gothic art in Austria or even in Europe: the St. Wolfgang altar piece in the parish church of St. Wolfgang in Upper Austria. It was created by Michael Pacher, a well known South Tyrolian

artist from Bruneck, between 1471 - which is the date of the contract
between Benedict, the abbot of Mondsee and the artist - and 1481 - the
date written on the exterior wings. It is a shrine altar[4] with sculptured
polychromed figures and painted wings. The altar consists of three main
parts: the shrine of the predella, the main shrine and the architectural
ornamentation, the "Gesprenge". The predella can be closed by one pair
of wings, the main shrine by two pairs of huge wings (386 x 164 cm). So
the altar-piece is a pentaptychon. It combines 72 large and small figures
and 31 panel paintings, the overall height is 12,16 (+ 0,7)m.

It is a changeable altar with a clear iconographical program, which covers
the whole ecclesiastical year. With closed wings, where the legend of
St. Wolfgang is depicted - this state is called the working day view -
the altar-piece serves for ordinary days, for Advent, Lent and funeral
services. When the outer main wings are opened the Sunday view is to be
seen, where the life of Christ is visualised in 8 paintings. This state
was used for Sundays and festa simplicia. Finally, the holy day view is
presented, when the inner main wings and the predella wings were opened,
where the life of Mary on the wing paintings is narrated together with
the open shrine, where the coronation of the Virgin by Christ with the
attendant figures of St. Wolfgang and St. Benedict is to be seen. The
predella shrine shows the adoration of the child by the Magi. The
glorious richness of this polychromed figures together with the paintings
of high quality is still a splendid and awe inspiring sight.

After the inspection of 1969, a conservation and investigation campaign
was planned, for some damage (fungi, anobia) was observed. As scientific
advisor acted the Central Laboratorium Voor Onderzoek van Voorwerpen van
Kunst En Wetenschap in Amsterdam and our laboratory, the Institute of
Color Chemistry and Painting materials. A few samples were taken. The
campaign provided a unique opportunity for a systematic investigation
of an extremely well preserved late Gothic retable of highest quality.
The used materials, the different techniques and the state of preservation
could be investigated in connex with stylistic problems of art historians,
because to the trained eye there are easily perceptible differences in
artistic quality of different parts of the altar piece to be observed.
The highest quality is presented in the sculptures of the shrine and the
paintings of the inner wings (Mary, Christ) of less quality are the
paintings of the outer wings and the back of the main shrine and the
sculptures of the "Gesprenge".

This means that different hands or artists could be distinguished. In close cooperation of the different disciplines a catalogue of questions and problems was composed, which implied for the scientific investigations the following questions:

a) What kind of materials are used in the different parts of the altar in its physical appearance. This implies the determination of wood, glue, painting materials like pigments, (organic lakes) precious metal foils, binding media, varnishes.

b) What are the techniques and tools for joinery, woodworking, gilding, painting and polychromy.

c) It is possible to make the different stages of the artistic genesis visible, like construction lines and the underdrawing or the economy of the actual craftsmanship like the use of stencils or tracings. (IR and X-ray methods would be the appropriate methods.)

d) Are there technological differences, which show different hands, like painting technique or style of underdrawing.

e) How was the original appearance of the altar, when it was erected by Pacher in St. Wolfgang and what kind of alterations by natural and artificial processes like restorations took place (blue, green, red lakes, zwischgold (saving technique) Florian armour).

f) How were special techniques like the imitation of the pearl embroidery or the appliqué relief brocade carried out.

g) How is the state of preservation and the climatic situation during the seasons of the year. What are the influences of the large number of visitors in summertime and the heating of the church in winter.

h) What are the materials used in different restoration processes. Are there temporal differences, which could be used for dating.

This catalogue gives a rough sketch of problems, which could be tackled with scientific investigations. Not mentioned are purely stylistic and aesthetic questions of history of art and conservation which bore much weight in this campaign.

The possibility of answering this questions depended mainly on two basic requirements, -provided the analytical methods were powerful enough-, a sufficient problem orientated sampling and the disassembly of certain parts of the altar. Both of these could be only met to a certain extent.

A complete disassembly of the altar was never intended and luckily the paintings of the wings were in such a good condition, that it would have been a sacrileg to take samples from well preserved, untouched areas. Only along the edges and near clefts and lacunae, sampling was allowed. The situation in case of the polychromy of sculptures, the architecture of the shrine and its paintings on the back was less severe, adequate samples could be taken.

It is not feasible to discuss now all the analytical methods, which were applied for solving these problems, the poster in the exhibition room compiled by the Institute for Analytical Chemistry (TU Wien) and the Institute for Colour Chemistry and Painting Materials gives some insight in this respect. It is also impossible to give a full and lengthy report of the results obtained and their interpretation. A monograph on this subject will be published hopefully in the Spring of 1981 [4].

Just a few remarks will be made, where problems could be solved and I will confine myself to the question of paintings.

Practically all samples taken,were embedded in the usual manner in an epoxy resin and a cross section, or a thin section was made to study the layer structure [5]. The first question concerning the used materials and the structure could be more or less completely answered [6] and gave much insight in the procedures and habits of the Pacher workshop and showed once more the elaborate and thorough craftsmanship [7]. The large wings were made of spruce (panels and frames) [8] and carefully lined with canvas about 70 cm broad [7]. The ground consisted of levigated chalk with 10 - 12% impurities. Under the microscop at high magnifications the typical discoidal platelets of coccoliths could be observed [9]. This type of material can also be found on the frames of the wings, the inner side of the shrine and the sculptures. Whereas the ground of the back of the shrine showed,additionally to the calcite,a high content (up to 50%) of dolomite and is void of fossiles. This indicates the use of ground limestone (coarse granular).

The palette was no real surprise.

Blue: azurite
Green: Verdigris basic and neutral, copper resinate and green lakes
Yellow: lead – tin yellow I and II [10], yellow ochres.
Red: vermilion, caput mortuum, red ochre, red lakes (caesalpinia sappan) [11,12]
Whites: lead white, cerussa [13], calcite (chalk)
Binding media: animal glue, drying oils, egg tempera, mastic.

On the ready prepared panels, which were partially gilded the underdrawing was executed with a fine hair brush and a grayblack water soluble colour and under the gilding, where the stroke would have been lost, incisions of the outlines made with a blunt needle were applied [7]. For repeating patterns like tendrils or textiles, stencils were used (e.g. carnation pattern on the frame), an economic way of doing it. There were two types: cut stencils and punctured stencils. Two types of gilding technique could be observed: bole and mordant gilding. For the first one a thin (2-5 μm) layer of red bole was applied on the perfectly smooth ground, which was accomplished by the use of a special tool: dry horse-tail. The application of the silver foil was done directly without bole on the perfectly prepared white ground. As mordant an oil-varnish was used.

The painting techniques of the different cycles (Wolfgang, Mary, Christ) and the back of the shrine show remarkable differences, not in the materials, but in the application of colour. The back was constructed in a more old fashioned way with a carefully planned layer structure, how it is observed in the netherlandish paintings of this time. This fact strengt hens the theory of the art historians [14], that a journeyman of the Pacher workshop, who served his apprenticeship in the Netherlands, could be the artist. A similar but less pronounced difference could be detected in the Wolfgang-legende, which is attributed to Friedrich Pacher, the brother of Michael [14].

There are strong discolourations by natural ageing processes in the panel paintings. All matt blues are blackened completely and irreversibly. The oil-bounded, glossy blue areas are shifted to a deep green, probably by formation of copper oleate. All greens are not recognisable as such any more, they show a deep embrownment which is a common and analytically unsolved problem in Gothic paintings, there are many theories about this phenomenon, but none of them could be proved yet analytically [15,16,17]. The layer-structure of the greens is always complicated:

Normally verdigris is mixed with lead white and lead tin yellow, often sap green is added [18], which was obtained from Rhamnus species (Buckthorn berries - complexing agent : alum). A deep brownish layer now hides these green areas, which is partly primary, partly definitely secondary and only soluble in dimethylformamide (DMFA).

In conjunction with changes of the metal foils (namely the "Zwischengold" and silver), these discolourations have altered the equilibrium of colour hues and their brightnesses to a great extent. That is of great importance for the evaluation of symbolism of colour. Instead of the four-fold harmony of the two pairs Blue-Gold and Red-Green, which is supposed to be the Gothic language of colour [19], only the triad of Gold, Red and Blue can be seen now. Also the fleshy parts turned out to be over-painted in a warmer hue, which is characteristic for the Baroque taste.

Only a few questions could be answered in respect to the special techniques. There are several possibilities for the production of the applique relief brocade, which are listed in the liber illuministarius (or Tegernseer Manuskript) in Munich. Prefabricated leafs were cut and applied with a special cement. To produce these foils a mould was carved out of hard wood or lead [20]. For the imitation of threads and the patterns up to 14 incisions /cm were applied. A tinfoil was pressed in the mould and filled with a hardening paste of wax and colophony or chalk in animal glue. The foils so obtained, were gilded afterwards and painted.

Not so clear is the procedure for the imitation of pearl embroidery. Probably it was done with a brush on the still wet priming with the same mass, which meant incredibly tedious and cumbersome work, if one considers the many thousand beads graded even in size. Clearly two greater restoration campaigns had taken place, one in the 17th century and one in 1857 by Adalbert Stifter, the first official conservator in Upper Austria. Technically and from the point of used materials they were easy to distinguish.

These two campaigns provide an interesting example how restoration is influenced by cultural attitude, philosophy and ideology. In the second half of the 19th century the theories of French architects like Violet le Duc, who preached the "unity of style", prevailed in Europe [21]. As a consequence baroque altars were taken out of Gothic churches and replaced by Neo-Gothic ones. Luckily Austria is a very traditional country and is always reluctant in accepting new ideas, so Stifter was hindered by the Rector of the Academy of Fine Arts in Vienna, Ruben, one of my predecessors, to do any overpainting and refreshing of altered coloured areas. Only minor details were newly added like, the gallery on the shrine. Very soon in the last two decades of the 19th century a new movement in the aesthetics of restoration arose in England by William Morris, who founded the Society for Protection of Ancient Buildings. His maxime was: Put protection in the place of restoration, or unitas in pluritate (Nicolaus Cusanus).

This is to a large extent still valid nowadays. So like Frankl pointed out this morning: "Do not forget there can be more dimensions" is true for works of art and for scientists working in this field too. Not sheer curiosity of applicability of methods should govern the doing, but the respect and care for the historically grown entity of a work of art.

References

1. Ch. Wolters: "Naturwissenschaftliche Methoden in der Kunstwissenschaft" in: "Enzyklopädie der Geisteswissenschaftlichen Arbeitsmethoden", 6. Lief. p. 69 ff. München-Wien, 1970.

2. R.D. Harley: "The Specialist Historian and Conservation." Maltechnik, Restauro, 81, 57-62 (1975).

3. H. Kühn: "Möglichkeiten und Grenzen der Untersuchung von Gemälden mit Hilfe Naturwissenschaftlicher Methoden." Maltechnik, Restauro, 80, 149-162 (1974).

4. M. Koller and N. Wibiral: "Der Pacheraltar in St. Wolfgang. Untersuchung und Restaurierung 1971-76." Schriften zu Denkmalschutz und Denkmalpflege, Bd. 11, Böhlau Wien, to be published spring 1981.

5. J. Plesters: "Cross-Sections and Chemical Analysis of Paint Samples." Stud. Conservation 2, 110-157 (1956).

6. F. Mairinger, G. Kerber and W. Hübner: "Analytische Untersuchungen der Gemälde des Pacheraltares." in: M. Koller and N. Wibiral, cf. 4 (Sec. II/8)

7. M. Koller: "Arbeitsweise und Werkstattbetrieb." in: M. Koller and N. Wibiral, cf. 4 (Sec. III/2).

8. M. Koller and R. Prandtstetten: "Hölzer, Holzbearbeitung und Holzkonstruktion." in: M. Koller and N. Wibiral, cf. 4 (Sec. II/4).

9. R. Gettens, R.W. Fitzhugh and R.L. Feller: "Calcium Carbonate Whites." Stud. Conservation 19, 157-184 (1974).

10. H. Kühn: "Lead-Tin Yellow". Stud. Conservation 13, 7-33 (1968).

11. J.H. Hofenk-de Graaf: "Natural Dyestuffs". ICOM Meeting Amsterdam, 15.-19.9.1969.

12. F. Mairinger: "Untersuchungen der Farb- und Bindemittel an gefaßten Skulpturen der Schwanthaler und ihres Kreises." Restauratorenblätter 2, 97-106 (1974) Bundesdenkmalamt Wien.

13. A.J. Pernety: "Dictionaire Portatif de Peinture, Sculpture et Gravure." Paris 1757. See p. 32 (Céruse).

14. N. Wibiral: "Probleme und Ergebnisse der bisherigen kunstgeschichtlichen Forschung." in: M. Koller and N. Wibiral, of. 4 (Sec. I).

15. F. Mairinger and G. Kerber: "Analytische Untersuchungen zur Grün-
        problematik."Lecture given at the ICOM-Symposion: "Der
        Pacheraltar in St. Wolfgang". 1975. St. Wolfgang, O.Ö.

16. C.M. Groen: "Towards Identification of Brown Discolouration on Green
        Paints", ICOM-Meeting, Venice (1975).

17. L. Kockaert: "Note on the Green and Brown Glazes of Old Paintings."
        Stud. Conservation, 24, 69-74 (1979).

18. Boltz v. Ruffach: "Illuminierbuch", 1549, Ed. by Benzinger, C.I., in:
        Sammlung Maltechnischer Schriften Vol. IV (E. Berger)
        München (1913).

19. E. Frodl-Kraft: "Die Farbensprache der gotischen Malerei". Wiener Jahr-
        buch f. Kunstgeschichte XXX/XXXI, 89-178, cit. by Koller M.,
        Sec. III/2, cf. 4.

20. B. Hecht: "Betrachtungen über Preßbrokate". Maltechnik, Restauro, 86,
        22-49 (1980).

21. N. Wibiral: "Methodische Überlegungen" in: Sec. IV, cf. 4

III. MICROCHEMISTRY IN LIFE SCIENCES

MICROSCALE ANALYSIS OF BIOMOLECULES BY TIME RESOLVED LASER SPECTROSCOPY

RUDOLF RIGLER

Department of Medical Biophysics, Karolinska Institutet,
Box 60400, S-104 01 Stockholm 60, Sweden

Abstract. Principles and applications of techniques for the study of dynamic properties of biomolecules in the time range $10^{-12} - 10^2$ sec are presented. They are based on the spectroscopic properties of laser radiation permitting extreme time and spectral resolution as well as sensitivity of detection. In this context instruments were developed in our laboratory for the measurement of fluorescence anisotropy decay, fluorescence correlations, fluorescence photobleaching recovery as well as dynamic laser light scattering. From the analysis of these spectroscopic properties information about the dynamics of the molecular architecture and the solution structure of biomolecules as well as on their interactions can be obtained.

1. Introduction

The advent of the laser as light source offering very high spectral and time resolution has led to the development of new spectroscopical techniques which have become very important for studying dynamic processes of biomolecules at the molecular level. The possibility to generate pulses of a few picoseconds pulsewidth has opened a time range in which fast primary chemical processes as well as fast dynamic structural modes of biopolymers can occur (Table I). The high sensitivity of laser induced fluorescence spectroscopy allows to detect a few hundred fluorescent particles in tiny volumes and to record fluctuations in the particle numbers as caused by diffusion processes and chemical reactions. The narrow band width of laser lines permits to investigate the motion of particles in solution by the line broadening of scattered laser light.

Table I: Time table of events in biological systems

	sec.	
primary events in visual processes and photo- synthesis	10^{-12}	
	10^{-11}	rotational motion of aromatic aminoacids and nucleotides
H-bonding	10^{-10}	internal rotations of antibodies,
	10^{-9}	tRNA
	10^{-8}	rotational motion of tRNA, bovine serum albumine
	10^{-7}	
α-helix-coil transition	10^{-6}	
base pairing	10^{-5}	rotational diffusion of rhodopsin in visual membranes, rotational diffusion of virus DNA
interaction between oligo- nucleotides and oligopeptides	10^{-4}	
base pariring in DNA and RNA. Intercalation of mutagenic dyes	10^{-3}	rotational diffusion of Ca^{2+}- ATPase in sarcoplasmic reticulum, of Bacteriorhodopsin in Brown membrane
Rate of protein and DNA synthesis	10^{-2}	
tRNA conformation transitions	10^{-1}	
Breathing of DNA helix	1	charging of tRNA
	10^{1}	cleavage of DNA by restriction endonucleases
	10^{2}	
lifetime of repressor operon complex	10^{3}	

As it is apparent from Table I, the time constants of biological events span from picoseconds to hours and require spectroscopic techniques of high time resolution and spectral selectivity. In what follows I shall demonstrate methods which are depending on or have been greatly improved by the use of laser sources and have provided new insight in the dynamic behaviours of biomolecules.

2. Structural dynamics of biomolecules

Natural or artificial fluorescent groups placed at known positions in the primary structure of a macromolecule are able to "see" the environment in their immediate vicinity (i.e. sense the interaction with the solvent and with other neighbouring groups) and also of more distant places if they can exchange radiation energy with fluorescent groups located there. All

these interactions compete with the natural emission processes and determine
the lifetime of the excited state of the probe (Figure 1). With probes at
strategic positions in a macromolecular structure each conformational state
will be characterized by its own lifetime of emission and can be analyzed
by measuring the lifetime (τ).

SCHEME OF LUMINESCENCE PROCESSES

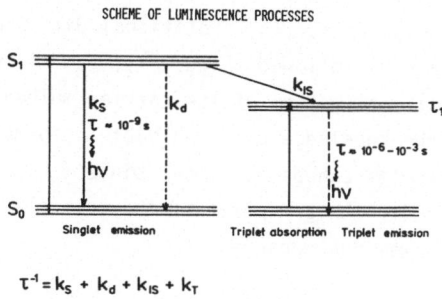

$$\tau^{-1} = k_S + k_d + k_{IS} + k_T$$

Chemical environment of probe (k_s)
Accessibility for external quenchers (k_d)
Interatomic distances (k_T)
Distribution of conformational states (a)
Brownian rotational motion and dynamics of structure (τ_ρ)

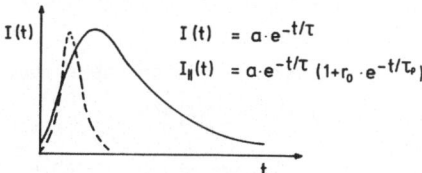

$$I(t) = a \cdot e^{-t/\tau}$$
$$I_{\parallel}(t) = a \cdot e^{-t/\tau} \, (1 + r_0 \cdot e^{-t/\tau_\rho})$$

Figure 1: Term scheme and pathways of decay from the excited state.
Information on lifetime τ and rotational relaxation times
(τ_ρ) is obtained from measuring the non polarized ($I(t)$)
and polarized intensities ($I_{\parallel}(t)$) emitted by the probe
after excitation with a short pulse of polarized light
(dashed curve)

The decay of the excited state yields information not only about processes influencing the lifetime of the fluorescent label, but also about the Brownian rotational motion of the label itself, if the sample is excited with a pulse of polarized light (Figure 1). Then an anisotropic distribution of fluorescence particles is generated in the excited state by photoselection. Due to Brownian rotational motion during the time of emission, the original anisotropic distribution is randomized as time goes on. By measuring the time course of polarized emission ($I_{||}(t)$) rotational diffusion times (τ_ρ) of fluorescent molecules can be followed (Figure 1). The fluorescent label then will "sense" the rotational motion of the carrier molecule as well as of localized parts. From the knowledge of τ_ρ valuable information on the solution structure of molecules can be obtained. Since τ_ρ varies with the molecular volume and thus with the 3rd power of the molecular radius, it is very sensitive to conformational changes.

A detailed theoretical basis for the analysis of rotational motion and internal flexibility of biopolymers has been developed (Chuang and Eisenthal,[1] Belford et al.,[2] Ehrenberg and Rigler[3]; Rigler and Ehrenberg[4]; Wahl[5]; Harvey[6]; Wegener[7] and has been applied to the study of biomolecules such as the Fab-fragments of antibody molecules (Yguerabide[8]; Munro et al.,[9]) DNA (Millar et al.,[10]), membrane proteins (Wahl et al.,[11]) and myosin (Highsmith et al.,[12]). Our own work is concerned with the study of dynamic apsects of the structure of transfer RNA. By labelling tRNA at known places with fluorescent structural probes we were able to demonstrate the existence of at least 3 tRNA conformations existing in a dynamic equilibrium in solution (Rigler et al.,[13]; Ehrenberg et al.,[14]). We are presently involved in investigating the biological role of these conformations of which one is very close to the crystal structure.

The use of synchronously pumped dye lasers which are able to generate pulses of pulse widths of a few ps and less has led to a considerable improvement of existing techniques (Ippen and Schank[16]; Flemming and Beddard[17]). An experimental set up, now used in our laboratory is shown in Figure 2. As exciting source a mode locked Krypton laser is used which pumps synchronously a dye laser leading to a ten fold pulse compression. From the train of pulses spaced by 13 nsec intervals single pulses are switched out by a cavity dumper and excite the sample. Single photon counting procedures are used to collect the decay data in a multichannel analyzer (Figure 2).

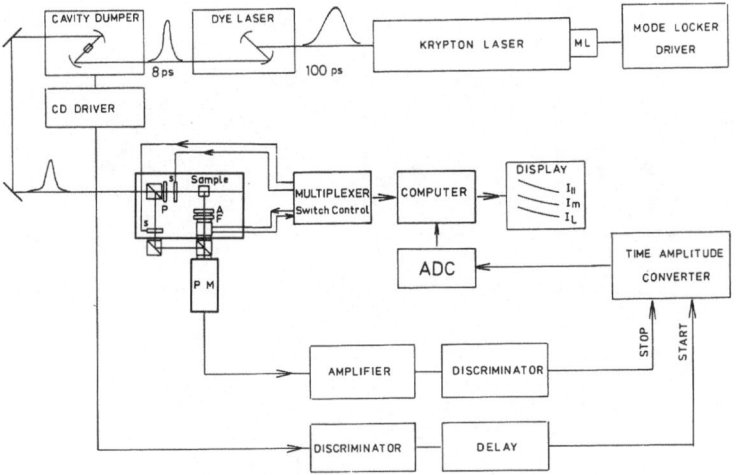

Figure 2: Set up for analysis of fluorescence lifetimes and rotational relaxation times. As excitation source a synchronously pumped dye laser is used. The laser pulse including the response of the photomultiplier (I_L) and the polarized ($I_{\parallel}$) and non polarized (I_m) intensities are measured by a split beam instrument by single photon counting techniques (Rigler and Ehrenberg [15])

The small pulse width of the laser together with the high repetition frequency permits to investigate dynamic processes in the picosecond domain such as rotational motion of individual aminoacids or nucleotides.

Measurements of this kind are of particular importance to verify the predictions of molecular dynamic calculations (McCammon et al., [18]).

3. Number fluctuation and diffusion

The availability of laser sources has initiated new techniques for the study of translational diffusion. Information on diffusion processes can be extracted from spontaneous fluctuations of the number of particles in a small volume element (Elson and Magde [19]). These fluctuations can be due to translational (Magde et al. [20]) or rotational (Ehrenberg and Rigler [21]) diffusion. Also chemical reactions (Magde et al., [20]) give rise to fluctuation if the change of the number of particles by e.g. dimerisation or by changes in the quantum yield (Figure 3). In order to extract the characteristic times from

the fluctuating signal the autocorrelation function G(t) has to be calculated
(Figure 3).

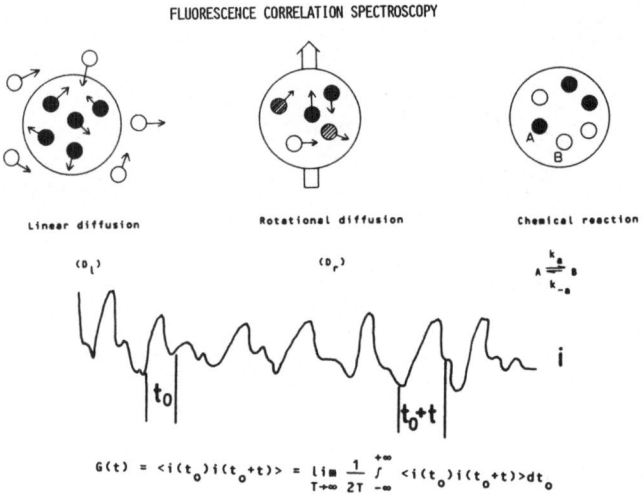

Figure 3: Cross section of the laser beam with excited particles undergoing translational and rotational diffusion as well as a
chemical reaction. They give rise to intensity fluctuations
i(t) for which the autocorrelation function G(t) is calculated

The autocorrelation function of fluorescence intensity fluctuations due to
translational and rotational diffusion of a biomolecule with spherical shape
is given by (Ehrenberg and Rigler [21])

$$G(t) = <i^2>\{1 + \frac{1}{N}(r_0 \cdot e^{-6D_R t} + \frac{1}{1+4D_T t\omega^{-2}})\}$$

with r_0=limiting anisotropy and ω=radius of laser beam.

The autocorrelation function contains the translational (D_T) and rotational
(D_R) diffusion coefficients as well as the number of particles in the observed volume element (N) as parameters. This possibility has been used to
determine molecular weights for high molecular substances from fluctuation
amplitude and the weight per volume of solute molecules (Weissman et al.,[22]).

The selective excitation of a fluorescent probe linked to a molecule permits
to follow its diffusion also in the presence of other molecules or in matrices
like membranes. For this reason fluorescence correlation spectroscopy (FCS)

has been used successfully to study diffusion processes in membranes. Similarly diffusion of nucleic acid specific proteins in the presence of DNA or RNA can be studied.

FLUORESCENCE   CORRELATION-SPECTROMETER.

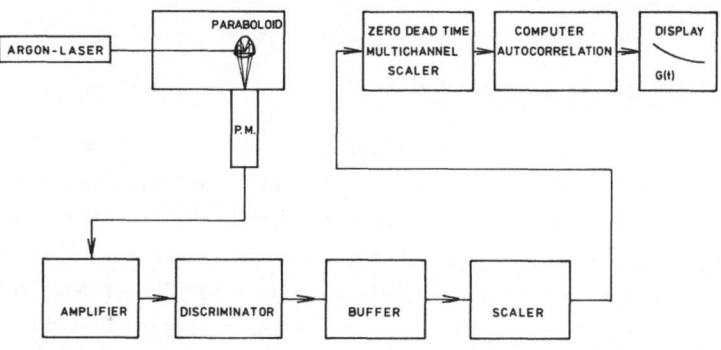

Figure 4: Block scheme of apparatus for fluorescence correlation measurements (Rigler et al. [23])

An instrument designed by us, consists of an argon ion laser focussed on the specimen with a focal spot of ca 7 μm radius (Figure 4). For maximum light collection a paraboloid cavity is used (Rigler et al. [23]). The photomultiplier signal is fed into a multichannel scaler and the autocorrelation function is calculated in a batch mode. Fluctuations of the laser source are compensated for by a reference circuit.

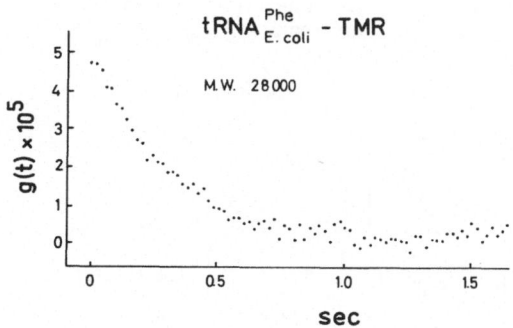

Figure 5: Autocorrelation function of fluorescence intensity fluctuations caused by E. coli tRNAPhe labelled with TMR and excited to fluorescence by the 514 nm Argon laser line

A typical result obtained with fluorescence correlation spectroscopy (FCS) is given in Figure 5 where the autocorrelation curve of E. coli-tRNA^Phe labelled with tetramethyl-rhodamin-isothiocyanate (TMR) at the X-base (Plumbridge et al., [24]) is shown. This experiment is part of an investigation in order to relate conformations and diffusion properties of the tRNA structure (Grasselli and Rigler, unpublished work). The high sensitivity of FCS allows to measure about 1000 molecules in a volume element of 10^{-12}l and thus very diluted solutions.

4. Photobleaching - an alternative

While photochemical destruction of fluorescent dyes is detrimental to correlation techniques it can be used to obtain information on slow diffusion processes in a time scale (sec and minutes), where FCS is unstable because of longterm drifts. In this approach one uses high intensity illumination in order to deplete the observed volume element from fluorescence particles by photobleaching. After the bleaching pulse the diffusion of unbleached dye molecules into the volume element is observed with low intensity illumination (Figure 6). From the time constant for the recovery of the original fluorescence intensity, the translational diffusion can be obtained.

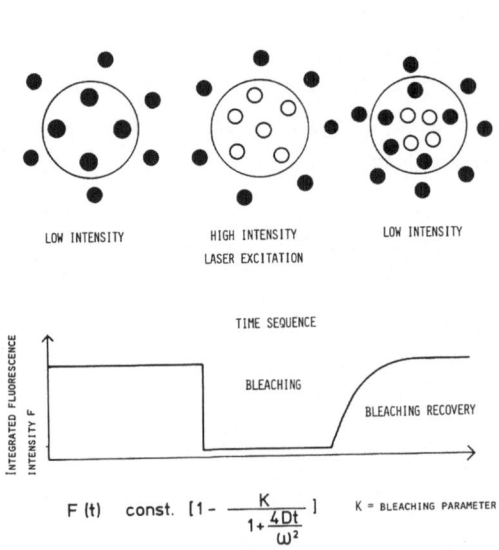

CROSS SECTION OF ILLUMINATED VOLUME

LOW INTENSITY   HIGH INTENSITY   LOW INTENSITY
                 LASER EXCITATION

TIME SEQUENCE

BLEACHING

BLEACHING RECOVERY

$$F(t) \quad const. \ [1 - \frac{K}{1 + \frac{4Dt}{\omega^2}}] \qquad K = BLEACHING \ PARAMETER$$

INTEGRATED FLUORESCENCE INTENSITY F

Figure 6: Principle of a photobleaching experiment

The translational diffusion of biological macromolecules observed by photobleaching recovery within an illuminated cross section of $\omega=10$ µm occurs in
a time ranging from 0.1 s (for a small enzyme) to about 10 sec (for a big
DNA molecule) or minutes (for a protein in a membrane). In this time scale
we can neglect the details of the photochemical reaction and describe it
as an irreversible first order reaction.

The theoretical description of the intensity recovery by diffusion has already been given by Axelrod et al. [25] For a laser beam with a Gaussian profile
of the intensity I, the solution for the time dependence of the fluorescent
signals F(t) has the form:

$$F(t) = F_0 \sum_{n=0}^{N} \frac{(-K)^n}{n!} \{1+n \frac{(1+2t)}{4Dt\omega^{-2}}\}^{-1}$$

with F_0=unbleached fluorescence level. The parameter $K=\alpha IT$ specifies the
amount of bleached dye under the specific conditions of laser intensity (I),
bleaching time (T) and bleaching sensitivity of the dye (α).

The principles of fluorescence photobleaching recovery (FPR) have been described in the literature and successfully applied to the determination of
the translational diffusion coefficient of labelled molecules in cell
membranes as well as in lipid bilayers (Peters et al., [26], Axelrod et al., [25];
Jacobson et al., [27]; Wolf et al., [28]; Schlessinger et al., [29]).

We have applied this technique to the study of diffusion of labelled macromolecules in solution. Since in this case the recovery times are much faster
than in cell membranes a precise characterization of the bleaching pulse and
a higher time resolution of the detection system are necessary.

Our basic set up for FCS measurements has been modified to perform FPR
measurements as shown in Figure 7. An acustooptical modulator provides
both the bleaching pulse as well as the attenuated illumination for the
detection of the recovery signal. After the bleaching time, which can be
selected over a broad range, the electric shutter of the photomultiplier
is opened as the laser intensity is attenuated 1000 fold, and the signal
is stored in a computer. For each measurement the sample is displaced by
a mechanical scanner to.expose a new area to the bleaching pulse. The
whole operation of the system is automatically timed by a control unit.
The present time resolution of the system is given by the opening time
of the electrical shutter in front of the multiplier, which in our case

is about 10 msec. This is sufficient for recovery times reflecting the translational diffusion of biological macromolecules.

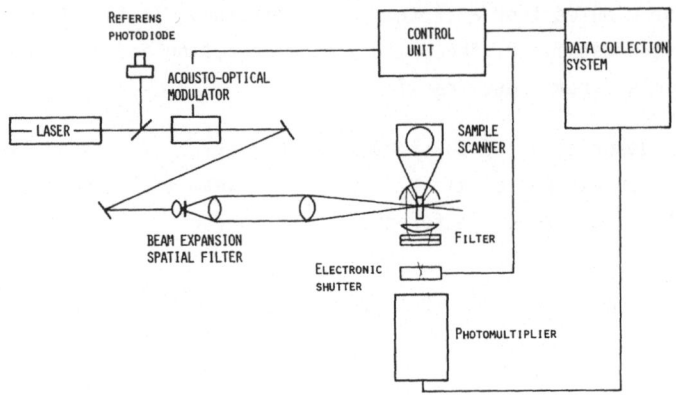

Figure 7: Set up of apparatus for fluorescence photobleaching experiments (Rigler and Grasselli [30])

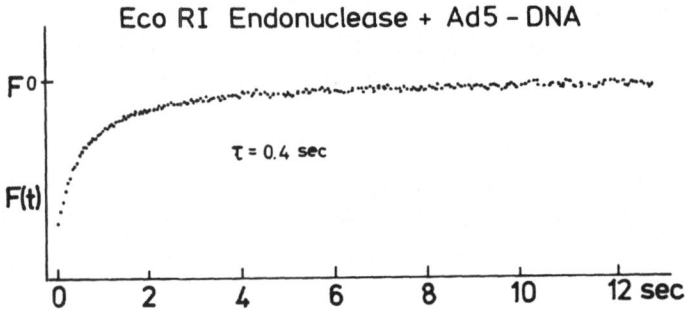

Figure 8: Photobleaching recovery curve for the diffusion of Eco RI endonuclease labelled with TMR in the presence of an excess of Adenovirus 5 DNA

As an example I would like to show the diffusion of Eco RI restriction endonuclease labelled with tetramethylrhodamine in the presence of Adenovirus DNA (Figure 8). The aim of this experiment is to find out to what extent diffusion processes notably along the DNA chain as proposed by Richter and

Eigen [31] might be involved in the cleavage of the DNA chain (Forsblom et al. [32]; Rigler and Forsblom, [33]).

5. Motion of living organisms

As last example I should like to mention another application of the laser in the spectroscopy of dynamic processes. Here information on diffusion and active motion of particles can be obtained from the Doppler shifted frequency of coherently scattered laser light (Cummins [34]). Intensity fluctuation of the light, scattered at a given angle are registered (Figure 9) and like in FCS the autocorrelation function is calculated. For particles undergoing Brownian diffusion with diffusion constant D and active motion with a velocity distribution P(V) the autocorrelation function can be calculated (Cummins [34]) (Figure 10).

Set up for measurement of dynamic light scattering

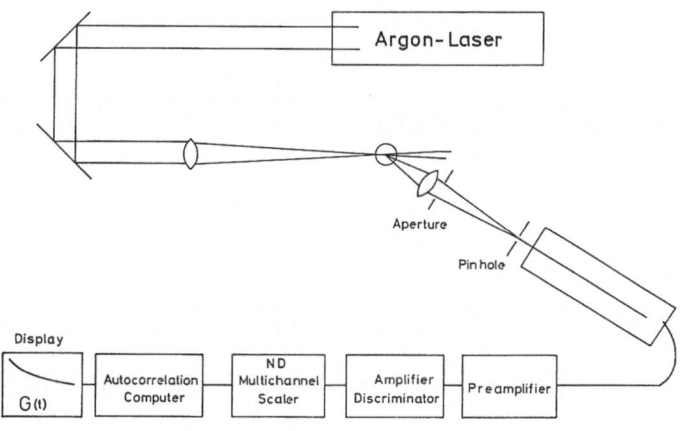

Figure 9: Set up for measurements of dynamic light scattering

From the measured autocorrelation curve of e.g. human sperms as done in our laboratory (Figure 11) detailed information on the swimming speed and the fraction of actively mobile sperms can be obtained. A more detailed analysis (Matsumoto et al., [35]) is able to separate also translational from rotational motion and is able to provide rather detailed information on the motion of living organisms.

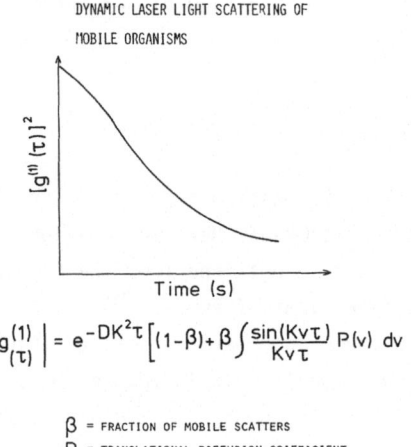

DYNAMIC LASER LIGHT SCATTERING OF
MOBILE ORGANISMS

$$\left| g^{(1)}_{(\tau)} \right| = e^{-DK^2\tau}\left[(1-\beta)+\beta \int \frac{\sin(Kv\tau)}{Kv\tau} P(v)\ dv\ \right]$$

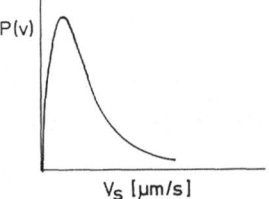

β = FRACTION OF MOBILE SCATTERS
D = TRANSLATIONAL DIFFUSION COIFFICIENT
P(v) = SWIMMING SPEED DISTRIBUTION

Figure 10: Autocorrelation function $g^{(1)}_{(\tau)}$ of particles with Brownian diffusion
(D) and active motion (P(V))

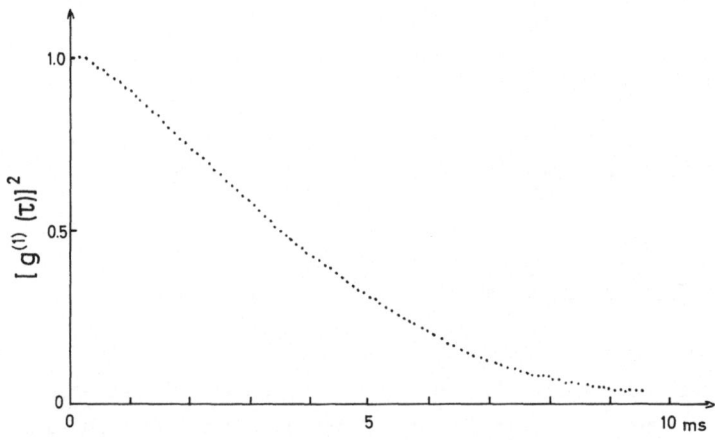

Figure 11: Autocorrelation function of the fluctuations of laser light
scattered by living human sperms

6. Chemical kinetics

A field, where the use of pulsed or continuous wave lasers has opened new experimental possibilities, is the study of the kinetics of chemical reactions.

The concept of chemical relaxation as developed by Eigen and De Maeyer [36] has led to the development of a variety of experimental procedures for perturbing the chemical equilibrium state by fast changes in temperature, pressure or electric field. This approach has been of big importance for studying the primary reaction steps in biological reactions and has given a detailed information on the molecular basis of the biological functions of proteins, nucleic acids or membranes (for a review see Pecht and Rigler [37]).

The generation of laser pulses in a spectral region where water absorbs allows to accomplish nsec temperature jumps (Flynn and Sutin [38]) and to study very fast biological reactions (Giannini and Grasselli [39]). Similarly high electric fields created by picosecond laser pulses could probably be used for fast dielectric relaxation experiments. The feasibility of fluorescence correlation measurements for the study of chemical reactions has been demonstrated by Magde et al. [20] and with further improvements of FCS avoiding photobleaching to occur, this technique could be used more generally. The possibility to observe tiny sample volumes when the laser is used as a spectroscopical tool is going to have implications for the development and improvement of perturbation procedures for kinetic studies.

Acknowledgement. The development of the instrumentation described was supported by grants from the Knut and Alice Wallenberg Foundation, the Swedish Natural Science Research Council and the Swedish Cancer Society.

References

1. T.J. Chuang and K.B. Eisenthal. (1972) J. Chem.Phys. 57, 5094.

2. G.G. Belford, R.L. Belford and G. Weber. (1972) Proc. Natl.Acad.Sci.USA 59, 1392.

3. M. Ehrenberg and R. Rigler. (1972) Chem.Phys.Lett. 14, 539.

4. R. Rigler and M. Ehrenberg. (1973) Quart.Rev.Biophys. 6, 139.

5. Ph. Wahl. (1975) Chem.Phys. 7, 220.

6. S.C. Harvey. (1979) Biopolymers 18, 1081.

7. W.A. Wegener, R.M. Dowben and V.J. Koester. (1980) J.Chem.Phys. 73, 4086.

8. J. Yguerabide, H.F. Epstein and L. Stryer. (1970) J.Mol.Biol. 51, 573.

9. I. Munro, I. Pecht and L. Stryer. (1979) Proc.Natl.Acad.Sci.USA 56, 76.

10. D.P. Millar, R.J. Robbins and A.H. Zewail. (1980) Proc.Natl.Acad.Sci.USA, in press.

11. Ph. Wahl, M. Kasai and J.P. Changeux. (1971) Eur.J.Biochem. 18, 332.

12. S. Highsmith, R.A. Mendelson and M.F. Morales. (1976) Proc.Natl.Acad.Sci. USA 73, 133.

13. R. Rigler, M. Ehrenberg and M. Wintermeyer. (1977) Mol.Biol.Biochem. and Biophys., vol. 24, pp. 219, Springer-Verlag.

14. M. Ehrenberg, R. Rigler and W. Wintermeyer. (1979) Biochemistry 18, 4588.

15. R. Rigler and M. Ehrenberg. (1976) Quart.Rev.Biophys. 9, 1.

16. E.P. Ippen and C.V. Shank. (1975) Applied Physics Letters 27, 488.

17. G.R. Fleming and G.S. Beddard. (1978) Optics and Laser Technology (Oct.) pp. 257.

18. J.A. McCammon, P.G. Wolynes and M. Karplus. (1979) Biochemistry 18, 927.

19. E.L. Elson and D. Magde. (1974) Biopolymers 13, 1.

20. D. Magde, E.L. Elson and W.W. Webb. (1974) Biopolymers 13, 29.

21. M. Ehrenberg and R. Rigler. (1976) Quart.Rev.Biophys. 9, 69.

22. M. Weissman , H. Schindler and G. Feher. (1976) **Proc.**Natl.Acad.Sci.USA 73, 2776.

23. R. Rigler, P. Grasselli and M. Ehrenberg. (1979) Physica Scripta 19, 486.

24. J.A. Plumbridge, H.G. Bäumert, M. Ehrenberg and R. Rigler (1980) Nucleic Acids Research 8, 827.

25. D. Axelrod, D.E. Koppel, I. Schlessinger, E.L. Elson and W.W. Webb. (1976) Biophys. J. 16, 1055.

26. R. Peters, J. Peters, K.H. Teur and W. Bähr. (1974) Biophys.Biochem.Acta 367, 282.

27. K. Jacobson, Z. Derzko, E.S. Wu, Y. Hou and G. Poste. (1976) J. Supramol. Structure 5, 565.

28. D.E. Wolf, J. Schlessinger, E.L. Elson, W.W. Watt, R. Blumenthal and P. Heukart (1977) Biochemistry 16, 3476.

29. I. Schlessinger, L.S. Barak, G.G. Hammes, K.M. Yamada, I. Pastan, W.W.Webb and E.L. Elson. (1977) Proc.Natl.Acad.Sci. USA 74, 2909.

30. R. Rigler and P. Grasselli. (1980) in Lasers in Biology and Medicine (C. Sacchi, ed.), Plenum Press, in press.

31. P.H. Richter and M. Eigen. (1974) Biophys. Chem. 2, 255.

32. S. Forsblom, R. Rigler, M. Ehrenberg, U. Pettersson and L. Philipson. (1976) Nucleic acids research 3, 3255.

33. R. Rigler and S. Forsblom. (1979) in FEBS 12th Meeting Dresden 1978, vol. 51, Gene Functions, S. Rosenthal et al.

34. H.Z. Cummins. (1977) in Photon Correlation Spectroscopy and Velocimetry, H.Z. Cummins and E.R. Pike (ed.), Nato Advanced Study Institutes Series - Series B: Physics, pp. 200. Plenum Press, New York 1977.

35. G. Matsumoto, H. Shimizu, J. Shimada and A. Wada. (1977) Optics Communications 22, 369.

36. M. Eigen and L. DeMaeyer. (1963) in Technique of Organic Chemistry (S.L. Friess, E.S. Lewis and A. Weissberger, eds.) vol.8, part II, pp. 895, J. Wiley, NY.

37. I. Pecht and R. Rigler. (1977) Chemical Relaxation in Molecular Biology, Springer-Verlag, Heidelberg.

38. G.W. Flynn and N. Sutin. (1974) in Chemical and Biochemical Application of Lasers (C.G. Moore, ed.), Academic Press, New York.

39. I. Giannini and P. Grasselli. (1976) Biochem. Biophys. Acta 445, 420.

MICROCHEMICAL REACTIONS IN NERVE IMPULSE TRANSMISSION

EBERHARD NEUMANN
Max-Planck-Institut für Biochemie,
D-8033 Martinsried,
Federal Republic of Germany

Abstract. The electrical signals in biological cells are generated in micro-reaction spaces of the cell membranes. The bioelectric signals and other bioelectric potential changes of nerve and muscle cells are caused by cross-membrane ion flows; and the most extensively studied ion flow gating system is the acetylcholine system. Basic features of elemental bioelectric signals are shown to be understandable in terms of recent relaxation kinetic data on isolated acetylcholine receptor and acetylcholinesterase. The conclusions from electrophysiological and physical chemical investigations can be summarized in a flow scheme for an essentially sequential processing of acetylcholine by receptor and esterase. The data of the acetylcholine system reveal striking similarities to basic features of the Na^+ ion gating system of axonal membranes of nerve and muscle cells. For example, the conducting conformation of both the acetylcholine receptor-channel and the Na^+-channel are of transient nature, they are intrinsically metastable compared to the thermodynamically stable inactivated (desensitized) states.

1. Introduction

The electric signals in living organisms are of ionic-chemical origin. More specifically, nerve impulses and other membrane potential pulses of nerve and muscle cells result from transient permeability changes in the excitable membranes to ions, in many cases selectively to Na^+, K^+, Ca^{2+}, or Cl^- ions. When a nerve impulse arrives at a nerve terminal of a contact site (synapse) to another nerve or a muscle cell, synaptic potential changes or a nerve

impulse are generated in the postsynaptic membrane of the target cells. A
typical neuromuscular junction is schematically represented in figure 1.

Figure 1: Schematic representation of a synapse connecting a nerve and a
muscle cell. The reaction products of acetylcholinesterase
activity are found along all excitable biomembranes, in the
synaptic cleft and in the tubular system. Acetylcholine receptor
as judged from reaction sites with α-bungarotoxin is accumulated
in postsynaptic membranes but is also found presynaptically. The
enlarged inserts models the location of acetylcholinesterase (O)
and of the membrane-bound receptor

The bioelectrical signal transmission between nerve cells and across neuromuscular synapses undoubtedly involves membrane permeability changes to ions.
The resulting local ion flows are chemically controlled by membrane-bound
gating proteins which form ion-selective channels when they are in the open,
permeable conformation. The gating reactions responsible for opening and
closing of the ion pathways in biomembranes appear to generally involve bimolecular reactions, but the molecular-mechanistic details are yet unknown
(Neumann,[1] Dorogi and Neumann[2]).

The most thoroughly studied ion flow gating is the rapid Na^+-K^+ ion transport mediated by the acetylcholine (AcCh) system. The transport gating by
this regulatory system appears to result from a direct interaction of AcCh
with two proteins: acetylcholine receptor (AcChR) and acetylcholinesterase
(AcChE), E.C. 3.1.1.7. According to Nachmansohn[3] a proper stimulus re-
leases AcCh from a storage site; the chemical activator then binds to AcChR
and induces a conformational change to an open, ion-conducting structure.
The enzyme AcChE rapidly hydrolyzes acetylcholine and thereby terminates
the receptor-mediated permeability increase. This gating concept is pictured
in figure 2.

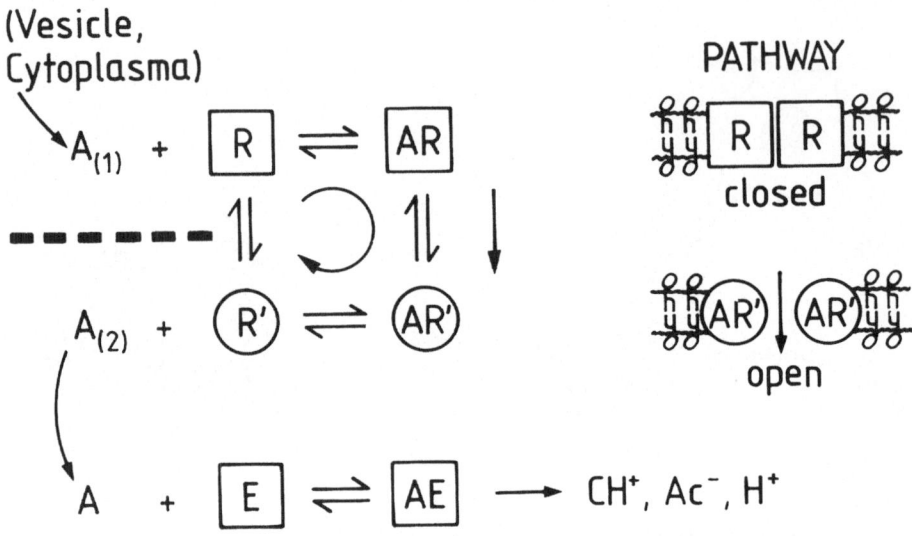

Figure 2: Functional concept for the interaction of acetylcholine, A, with receptor (R, closed conformation and R', open conformation) and esterase. On the right is a schematic view of the receptor-mediated permeability change in the membrane

The reactions of AcCh with AcChR and AcChE occur in microreaction spaces near the membranes. In figure 1 the so far most extensively studied reaction space of AcCh is enlarged, showing that the receptors are accumulated in postsynaptic parts, but are also present in presynaptic parts of excitable membranes as far as can be judged from the α-bungarotoxin binding believed to be specific for AcChR. Reaction products of esterase activity are found along all excitable biomembranes and in particular in the synaptic gap. Therefore, in figure 1 the enzyme localization is modelled within the cleft space close to the membranes (see, e.g., Neumann et al. [4]).

The reactions of AcCh in this reaction space cause the wellknown electrical epiphenomena: miniature end-plate potentials, end-plate potentials and action potentials (or nerve impulses). The molecular mechanisms of the AcCh reactions with AcChR and AcChE are therefore an essential part of the complicated mechanism of the signal transmission in neuronal systems. In the following assay it will be shown how the results of microchemical experiments with the isolated AcCh-proteins permit a deeper insight in the special type of electrical epiphenomena and in the possible molecular interaction mechanisms involved in bioelectrical signal generation.

2. Elemental bioelectrical signal

The probably most elemental epiphenomenon resulting from AcCh action is the
spontaneous miniature end-plate current, mepc (see, e.g. Gage [5]). As seen in
figure 3, where such a mepc is redrawn, a rapid growth phase is followed by
a slower decay phase. Gage and McBurney [6] explicitly state that this decay
is exponential from the peak, no rounding is observed. The growth phase
probably reflects the AcCh-induced conformational change; the decay phase
is determined by AcChE activity, but not in a rate-limiting manner.

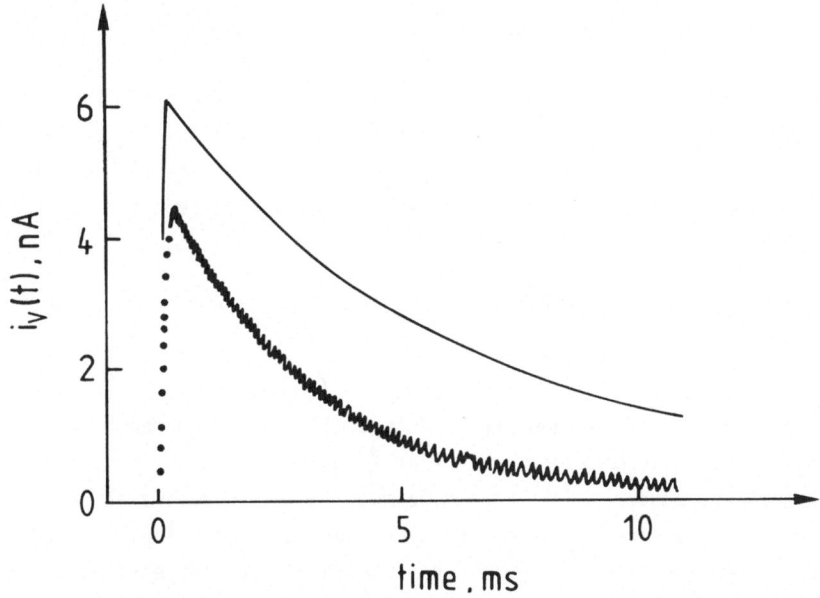

Figure 3: Miniature end plate current, mepc, in toad neuromuscular junction
(redrawn from Figure 8 of Gage and McBurney [6]; and Figure 2 of
Gage [5]); clamp potential - 70 mV, 293K in standard Ringer solution.
Upper trace recorded after 30 min exposure to 1mg/l neostigmine in
Ringer solution, a condition assumed to completely inhibit esterase
activity: the slow decay reflects slow diffusion of acetylcholine
in the junction (and probably also further receptor desensitization)

The voltage dependence of the decay time constant as well as the average
life-time t_0 of an open channel is not affected by the presence of esterase
inhibitors; time constants of mepc and of end-plate currents are the same
and equal to t_0 (see, e.g. Gage [5], Stevens [7]). It thus appears that the
closure phase of the AcCh-activated channels is practically independent
of AcCh; it is rate-limited by the return of AcChR to the closed conformation
(Stevens [7]). This observation requires, however, that AcCh is much faster
removed than it can return to AcChR; effective association to AcChE should

therefore be faster than to AcChR or in terms of association rate constants:

$$k_1(eff, R) < k_1(eff, E). \tag{1}$$

Since practically complete inhibition of AcChE with neostigmine causes an increase of the peak amplitude of mepc by a factor of 1.4 and a prolongation of the decay phase by a factor of about 2 (see Figure 3 and Gage and McBurney [6]), the receptors in a normal mepc appear not to be saturated.

3. Flow concept of AcCh-mediated gating

In order to model AcChE inhibitor effects on mepc consistent with the inequality (1), the elementary conductance increase must involve at least two AcCh molecules (Neumann, Rosenberry and Chang [4]; Rosenberry [8]). Using several assumptions on receptor density in the synaptic cleft and applying the reactions

$$2A + R \rightleftharpoons ARR' + A \rightleftharpoons (AR')_2 \tag{2}$$

$$A + E \rightleftharpoons AE \rightarrow E + P \tag{3}$$

in a competitive model, Rosenberry [8] successfully simulated mepc; for equation (2) see also Sheridan and Lester [9]. However, the sharp peak of mepc was not reproduced and it cannot be modelled by any competitive model, even if the time course of the AcCh concentration, $[A]$, is a delta function during the growth of mepc. Thus the particular shape of mepc indicates an essentially sequential processing of AcCh: the main part of neurally evoked AcCh binds first with AcChR in a microreaction space 1. Because of the inequality (1) it appears that initially in the growth phase of a mepc the AcCh concentration close to activatable receptors, $[A]_1$, is larger than the AcCh concentration $[A]_2$ in a second microreaction space where AcChE fully competes with AcChR for activator. To fulfill the initial mepc condition

$$[A]_1 > [A]_2 \tag{4}$$

a partial diffusion barrier between the local microreaction spaces 1 and 2 appears to be necessary.

These conclusions from purely electrophysiological data have been summarized in a flow scheme (Neumann and Bernhardt [10], Neumann, Rosenberry and Chang [4]). The present form of the AcChR-gating cycle for neurally triggered AcCh is shown in figure 4.

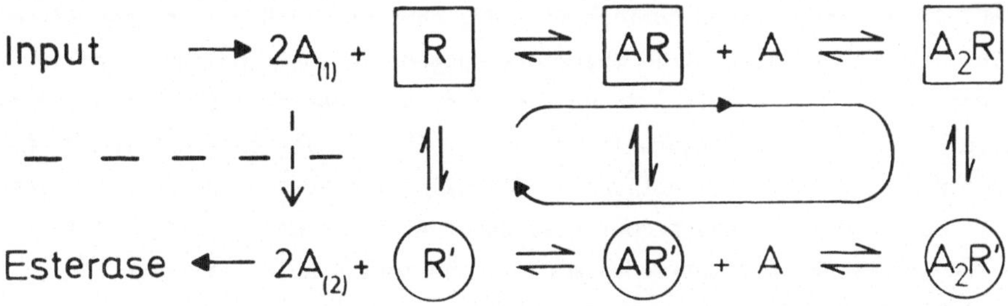

Figure 4: Flow scheme (AcChR-gating cycle) for neurally triggered acetylcholine (input), which reacts essentially in a sequential manner
with receptor and esterase. The curved arrow indicates the flow
of acetylcholine A from a microreaction space 1 through the
closed (R) and open (R') receptor states to a separate microreaction space 2 where the esterase has full competitive access
to acetylcholine $A_{(2)}$. The dashed line represents a partial
diffusion barrier such that initially in a mepc the condition
$[A]_1 > [A]_2$ holds

It is readily realized that, once AcCh is more rapidly removed, i.e. $[A]_2 \approx 0$,
closure of a channel can occur solely along intramolecular pathways, probably
mainly via the $R' \rightleftharpoons R$ step and less via the $A_2R \rightleftharpoons A_2R'$ step; both pathways are
consistent with first order decay of a mepc.

When AcCh is of non-neural origin, for instance applied artificially using a
micropipette, AcChRs are multiply activated (Katz and Miledi [11]) and the
decay phase of a mepc is prolonged. The inequality (4) may no longer hold
for bath application of AcCh and the opening-closure kinetics is mainly
determined by the kinetics of the $A_2R \rightleftharpoons A_2R'$ step, as suggested by experiments
(see Neher and Sakmann [12], Stevens [7]; Sheridan and Lester [9], Stevens [13]).

Desensitization. Prolonged exposure to AcCh indicated additional reaction
pathways for the closure of AcCh-activated channels. Longer bath application
of activators causes inactivation or pharmacological desensitization of
AcChRs, according to a cyclic scheme first proposed by Katz and Thesleff [14]:

$$
\begin{array}{ccc}
A & + & R \rightleftharpoons AR' \\
& & \updownarrow \quad \updownarrow \\
A & + & R'' \rightleftharpoons AR''
\end{array}
\qquad (5)
$$

where R" is one of probably more, desensitized receptor states (see also
Rang and Ritter [15]).

A direct consequence of the electrophysiological data describable with scheme

(5) is that, even in the absence of AcCh, a certain fraction of AcChR exists a priori in the inactivated conformation R", characterized by a higher affinity to AcCh than the R-state. It is therefore suggestive that the 'non-depolarizing receptors (Katz and Thesleff [14]) and the 'diffusion barrier' (De Motta and del Castillo [16]) are just receptors in the high AcCh-affinity desensitized R"-form. Furthermore, AcCh-induced inactivation following the activation phase suggests that the activated, permeable conformation of AcChR is metastable; the inactivated states AR" are thermodynamically the most stable states in the presence of activator.

Ca^{2+} -ions. The permeability changes induced by AcCh in many biomembranes are cation selective. Corresponding to the high concentrations of external Na^+ ions and internal K^+ ions the alkali metal ions predominantly contribute to the ion flows. Since Ca^{2+} selectively inhibits the Na^+-flow there appears competition for sites and thus transient, short-lived binding of Na^+ and Ca^{2+} ions in the Na^+ pathway (Takeuchi, A. and Takeuchi, N. [17]; see also Gage and van Helden [18])

The detailed role of Ca^{2+} ions in activation and inactivation is not yet molecularly understood. In frog muscle increased external Ca^{2+} concentration slightly decreases mepc decay time constants and life times of open channels (see Table 3 of Magleby and Weinstock [19]). Thus Ca^{2+} ions appear to facilitate the closure reaction and to stabilize the R-conformation of AcChR. In addition, increased Ca^{2+} concentration appears to accelerate the activator-induced desensitization (see, e.g., Nastuk and Parsons [20], Manthey [21]; Fiekers, Spannbauer, Scubon-Mulieri, and Parsons [22]).

In summary, AcCh-induced rapid activation eventually releasing Ca^{2+} ions followed by slower inactivation processes accompanied by uptake of Ca^{2+} ions appears to be a fundamental characteristics of functionally intact AcChRs.

4. Kinetics of isolated AcChR and AcChE

As discussed in chapter 2, the rate constants $k_1(R)$ and $k_1(E)$ of effective binding of AcCh to the two gating proteins AcChR and AcChE according to the overall reactions

$$A + R \underset{k_{-1}}{\overset{k_1}{\rightleftharpoons}} AR \rightleftharpoons .... \qquad (6)$$

$$A + E \overset{k_1}{\rightleftharpoons} E \cdot A \rightleftharpoons .... \qquad (7)$$

are of fundamental importance; see the inequality (1). The constants may be derived from kinetic studies with the isolated proteins. One might anticipate that both bimolecular reactions are close to diffusion controlled; that is, nearly every collisional encounter between acetylcholine and an R or E binding site leads to binding. In practical terms this implies that the bimolecular reactions are very rapid (in the μsec or msec time range) and require special rapid kinetic methods. Furthermore, the analysis of rapid kinetic reactions is greatly facilitated when the reaction systems are homogeneous. Thus we have so far applied rapid kinetic techniques only to solubilized receptor or esterase of high purity.

A powerful method for the determination of rate constants of rapid chemical reactions is chemical relaxation spectrometry (Eigen [23]). The principle of this method is the perturbation of a chemical equilibrium or a steady-state by a rapid change in a physical variable, for example, temperature, followed by a measurement of a rate of concentration change as the components adjust to their new equilibrium or steady-state concentrations. Generally the concentration of one component is monitored by spectrophotometric or spectro-fluorometric techniques. For small physical perturbations, the concentration change associated with a single elementary reaction such as those in equations (6) and (7) has an exponential time dependence and is thus characterized by a relaxation time and a relaxation amplitude. A simple theoretical analysis (Neumann and Chang [24]) predicts that this relaxation time τ for the reaction in equation (6), for instance, will depend on the reactant concentrations according to equation (8),

$$\tau = \{k_1^{(R)} ([A] + [R]) + k_{-1}^{(R)}\}^{-1} \tag{8}$$

where $[A]$ and $[R]$ are equilibrium concentrations. Usually $[A]$ and/or $[R]$ cannot be directly measured; and it is convenient to express equation (8) in terms of the total concentrations $[A^o]$ and $[R^o]$, where $[A^o] = [A] + [AR]$ and $[R^o] = [R] + [AR]$, as shown in

$$\tau = \{k_1^{(R)} [R^o] \sqrt{(1 + p + [A^o]/[R^o])^2 - 4[A^o][R^o]}\}^{-1}, \tag{9}$$

where $p = K/[R^o]$ and $K = k_{-1}^{(R)}/k_1^{(R)}$. Equation (9) is readily rearranged to

$$\tau^{-2} = \alpha + \beta[A^o] + \alpha[A^o]^2 \tag{10}$$

where $\alpha = (k_1^{(R)})^2(K + [R^o])^2$, $\beta = 2(k_1^{(R)})^2 (K - [R^o])$, and $\gamma (k_1^{(R)})^2$ (Rosenberry and Neumann [25]). Equations 9 and 10 have been written with apparent asymmetry of $[A^o]$ and $[R^o]$ in anticipation of experimental measurements of τ in which $[R^o]$ is constant while $[A^o]$ is varied. Under these conditions it is readily shown that τ has a maximum value (τ^{-2} a minimum) when $[A^o] = [R^o]$ - K, provided that $[R^o] > K$, i.e., p < 1, (Neumann and Chang [24]).

While measurement of the reaction rate constants of acetylcholine with receptor and esterase defined in equations (6) and (7) by means of equations (9) and (10) thus would appear straightforward, direct measurements have not yet been possible. In the receptor case, no suitable optical signal for monitoring the reaction in equation (6) itself has been found. However, this reaction can be coupled to Ca^{2+}-binding equilibria involving the receptor. Recently, a relaxation spectrum in this coupled system has been observed, and corresponding extensions of equation (9) permitted estimates both of several rate constants including $k_1^{(R)}$ in equation (6) and of receptor site normalities. In the esterase case, the complex AE in equation (7) is an intermediate which can also react along a hydrolytic pathway with such speed that equilibrium relaxation measurements are not possible.

However, a minimum estimate of $k_1^{(E)}$ in equation (7) is available from steady-state kinetic data.

For both receptor and esterase, the relaxation kinetics of specific fluorescent ligand binding at the active site have been studied. The fluorescent ligands used in these studies are shown in figure 5.

They are highly fluorescent when free in solution but totally quenched on binding to the active site, and this difference in fluorescence intensity allows monitoring of concentration changes during temperature-jump relaxation studies. These ligands appear to act as specific analogs of acetylcholine and have provided valuable information about bimolecular reaction rate constants for both proteins. AcChE and AcChR were isolated from electric fish (see, e.g., Neumann et al., [4]).

Acetylcholinesterase

The essential role of AcChE in bioelectrogenesis is manifested in a very high turnover number $k_{cat} = 1.6 \times 10^4$ s^{-1} at 0.1 M NaCl, pH 8 and 298 K. The isoelectric point of the eel protein is pI=4.5; thus, under normal conditions of pH 7 the protein is anionic. In order to explore ionic-electrostatic contributions to substrate binding fluorescent non-substrates were used, which bind specifically to the catalytic sites and are cations like acetylcholine, but are not hydrolyzed.

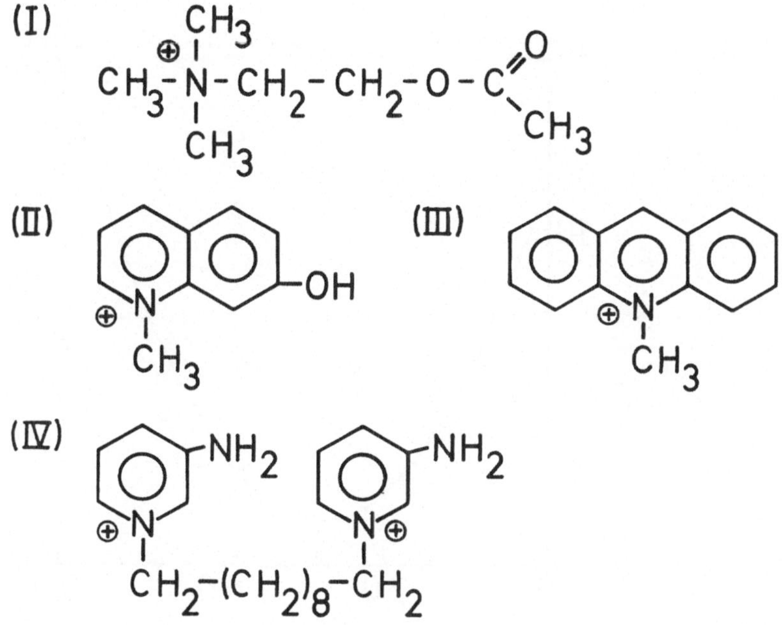

Figure 5: Chemical formula of acetylcholine (I), I-methyl-7-hydroxyquinolinium (II), N-methylacridinium (III), and bis(3-aminopyridinium)-1,10--decane (IV)

Two cationic ligands which have been used in temperature jump studies of AcChE are N-methylacridinium and 1-methyl-7-hydroxyquinolinium. For instance, the interaction of N-methylacridinium with acetylcholinesterase was analyzed according to equation (10) and is shown in figure 6. A key result of the relaxation kinetic studies is that the overall relaxation of AcChE and N-methylacridinium (and 1-methyl-7-hydroxyquinolinium) is bimolecularly controlled (Rosenberry and Neumann [25]). The reaction rate constants and equilibrium dissociation constants for both ligands are given in Table 1.

The observed bimolecular rate constants between 10^{10} and 10^{9} $M^{-1}s^{-1}$ are unusually high for enzyme ligand interactions. In addition, the association rate constants $k_{12} = k_1^{(E)}$ are very strongly dependent on the ionic strength, I_c, of the solution (Nolte, Rosenberry, Neumann [26]). Virtually the same strong I_c-dependence has been observed for a catalytic parameter proportional to k_{12} of acetylthiocholine, k_{cat}/K_{app}, a substrate whose structure and kinetic properties are very similar to those of acetylcholine.

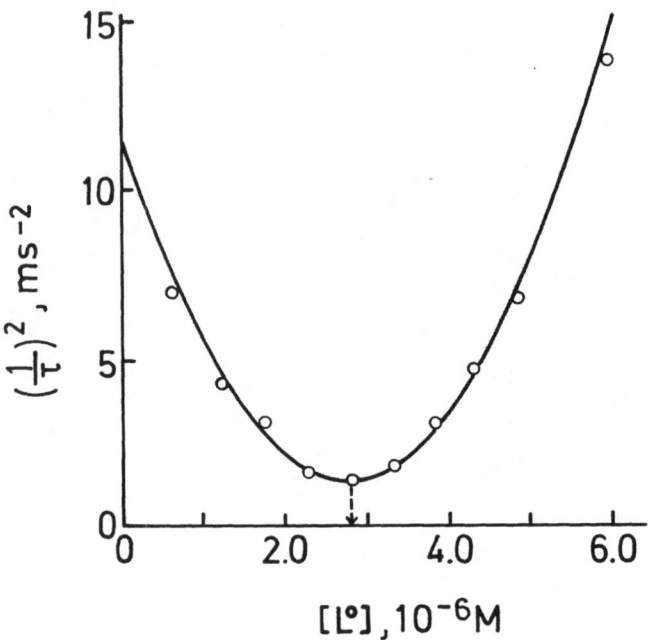

Figure 6: Dependence of the mean values of the relaxation times τ, observed with acetylcholinesterase and N-methylacridinium on total ligand concentration, [L°], in 0.1 M sodium phosphate at pH 8.0 and 23° C. Initial concentration of esterase active sites [E°] = 2.92 x 10⁻⁶ M. Data are plotted according to equation (10). The solid line was calculated from a weighted least-squares analysis based on a second-order polynomial in [L°]. From the position of the minimum, the value of [E°] = [L°]$_{min}$ - K = 2.94(±0.04) x 10⁻⁶ M can be obtained independently from the kinetic titration; it is consistently the same as the initial concentration of esterase sites determined by thermodynamic techniques (25)

Table 1: Equilibrium dissociation constants K and rate constants for
the interaction of acetylcholinesterase and cationic ligands.

Compound	$k_1(E)$, $M^{-1}s^{-1}$	K, M	$k_{-1}(E)$, s^{-1}
(I)	$1.6(\pm0.1)\times10^8$		
(II)	$2.18(\pm0.15)\times10^9$	$2.03(\pm0.3)\times10^{-7}$	4.4×10^2
(III)	$1.18(\pm0.03)\times10^9$	$1.49(\pm0.03)\times10^{-7}$	1.8×10^2

(I)   Acetylcholine in 0.1 M ionic strength at 25°C, pH 8.0;

(II)   1-Methyl-7-hydroxyquinolinium in 0.1 M sodium phosphate at 23°C, pH 7.0;

(III)   N-methylacridinium in 0.1 M sodium phosphate at 23°C, pH 8.0 (Rosenberry
and Neumann [25]).

The I_c-dependencies have been analyzed in terms of a Brønsted-Debye-Hückel
relationship as given in figure 7.

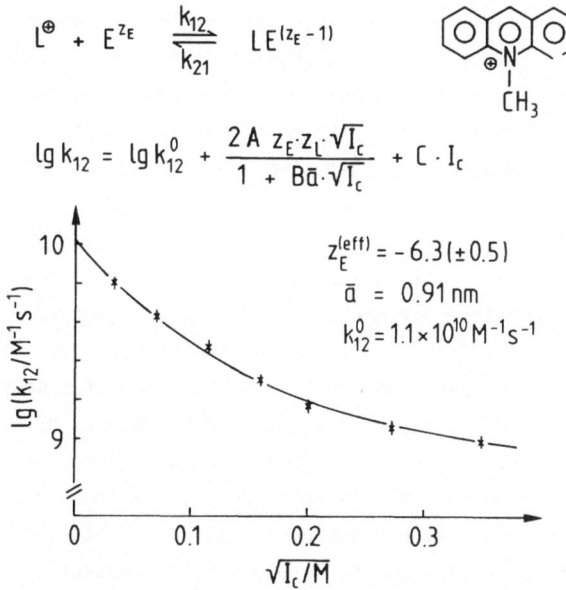

$$L^\oplus + E^{z_E} \xrightleftharpoons[k_{21}]{k_{12}} LE^{(z_E-1)}$$

$$lg\,k_{12} = lg\,k_{12}^0 + \frac{2A\,z_E\cdot z_L\cdot\sqrt{I_c}}{1 + B\bar{a}\cdot\sqrt{I_c}} + C\cdot I_c$$

$z_E^{(eff)} = -6.3(\pm0.5)$

$\bar{a} = 0.91\,nm$

$k_{12}^0 = 1.1\times10^{10}\,M^{-1}s^{-1}$

Figure 7: Ionic strength (I_c) dependence of the association rate constant
k_{12} of the equilibrium of electric eel AcChE and N-methylacridinium,
analyzed in terms of the semiempirical Brønsted-Debye-Hückel
equation giving $k_{12}(I_c\rightarrow0) = k_{12}^0$, the effective point charge
equivalent z_E of an enzyme active site and the approach distance
$\bar{a}$, (26)

The extended Debye-Hückel equation for the activity coefficient product involved in this relationship, contains the bulk dielectric constant ε of H_2O (at 298K, $\varepsilon=79$) and the charge numbers Z_i of the interacting ionic species; it has been found to describe not only the experimental data but also the rigorous Monte-Carlo results of electrolytes even with higher charge number (Pitzer [27]). We have chosen a Brønsted-Debye-Hückel relationship to evaluate the effective charge number Z_E as the point charge equivalent of the enzyme active site. The analysis results in $Z_E = -6.3(\pm 0.5)$. Thus at least six to seven monovalent anionic groups contribute to the comparatively large values k_{12} for both the non-substrate N-methylacridinium and acetylthiocholine ($k_{12}^6 = 0.42 \times 10^{10}$ $M^{-1}s^{-1}$), suggesting that an enzyme surface area larger than the ligand binding site itself is effective in trapping a ligand in encounter complexes. This larger surface area might include peripheral anionic sites from which ligand would move to the active site by surface diffusion. The high effective charge number supports this concept. The charged groups contributing to Z_E would be expected to be dispersed over an enzyme surface greater than the immediate catalytic site.

In summary, we may conclude that the high bimolecular association rate constants and the unusually strong ionic strength dependence of kinetic and thermodynamic parameters have its physical origin in a dominantly anionic surface structure of this enzyme. Physiologically, the polyionic enzyme acetylcholinesterase appears to be a powerful electrostatic sink for trapping and decomposing the acetylcholine cation.

Acetylcholine receptor

The smallest, functionally intact acetylcholine binding protein which can be decoupled under gentle experimental conditions from Torpedo fish, minimizing chemical modifications, is the H-form of receptor rich membrane fragments. According to Chang and Bock [28] this H-form of a molecular weight of roughly 400.000 d (Torp. cal.) is a dimer of two probably not identical monomers: the L- and the L'-form which are linked by intersubunit disulfide bond (see also Suarez-Isla and Hucho [29]; Hamilton, McLaughlin, and Karlin, [30,31]). Provided that endogenous lipids remain attached to the isolated receptor-lipid complex, the H-form exhibits essentially the same positive-cooperative acetylcholine binding isotherms as biomembrane fragments (Chang and Bock [32]).

Ca^{2+} and $AcCh^+$ binding. The isolated AcChR in detergent solution (probably largely in the L-forms) was found to bind large amounts of Ca^{2+} ions associated with high (µM) and low (mM) affinity sites (Eldefrawi, M.E., Eldefrawi, A.T., Penfield, O'Brien and van Campen [33]; Chang and Neumann [34]; Rübsamen, Hess, Eldefrawi [35]); see Table 2.

Table 2: Apparent equilibrium dissociation constants K_{Ca} and maximum number B° of Ca binding sites per 380,000 ($\pm$20,000) daltons of the isolated acetylcholine receptor from Torpedo california in 0.1 M NaCl, 0.05 M Tris.HCl, 0.1% Brij, pH 8.5 at 20°C.

Region	K_{Ca}, M	B°
(1)	$3.3(\pm 0.3) \times 10^{-4}$	$44(\pm 4)$
(2)	$2.5(\pm 0.5) \times 10^{-6}$	$34(\pm 4)$

The data were obtained from a Scatchard plot where two extended linear regions suggest at least two types of independent Ca^{2+} binding sites. The overall acetylcholine equilibrium constant of the receptor was 10^{-6}M (Neumann, Rosenberry, and Chang [4]).

There is competition of Ca^{2+} binding with other cations (see also Raftery, Vandlen, Reed, and Lee [36]). This competition has been used to estimate some thermodynamic and kinetic constants of acetylcholine binding to AcChR. It was found that upon binding of one AcCh-ion about 2 to 3 Ca^{2+} ions are released; subsequent binding of α-bungarotoxin causes reuptake of Ca^{2+} (Chang and Neumann [34]; see also Rübsamen, Hess, Eldefrawi [35]). See figure 8;

Relaxation kinetic studies have provided estimates for the (effective) rate constants of acetylcholine and of Ca^{2+} ion binding; figure 9 and Table 3. As seen, $k_1^{(R)} = 2.4 (\pm 0.5) \times 10^7$ $M^{-1}s^{-1}$ for AcCh suggests that the measured association rate constant is probably an effective, complex rate parameter involving several, more rapid elementary steps. In the simplest case of very rapid pre-equilibria the effective association rate constant for the reaction model in figure 10 would be $k_1(\text{eff, R}) = k_3 \cdot k_1 \cdot k_2 / ((k_{-1} \cdot k_{-2})$. The life time of the effective association is $(k_{-1}^{(R)})^{-1} = 7$ ms; see Table 3. The kinetic constants for acetylcholine so far characterize a receptor preparation with overall dissociation equilibrium constant $\bar{K}_A = 10^{-6}$ M (at 296 K, pH 8.5 and 0.1 M NaCl). This value is close to the acetylcholine concentration which causes the 'electrical half-response'. On the other hand, K_A for crude extracts, membrane fragments and recent receptor preparations where chemical modification could be largely reduced (Chang and Bock [32]), is between 10^{-8} and 10^{-9} M, most likely representing the acetylcholine affinity to the inactivated receptor. Moreover, the rate constants for the bimolecular overall reaction (Table 3) compare well with data from electric current relaxations (Sheridan and Lester [9]): $k_{on} \cong 10^7$ $M^{-1}s^{-1}$ and $k_{off} \cong 10^2$ to 10^3 s^{-1} depending on membrane potential. The electrophysiological data further indicate that at least two acetylcholine ions must bind in order to open a single permeation site. This may be related to the fact that the H-form of the isolated receptor protein has two binding sites for acetylcholine.

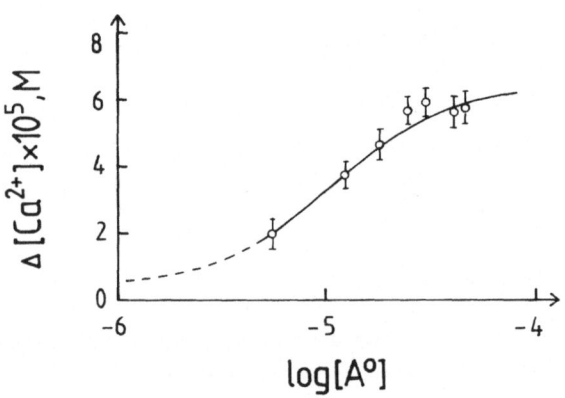

$$nA + RCa_x \rightleftharpoons A_nR + xCa$$
$$\alpha \equiv A_b/A_b^o = \Delta Ca/\Delta Ca^o \quad ; \; K_A = 10^{-6}M$$
$$\alpha[R_A^o] + \frac{\alpha}{1-\alpha} K_A = [A^o] \quad ; \; n = 2$$
$$x = 5(\pm 1)$$

Figure 8: The change in concentration of free Ca^{2+} ions reflecting release of bound Ca^{2+}, Ca_b^{2+} ($\Delta[Ca^{2+}] = -\Delta[Ca_b^{2+}]$), from isolated acetylcholine receptor of Torpedo cal. (in 0.1 M NaCl, 0.05 M Tris·HCl, 0.1% Brij, 0.0012 M Ca, pH 8.5, at 20°C) as a function of the total acetylcholine concentration $[A^o]$. The receptor concentration is 2.6 mg/ml

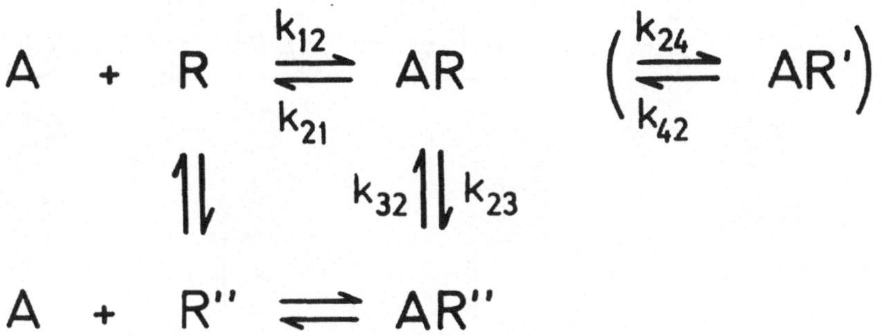

$$A \;+\; R \;\underset{k_{21}}{\overset{k_{12}}{\rightleftharpoons}}\; AR \qquad \left(\underset{k_{42}}{\overset{k_{24}}{\rightleftharpoons}}\; AR'\right)$$

$$\big\updownarrow \qquad\quad k_{32}\big\Updownarrow k_{23}$$

$$A \;+\; R'' \;\rightleftharpoons\; AR''$$

Figure 9: Reaction scheme for the interaction of acetylcholine with isolated
receptor from Torpedo cal. suggested by kinetic studies in which
the release of Ca^{2+} ions is used as an indicator for acetylcholine
binding. The observed main reaction path is $A + R \rightleftharpoons AR \rightleftharpoons AR''$, with
$[R'] \ll [R]$. The $AR \rightleftharpoons AR''$ relaxation is slow and well separated
from the initial binding step. The bimolecular rate constant
k_{on} is k_{12}, equivalent to $k_1(R)$ in equation (6); the dissociation
rate constant k_{off} is k_{21} equivalent to $k_{-1}(R)$ in equation (6)
or, if a step $AR \rightleftharpoons AR'$ is rapidly coupled to the observed bimolecular reaction, $k_{off} = k_{21}k_{42} (k_{24} + k_{42})^{-1}$ (see Neumann
and Chang [24])

Table 3: Interaction parameters for the isolated acetylcholine receptor from
Torpedo californica and acetylcholine and Ca ions in terms of a
direct competition of both ligands, for the main reaction path
$A + R$ AR AR^* (Neumann and Chang [24]; Neumann, Rosenberry and
Chang [4]).

$A + R \rightleftharpoons AR$	$AR \rightleftharpoons AR^*$	$Ca + R \rightleftharpoons CaR$
$k_1 = 2.4(\pm 0.5) \times 10^7 \; M^{-1}s^{-1}$	$k_2 = 43.5 \; s^{-1}$	$k_o = 10^8 \; M^{-1}s^{-1}$
$k_{-1} = 140 \; s^{-1}$	$k_{-2} = 6.5 \; s^{-1}$	$k_{-o} = 10^5 s^{-1}$
$K_1 = 0.6 \times 10^{-5} \; M$	$K_2 = 6.7$	$K_0 = 10^{-3} \; M$

Solvent is 0.1 M NaCl, 0.1% Brij, 1 mM Ca^{2+}, 0.05 M Tris·HCl, pH 8.5 at 296 K.
The overall equilibrium constant for acetylcholine is given by $\bar{K}_A =$
$K_1(1+K_o^{-1} \cdot c_{Ca}) (1 + K_2)^{-1} = 10^{-6}$ M, where $K_1 = k_{-1}/k_1$, $K_2 = k_2/k_{-2}$ and for
the Ca^{2+}-binding $K_0 = k_{-o}/k_o$.

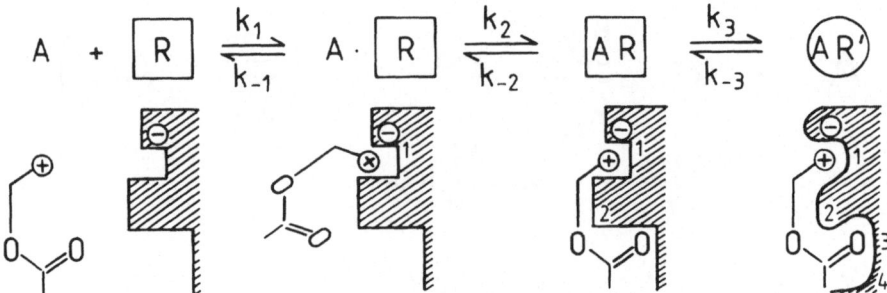

Figure 10: Chemical model for the induced conformational change in AcChR:
step 1, encounter ion pairing with $K_1 = k_{-1}/k_1$ 0.1 M,
$k_1 \geq 10^8$ $M^{-1}s^{-1}$ and $k_{-1} \cong 10^7$ s^{-1}; step 2, contact with a
second site in the life-time of the ion-pairing; step 3, conformational change to the permeable state induced by steps 1
and 2. The state AR' may involve a 'distorted' AcCh

Among the various points of comparisons, in particular the similarity between
$k_{on} \cong 10^7$ $M^{-1}s^{-1}$ from current relaxations and $k_1^{(R)} = k_1(eff) = 2.5(\pm 0.5) \times$
10^7 $M^{-1}s^{-1}$ from studies on isolated receptors suggests that the rate of
coupled Ca^{2+} release (upon binding of acetylcholine) not only reflects the
overall rate of effective acetylcholine binding to isolated receptor, but
is also characteristic for the rate-limiting step in the conductivity increase of the membrane. Thus, stoichiometry of acetylcholine binding,
equilibrium and rate constants suggest the low affinity receptor with
$K_A \cong 10^{-6}$ as a candidate for the in vivo metastable, conducting receptor
conformation, which by chemical modification (sulfhydryl-disulfide redox
reactions) during isolation may be stabilized in detergent solution.

In contrast to the relatively low value for acetylcholine, the measured
association rate constant for the binding of the dicationic inhibitor
bis(3-aminopyridinium)-1,10-decane (DAP) to isolated Torp.-marm. AcChR
is $k_{ass} = 1.2 \times 10^8$ $M^{-1}s^{-1}$ in 0.1 M ionic strength, pH 7.0 and 293 K.
The ionic strength dependence of the association rate of k_{ass}(DAP) suggests
an effective charge of $-3(\pm 1)$ on the binding site of the protein (Neumann
et al. [4]). This value is somewhat less negative than that indicated for the
esterase. But it seems that in both proteins of the permeability control
system there are larger electrostatic contributions to the rate with which
cationic ligands like acetylcholine are bound.

The estimates for biomolecular rate constants k_1(eff) of AcChR-ligand binding appear to depend strongly on the type of ligand used; for decamethonium 2×10^8 $M^{-1}s^{-1}$ and for carbamylcholine and acetylcholine $\cong 10^7$ $M^{-1}s^{-1}$ (Sheridan and Lester [9]) for suberyldicholine 0.98×10^7 $M^{-1}s^{-1}$ (Barrantes [37]), for NBD-5-acylcholine $\cong 10^8$ $M^{-1}s^{-1}$ (Jürss, Prinz and Mealicke [38]).

AcChR-lipid complexes. Studies with the isolated AcChR-lipid complexes from Torp. cal. confirm that the Ca^{2+}-binding isotherm is essentially two-phasic with equilibrium constants in the μM and mM range (Chang and Neumann [34], Neumann et al. [4]) suggestive for intracellular and extracellular Ca^{2+} sites in AcChR. If this complex is incorporated in lipid vesicles and then trans-ferred via a surface monolayer into planar lipid bilayer, 'specific' con-ductivity changes inclusively inactivations are evoked upon addition of carbamylcholine. Inhibitors like d-tubocurarine and α-bungarotoxin block the membrane (Spillecke, Schindler and Neumann, 1981). It thus appears that the H-form of AcChR (Trop. cal.) with its subunits α(40,000d), β(48,000d), γ(59,000d) and δ(67,000d) contains not only the AcCh binding subunits (Karlin, Weill, McNames, Valderama [39]) but also the subunits forming, after structural changes, the ion transporting channel.

Ca^{2+} ions may be involved to preserve stability of the protein-lipid complex and may also bind to the anionic groups of the channel subunits. A large number of anionic, probably carboxylate, groups are suggested by the large Ca^{2+} ion binding capacity of AcChR. Provided that the density of the anionic charges exceeds a certain value, divalent ions like Ca^{2+} are preferentially bound. If AcCh induces a structural change which increases the average distance between the charged groups, this polyelectrolyte preference for Ca^{2+} ions would be lost and ion exchange with, say, Na^+ ions could occur [40]. The 'indicator' of 'effective binding' of AcCh to AcChR would therefore be the allosteric release of bound Ca^{2+}. However, prolonged exposure to receptor activators leads to definitely allosteric uptake of Ca^{2+}-ions; see also Sugiyma and Changeux [41]).

Transient release of Ca^{2+} ions followed by uptake of Ca^{2+} upon addition of activators to AcChR suggest at least one metastable state in vitro, parallel to the electrophysiologically indicated in vivo, metastability for the con-ducting channel configuration, and parallel to the metastability suggested for the permeable state in sealed biomembrane vesicles (Bernhardt and Neumann [42], and Neumann [1]).

Asymmetry in AcChR. Bulger and Hess [43] found that two types of binding sites
in membrane-bound AcChR are apparent; this non-equivalence is induced by
α-bungarotoxin and leads to interconversion of the sites (Bulger, Fu, Hindy,
Silberstein and Hess [44]), which in unbound state must not necessarily be of
different type. Binding studies with DAP and α-bungarotoxin have also been
interpreted in terms of two classes of binding sites (Raftery, Vandlen, Reed
and Lee [36]). In the same line are data of Damle et al. [45] and of Delegeane
and McNamee [46]. Structurally, there appears a distinct asymmetry in the results
of digital image processing of electron micrographs of Torp. marm. membrane
fragments (Zingsheim, Neugebauer, Barrantes and Frank [47]).

These data suggestive for intrinsic asymmetry of AcChR structure and function
invites speculation on a possible functional role. If the 'monomers' of the
dimeric H-form have asymmetric subunit compositions, say an L-form with $\alpha_2\beta\gamma\delta$,
and an L'-form with $\alpha_2\gamma\beta\delta$ stoichiometry, a molecular weight of about 250,000 d
for each monomer and of about 500,000 d for the asymmetric H-form would result,
a value not inconsistent with recent estimates (Raftery, Vandlen, Reed and
Lee [36]; Reynolds and Karlin [48]), where, however, the 43,000 d chain appears
to be included (see e.g. Wennogle and Changeux [49]). Functionally, the L and
L'-forms are candidates for separate channels for Na^+ and K^+, both, however,
controlled by the binding of at least two acetylcholine ions.

These characteristic features of the dimeric AcChR gating protein are only
covered by more complicated reaction schemes. In the light of the previous
arguments of a dimeric channel structure, it appears appropriate to formally
express the experimental complexity in terms of the AcChR dimer, see figure 11;
the scheme is analogous to a general scheme developed by Eigen [23] for a
trametic subunit system.

To the extent to which data on isolated proteins can be used to extrapolate
to the cellular level, it is very tempting to compare the effective association
rate constant of the receptor-acetylcholine interaction $k_1(eff)$ = 2.4($\pm$0.5) x
10^7 $M^{-1}s^{-1}$ with that of the enzyme k_1 (eff) = k_{12} ⩾ 2 x 10^8 $M^{-1}s^{-1}$, probably
10^9 $M^{-1}s^{-1}$. It is readily recognized that the inequality (1) suggested by
electrophysiological data is parallel to $k_1(eff,R) < k_1(eff,E)$ found for the
isolated proteins. The physical reason for the lower k-value of AcChR may
reside in the pre-equilibria preceding the step $(AR)_2 \rightleftharpoons (AR')_2$; see figures
10 and 11. Since in the minimum reaction scheme for the enzyme acylation (see,
e.g. Neumann et al.[4])

$$A + E \underset{k_{21}}{\overset{k_{12}}{\rightleftharpoons}} E \cdot A \overset{k_2}{\longrightarrow} E \cdot acetyl + CH \rightarrow \ldots..$$

$k_2 \cong 10^5$ s^{-1} and $k_{21} \cong 10^4$ s^{-1} thus $k_2 > k_{21}$, the estimation of $k_{12} = k_1$(eff,E) is adequate for a comparison with k_1(eff,R).

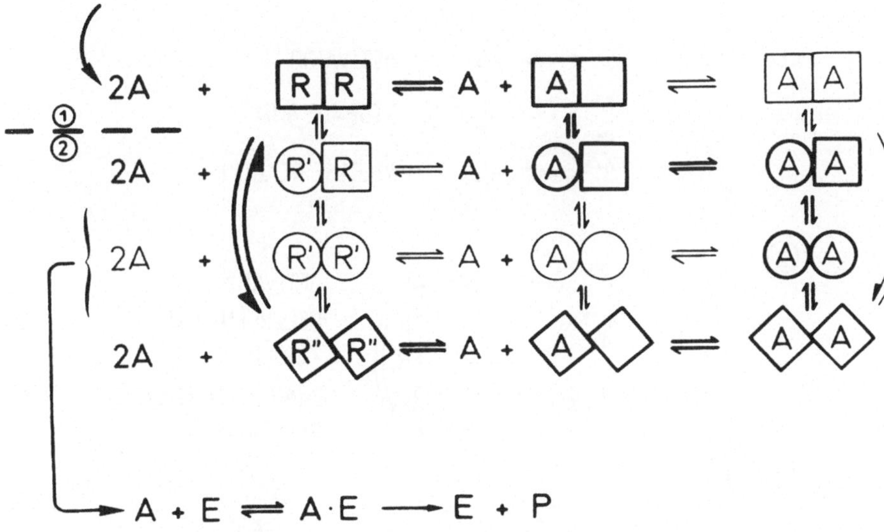

Figure 11: Chemical representation of acetylcholine (AcCh)-induced activation and inactivation of acetylcholine receptors (AcChR) in terms of a basically sequential processing of AcCh from a microreaction space 1 to a microreaction space 2 where acetylcholinesterase (AcChE) has fully competitive access to AcCh. The receptor states, in terms of activator binding sites, correspond to R_2(closed), R_2' (open) and R_2'' (desensitized). The main reaction pathways are drawn in thick symbols. For neurally evoked AcCh the changes along the desensitization pathways $R_2 \rightleftharpoons R''_2$ and $(AR')_2 \rightleftharpoons (AR'')_2$ appear to be uninvolved

5. Axonal gating

The existence of basic similarities between synaptic and axonal parts of excitable membranes has been frequently discussed; see, e.g., Nachmansohn and Neumann [50], Neumann and Bernhardt [10]. In particular, the conducting conformation of the Na$^+$ ion channel is clearly a metastable, short-lived state as appears to be the case with the permeable AcChR conformation (Neumann [1]). In a recent study it was found that kinetic models which can successfully simulate the ion-permeability features of axonal Na$^+$ channels, suggest the presence of bimolecular reaction steps in the activation of channels (Dorogi and Neumann [2]). The implied chemical formalism is highly suggestive of an activator-controlled gating system with strong similarities to the acetylcholine regulated ion transport systems; see figure 12.

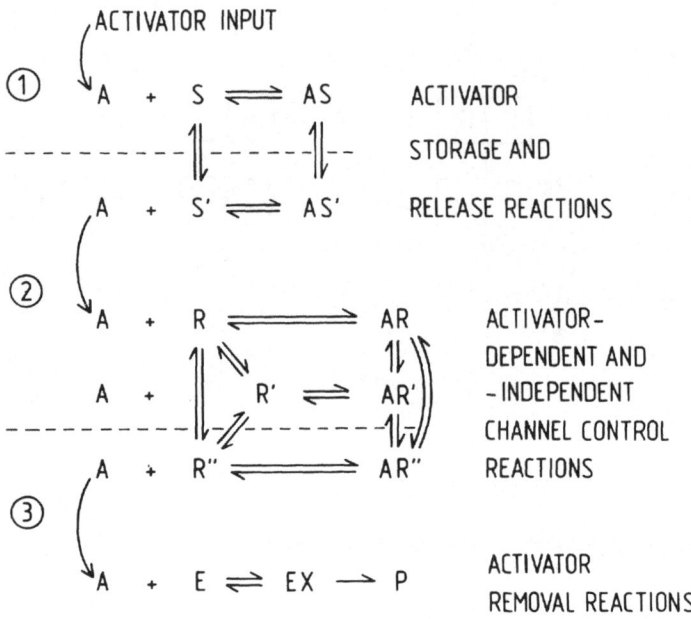

Figure 12: Overall chemical gating model for the axonal Na$^+$ channel, showing an essentially sequential processing of activator through three reaction spaces during a maintained depolarization. R is the activatable-closed state, R' and AR' are activated-conducting states, and R" and AR" correspond to inactivated-closed states. During a normal action potential the reactions in the micro-reaction space 3 appear uninvolved, providing minimum dissipation for the activator (Dorogi and Neumann [2])

Conformational changes which underlie the ion conductance changes are suggested to possess a greater sensitivity to the membrane field in axonal than at synaptic parts of excitable membranes. This allows axonal permeability changes to be regulated energetically more conservatively. Axonal K$^+$ channels with delayed activation kinetics would serve to reverse the increase in membrane permeability to Na$^+$ ions with a minimum of chemical dissipation (Dorogi and Neumann [2]).

In summary, an AcCh-mediated or fundamentally similar, chemical system is proposed as a plausible candidate for the regulation of axonal permeability changes leading to the action potential. Thus rapid bioelectrical signals based on transient permeability changes in axonal and synaptic parts of excitable biomembranes appear to be specialized cases of a more general chemically dissipative control principle involving activator-receptor interactions and structural metastability for the activated, ion conducting state.

Acknowledgement. The financial support of the Deutsche Forschungsgemeinschaft, grant NE 227, is gratefully acknowledged.

References

1. E. Neumann, In Balaban, M. (Ed.) Molecular Mechanisms of Biological Recognition. Elsevier, Amsterdam, 1979, pp. 449 - 63.

2. P.L. Dorogi and E. Neumann. Proc. Natl. Acad. USA, 77.

3. D. Nachmansohn. Chemical and Molecular Basis of Nerve Activity. Academic Press, New York, 1959, pp. 235.

4. E. Neumann, T.L. Rosenberry, and H.W. Chang, In A. Karlin, V.M. Tennyson, and H.J. Vogel. (Eds.) Neuronal Information Transfer. Academic Press, New York, 1978, pp. 183 - 210.

5. P.W. Gage. Physiol. Rev., 56, 177 - 247 (1976).

6. P.W. Gage and R.N. McBurney. J. Physiol., Lond., 244, 385 - 407 (1975).

7. Ch.F. Stevens. Cold Spring Harbor Symp. Quant. Biol., 40, 169 - 173 (1976).

8. T.L. Rosenberry. Biophys. J., 26, 263 - 289 (1979).

9. R.Z. Sheridan and H.A. Lester. J. Gen. Physiol., 70, 187 - 219 (1977).

10. E. Neumann and J. Bernhardt. Ann. Rev. Biochem., 46, 117 - 41 (1977).

11. B. Katz and R. Miledi. J. Physiol., Lond., 231, 549 - 74 (1973).

12. E. Neher and B. Sakmann. Proc. Natl. Acad. Sci. 72, 2140 - 45 (1975).

13. Ch.F. Stevens. Ann. Rev. Physiol., 42, 643 - 53 (1980).

14. B. Katz and S. Thesleff. J. Physiol., Lond., 138, 63 - 80 (1957).

15. H.P. Rang and J.M. Ritter. Mol. Pharmacol., 6, 357 - 382 (1970).

16. G.E. DeMotta and J. del Castillo. Nature, 270, 178 - 180 (1977).

17. A. Takeuchi and N. Takeuchi. Adv. in Biophys., 3, 45 - 95 (1972).

18. P.W. Gage and D. van Helden. J. Physiol., Lond., 288, 509 - 28 (1979).

19. K.L. Magleby and M.M. Weinstock. J. Physiol. Lond., 299, 203 - 18 (1980).

20. W.L. Nastuk and R.L. Parsons. J. Gen. Physiol., 56, 218 - 49 (1970).

21. A.A. Manthey. J. Membrane Biol., 9, 319 - 40 (1972).

22. J.F. Fiekers, P.M. Spannbauer, B. Scubon-Mulieri and R.L. Parsons. J. Gen. Physiol., 75, 511 - 529 (1980).

23. M. Eigen. Nobel Symp., 5, 333 - 367 (1967).

24. E. Neumann and H.W. Chang. Proc. Natl. Acad. Sci. USA, 73, 3994-98 (1976).

25. T.L. Rosenberry and E. Neumann. Biochemistry, 16, 3870 - 78 (1977).

26. H.-J. Nolte, T.L. Rosenberry and E. Neumann. Biochemistry, 19, 3705-11, (1980).

27. K.S. Pitzer. Acc. Chem. Res. 10, 371 - 77 (1977).

28. H.-W. Chang and E. Bock. Biochemistry 16, 4513 - 20 (1977).

29. B.A. Suarez-Isla and F. Hucho. FEBS Lett., 75, 65 - 69 (1977).

30. S.L. Hamilton, M. McLaughlin and A. Karlin. Biochemistry 18, 155-163 (1979).

31. S.L. Hamilton, M. McLaughlin and A. Karlin. Biochem. Biophys. Res. Commun., 79, 692; (1977).

32. H.-W. Chang and E. Bock. Biochemistry, 18, 172 - 79 (1979).

33. M.E. Eldefrawi, A.T. Eldefrawi, L.A. Penfield, R.D. O'Brien and E.van Campen. Life Sciences, 16, 925 - 36 (1975).

34. H.-W. Chang and E. Neumann. Proc. Natl. Acad. Sci. USA, 73, 3364-68 (1976).

35. H. Rübsamen, G.P. Hess, A.T. Eldefrawi and M.E. Eldefrawi. Biochem. Biophys. Res. Commun., 68, 56 - 63 (1976).

36. M.A. Raftery, R.L. Vandlen, K.L. Reed and T. Lee. Cold Spring Harbor Symp. Quant. Biol., 40, 193 - 202 (1976).

37. F.J. Barrantes. J. Mol. Biol., 124, 1 - 26 (1978).

38. R. Jürss, H. Prinz and A. Maelicke. Proc. Natl. Acad. Sci. USA, 76, 1064 - 68 (1979).

39. A. Karlin, C.L. Weill, M.G. McNamee and R. Valderrama. Cold Spring Harbor Symp. Quant. Biol., 40, 203 - 210 (1976).

40. E. Neumann, A. Katchalsky and D. Nachmansohn. Proc. Natl. Acad. Sci., USA, 70, 727 - 31 (1973).

41. H. Sugiyama and J.-P. Changeux. Eur. J. Biochem., 55, 505 - 15 (1975).

42. J. Bernhardt and E. Neumann. Proc. Natl. Acad. Sci. USA, 75, 3756 - 60 (1978).

43. J.E. Bulger and G.P. Hess. Biochem. Biophys. Res. Commun., 54, 677 - 84 (1973).

44. J.E. Bulger, J.L. Fu, E.F. Hindy, R.L. Silberstein and G.P. Hess. Biochemistry, 16, 684 - 692 (1977).

45. V.N. Damle, M. McLaughlin and A. Karlin. Biochem. Biophys. Res. Commun. 84, 845 - 51 (1978).

46. A.M. Delegeane and M.G. McNamee, Biochemistry, 19, 890 - 95 (1980).

47. H.P. Zingsheim, D.-Ch. Neugebauer, F.J. Barrantes and J. Frank. Proc. Natl. Acad. Sci. USA, 77, 952 - 56 (1980).

48. J.A. Reynolds and A. Karlin. Biochemistry. 17, 2035 - 38 (1978).

49. L.P. Wennogle and J.P. Changeux. Eur. J. Biochem., 106, 381 - 93 (1980).

50. D. Nachmansohn and E. Neumann. Chemical and Molecular Basis of Nerve
     Activity. Rev. Academic Press, New York, 1975, pp. 403.

X-RAY FINE STRUCTURE ANALYSIS OF BIOPOLYMERS

OTTO   KRATKY

Institute for X-Ray Fine Structure Research,
Austrian Academy of Science,
Steyrergasse 17, A-8010 Graz, Austria

Abstract. Three kinds of systems were studied:
1. Homodisperse solutions
   From the particulate scattering at very small angles important parameters
   of the macromolecules can be derived. A survey of the theory and selected
   examples for the application are given.
2. Crystals
   Supposed the availability of suitable crystals the reward of a crystal
   structure determination is considerable. The utmost is to get the whole
   of the crystal at essentially atomic resolution.
3. Fibers
   Biological fibers consist of parallel, chain-like molecules. They can be
   fully extended or have a complex shape, for example a helix.

Introduction

The subject of this lecture will be the study of three kinds of systems.

1. Homodisperse solutions of corpuscular biological macromolecules.

2. Crystals of macromolecules, in particular of proteins.

3. Fibers of chain-like biological molecules.

The resolution which can be achieved in the determination of a molecular
structure by scattering experiments depends essentially on the degree of

order of the system. In this respect, the best conditions are of course
those of the cristalline state.

Unfortunately a large group of biological macromolecules cannot be crystallized, whereas they can be obtained in the form of a homodisperse solution.
This state offers also the possibility to study problems connected with the
solute state, like dissociation and association phenomena, reactions with
other molecules, changes of the conformation by solvent or temperature, etc.
Therefore, the investigation of the dissolved state is of increasing importance, and we shall begin a treatment of structural studies on solutions of
macromolecules.

1.Homodisperse solutions of corpuscular biological macromolecules [1-10]

In all scattering phenomena, the observed scattering angles are inversely
related to the dimensions of the scattering particles. As macromolecules
are large compared with the generally used CuK_α wavelength ($\lambda=1,54$ Å),
we observe the scattering of such particle at correspondingly small angles,
we observe "Small Angle Scattering".

In diluted solutions, the scattered intensities from individual molecules
simply add up, so that the scattering curve, which is a plot of the scattered
intensity I versus the scattering angle 2θ, in homodisperse solutions charac-
terizes the individual particle; we speak of particulate scattering.

We will now try to explain, what kind of parameters can be derived from
this scattering.

According to Guinier,[1] the innermost part of the scattering curve has always a gaussian shape. It follows that a plot of ln I versus the square of
the scattering angle $(2\theta)^2$ (Guinier-plot) gives a descending line, the slope
of which equals $- KR^2$, R being the radius of gyration. ($K = \dfrac{4\pi^2}{\lambda^2} \cdot \dfrac{1}{3}$) R
is defined as the root mean square distance of all electrons from the center
of gravity, and it corresponds formally to the radius of inertia in mechanics.
Generally speaking, it is a measure for the extension of the particle in space.

Frequently, R alone can yield important information. This, for example, is
the case in Phenylalaninspecific t-RNA. (Figure 1) We see that, with increasing temperature, the radius of gyration increases, beginning at 40° C.
This increase in R is due to the fact, that the bonds of the tightly folded
particle begin to open. We say the molecule melts. At the same time the

absorption coefficient also increases, which is frequently used as a qualitative indication for the same phenomenon. The small angle measurement shows quantitatively to what extent the opening process has progressed.

Another important relationship is the one of Porod: [1,2] The volume V of a dissolved partice is given by

$$V = K \frac{I_o}{\int_0^{\infty} I(2\theta)^2 d(2\theta)} \qquad (K = \frac{\lambda^3}{4\pi}) \qquad (1)$$

The volume too can occasionally give a valuable information alone.

An interesting example is the saturation of the apoenzyme of yeast glyceraldehyde-3-phosphate dehydrogenase with the coenzyme NAD. It was found that complete saturation causes a contraction of the volume by 7 % (Figure 2).

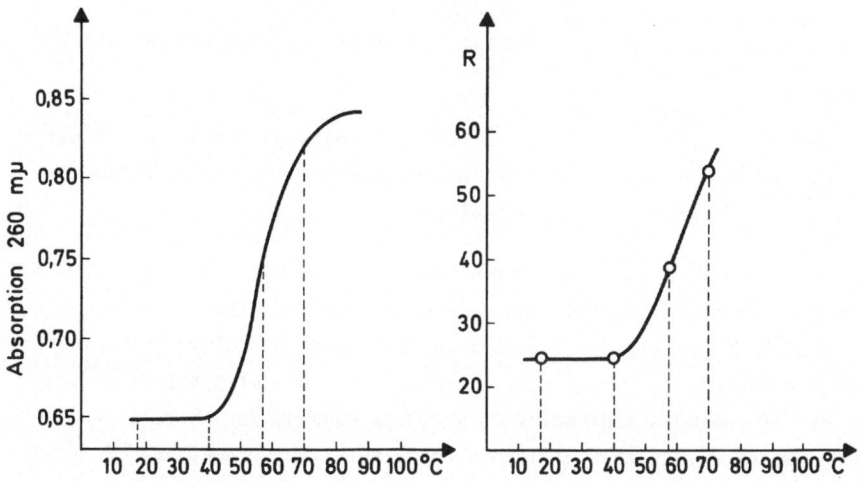

Figure 1: Melting curve of t-RNA [Phe] (yeast). Dependance of the radius of gyration (right) and of the absorption (left).
(I. Pilz, O. Kratky, F. Cramer, F. v.d. Haar, E. Schlimme, Eur. J. Biochem. 15, 401 (1970))

We see, that there is not a linear relationship between the degree of saturation and the volume contraction. This result clearly supports an allosteric mechanisms, and disfavours a sequential one.

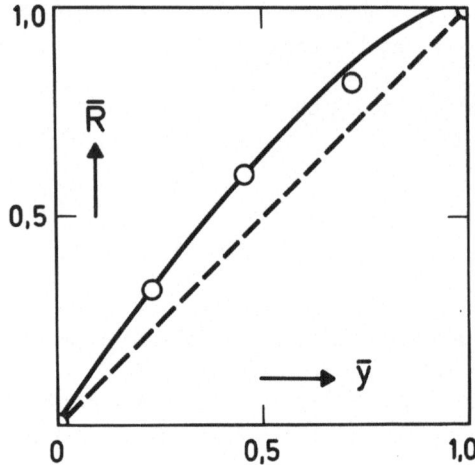

Figure 2: Volume contraction of glyceraldehyde-3-phosphate dehydrogenase
as a function of saturation with NAD. $\bar{y}$ = degree of saturation
with NAD; $\bar{R}$= degree of relative volume contraction.
(H. Durchschlag, G. Puchwein, O. Kratky, I. Schuster, K. Kirschner,
Eur. J. Biochem. 19, 9 (1971))

We see, that there is not a linear relationship between the degree of satura--
tion and the volume contraction. This result clearly supports an allosteric
mechanisms, and disfavours a sequential one.

From a comparison of R and V we obtain a measure for the anisotropy, similar
the unsymmetry factor obtained from ultracentrifuge studies.
The ellipsoids of rotation in Figure 3 may all have the volume determined
according to formula 1. The arrows indicate the radii of gyration of these
ellipsoids. If the measured radius of gyration equals, for example, that of
Figure 3b, we know that this ellipsoid represents the correct degree of anisotropy.
A better approach to obtain the overall shape is as follows: One compares
the experimental scattering curve with a collection of theoretically calculated curves of various three-axial bodies and selects the one which fits
best. The selected shape is considered as a reasonable approximation, and
we say the particle is equivalent in scattering. (Figure 10 shows such an
example.)
The small angle method can be successfully applied to weigh particles.
From the quotient of the scattered intensity extrapolated to zero angle

and the primary energy we obtain the particle molecular weight.

$$\frac{I_o}{P_o} = K M \tag{2}$$

The constant K contains several experimental quantities, such as the concentration, the partial specific volume etc.

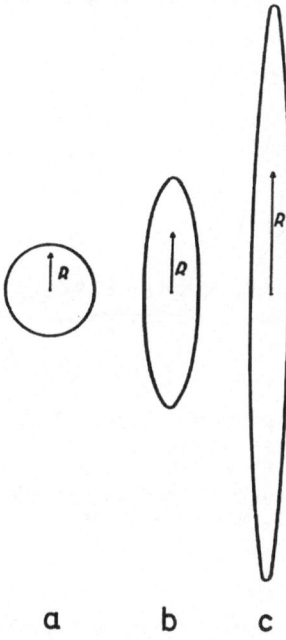

a   b   c

Figure 3: Sphere and ellipsoids of rotation of equal volume. The radius of
gyration R is enhanced with increasing anisotropy.
(O. Kratky, Ang. Chem. 72, 467 (1960))

An example for which these possibilities were combined is a Haemocyanin from
the blood of the snails. The molecular weight is M = $9 \cdot 10^6$, the shape analysis
yields to a hollow cylinder. The intensity ratio of the first side maximum
relative to the main maximum yielded, according to Figure 4, the ratio between
inner and outer radius to ∿ 0.45.

The determination of the substructure in based on the following fact: When a
particle is composed of many sphere-like subunits of the same size, the
distances of neighbouring units can be found from the position of certain
minima. For theoretical reasons, these minima are expected at relatively

large angles.

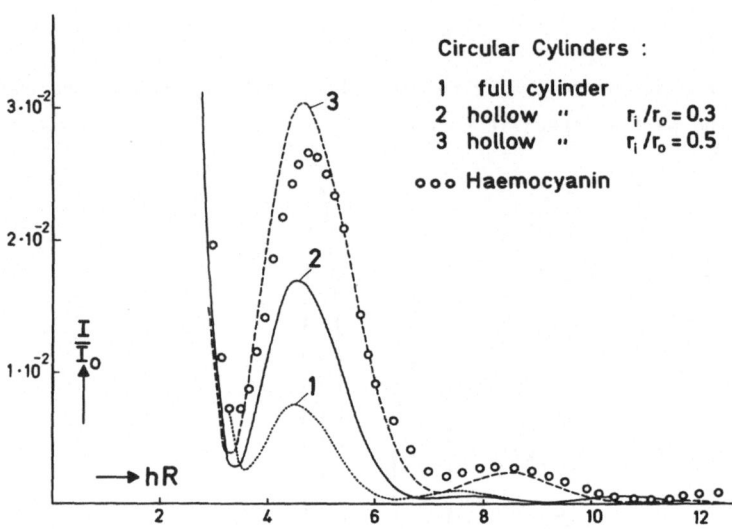

Figure 4: Outer portion of the scattering curve of haemocyanin helix pomatia
   compared with theoretical curves of circular cylinders. Curve 1,
   full cylinder; curve 2, hollow cylinder (inner radius to outer
   radius r_i/r_a = 0,3); curve 3, hollow cylinder (r_i/r_a= 0,5).
   (I. Pilz, O. Kratky, I. Moring - Claesson, Z. Naturfschg. 25b,
   600 (1970))

In the case of Haemocyanin the minima in Figure 5 yield a diameter of the
sphere-like subunits of 40 $\overset{o}{A}$.

The resulting model consists of 360 sphere-like subunits, as shown in
Figure 6. Figure 7 shows the good agreement of the theoretical scattering
curve with the experimental one, which supports the validity of this
model.

Many interesting macromolecules are of elongated shape. In this case the
scattering curve can be split into two factors, the cross-section factor
I_c and a factor 1/ 2θ

$$I = I_c \cdot \frac{1}{2\theta} \tag{3a}$$

1/ 2θ is the scattering curve of a gas of infinitely long and infinitely thin
needles. If the needles are not infinitely thin, 1/2θ has to be multiplied
with a factor I_c, which depends on the cross-section.

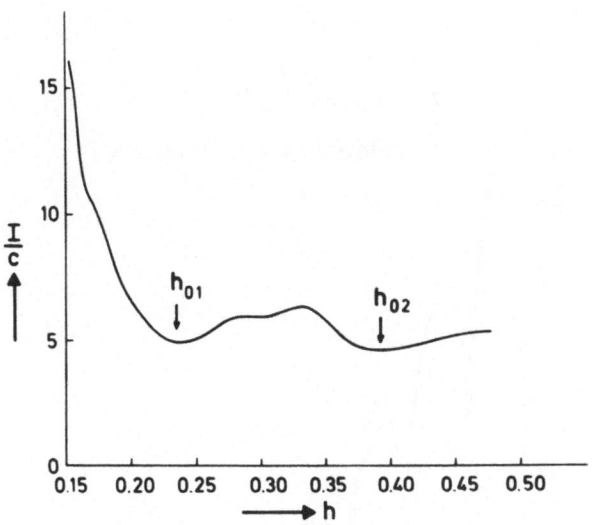

Figure 5: Tail-end of the scattering curve of haemocyanin helix pomatia.
(I. Pilz, O. Glatter and O. Kratky, Z. Naturfschg. 27b, 518 (1972))

Figure 6: Model for haemocyanin helix pomatia, composed of 360 identical
sherical subunits.
(I. Pilz, O. Glatter, O. Kratky, Z. Naturfschg. 27b, 518 (1972))

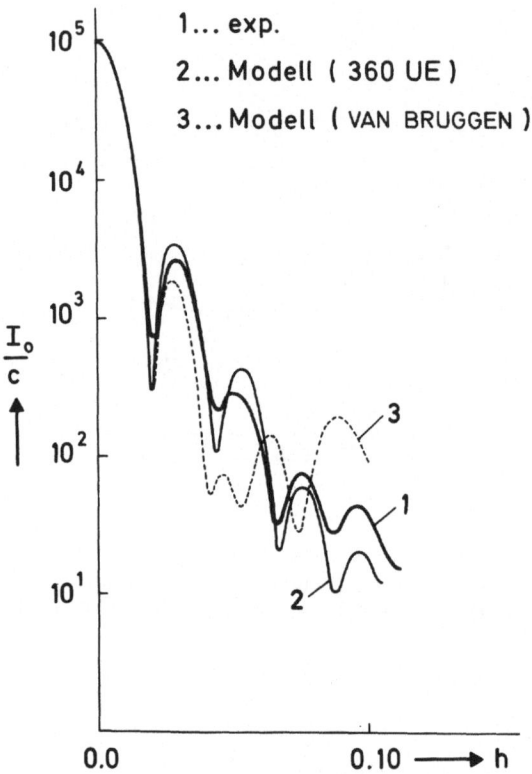

1... exp.

2... Modell (360 UE)

3... Modell (VAN BRUGGEN)

Figure 7: Comparison of the scattering curve of haemocyanin helix pomatia
(1, thick line) and theoretical curves (2 and 3). Curve 2 (thin
line) is the scattering curve of the model shown in Figure 6.
(I. Pilz, O. Glatter, O. Kratky, Z. Naturfschg. 27b, 518 (1972))

It can be determined by multiplication of (3a) with θ:

$$I_c = I \cdot (2\theta) \qquad\qquad (3b)$$

The slope of a Guinier plot of I_c yields the radius of gyration of the cross
section

$$R_c \left(\text{slope} = -\frac{2\pi^2}{\lambda} \cdot \frac{1}{2} \cdot R_c^2\right)$$

Moreover we find the weight per unit length $M/1\text{Å}$ by substituting I_p by $(I \cdot \theta)_0$
in equation (2). The small angle method is the only technique which can
"weigh" the mass per unit length of a rod-like particle.

A nice example illustrating the application of this relationship is the Glutamate Dehydrogenase. It has been known that, with increasing concentration, monomeric particles of this enzyme form oligomers. This reaction can be thoroughly characterized by the following three steps.

a) From the outer portion of the scattering curves we find a concentration-independent cross-section factor corresponding to a $R_c \approx 30$ Å. (Figure 8) From the absence of a concentration-dependance we conclude that the association occurs in the longitudinal direction.

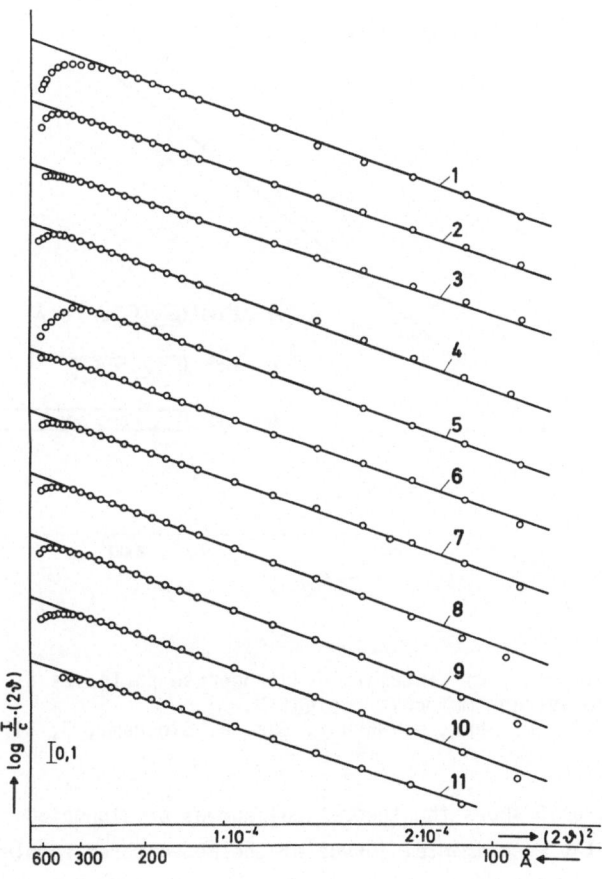

Figure 8: Cross-section factors of glutamate dehydrogenase. The concentration varies from 1 mg/ml (curve 1) to 33 mg/ml (curve 11).
(I. Pilz, H. Sund, M. Herbst, Eur. J. Biochem. 7, 517 (1969))

b) On the other hand, we observe from the inner portion of the scattering curve an increasing radius of gyration with increasing concentration. From R and R_c we can calculate the length L of the particles according to

$$L = \sqrt{12(R^2 - R_c{}^2)} \qquad\qquad (4)$$

The direct proportionality between this length and the independently determined molecular weight is shown in Figure 9. Again, this confirms the idea, that the particles associate in the longitudinal direction. The individual particles studied have length ratios and molecular weight ratios of 1 : 2 : 4

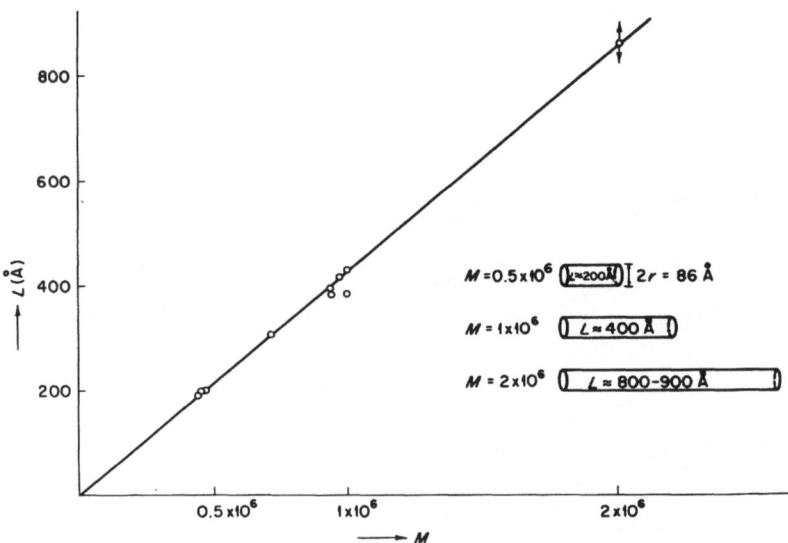

Figure 9: Mean length L of glutamate dehydrogenase particles as function of the average molecular weight M.
(I. Pilz, H. Sund, M. Herbst, Eur. J. Biochem., 7, 517 (1969))

c) Finally Figure 10 shows the theoretical curves of elongated cylinders compared to the experimental curves of the above three samples. The axial ratios derived from this comparison are 2,4, 4,8 and 10. It is very gratifying that the curve fitting yields exactly the same lengths ratios 1 : 2 : 4 found from application of (4).

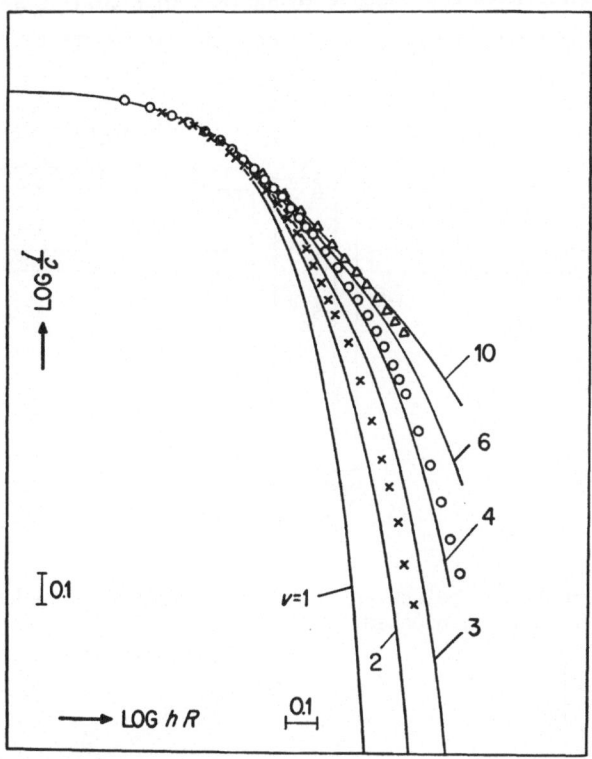

Figure 10: Theoretical scattering curves of circular cylinders (full lines).
The ratio v of length (L) to the diameter (2r) of the cylinder
(v = L/2r) is 1, 2, 3, 4, 6 and 10. Experimental scattering curves
of glutamate dehydrogenase particles of the following approximate
molecular weight: $0,5 \times 10^6$ (X), 1×10^6 (0), 2×10^6 ($\triangle$).
(I. Pilz, H. Sund, M. Herbst, Eur. J. Biochem. 7, 517 (1969))

So far, we have discussed particles whose shape can be approximated with a
two - or triaxial body. Now we give an example for a more complex shape:

Myeloma-Globulin is an antibody belonging to the γ-globulins. Its cross-
section factor curve shows a distinct break. Starting from the chemical
evidence given by Edelman (Figure 11), we have calculated theoretical
scattering curves of the bodies, shown in Figure 12.

We see, that a molecule with the two "wings" running in one direction fits
the experimental scattering curve best. This is in agreement with Edelman,
but in contradiction to electron microscopic results, which yielded a
Y-shaped body. Later, a crystal structure analysis of the same molecule was

carried out (Figure 13). The "wings" from the crystal structure are shorter than in our model. There are reasons to believe that this is a real difference in conformation between the crystalline and the solute state.

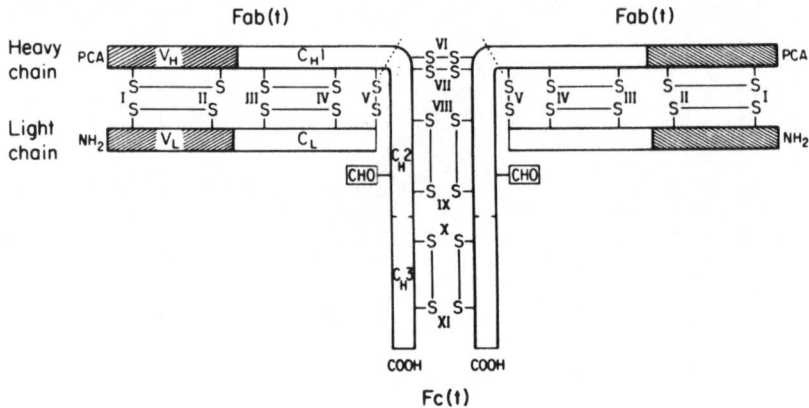

Figure 11: Scheme of the molecular structure of Myeloma-Globulin.
(G.M. Edelman, Biochem. 9, 3197 (1970))

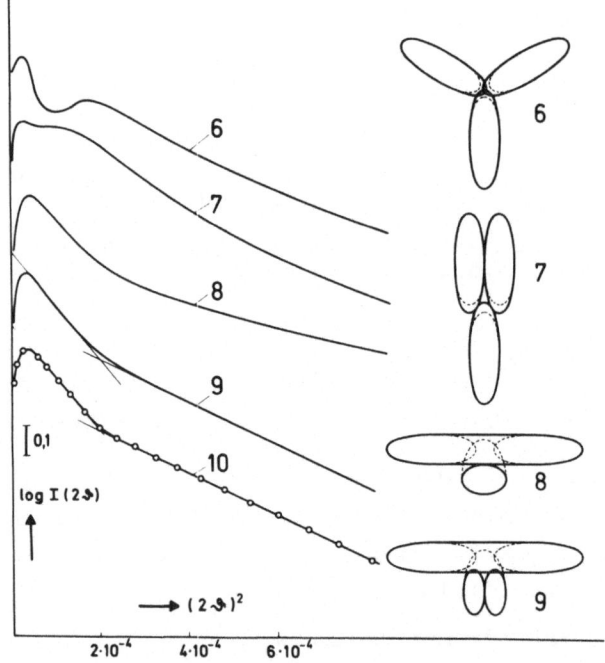

Figure 12: Cross-section curves of the γ-and T-shaped models shown in the Figure compared with the experimental curve of an IgG-molecule in solution. The T-shaped model 9 shows the best agreement with the experimental curve. (I. Pilz, G. Puchwein, O. Kratky, M.Herbst, O. Haager, W.E.Gall, G.M.Edelmann, Biochemistry N.Y.9,211 (1970))

Figure 13: Model deduced from X-ray crystallographic studies on an IgG-molecule.
(R. Sarma, E.W. Silverton, D.R. Davies and W.D. Terry, J. biol. Chem.
246, 3753 (1971))

Often, the overall shape is not as complicated as in the myeloma-example, but
the electron density within the particle shows strong inhomogeneities. This,
for example is the case in

Lipoproteins. [10]

Their physiological importance for the transport of lipids is well known.
From the numerous sidemaxima and the deep minima between them (Figure 14)
we conclude, that the particles have approximately spherical shape. In the
case of spherical symmetry, a Fourier inversion of the amplitudes yields
the radial electron density distribution, as shown in Figure 15. Considering
the relative amounts of the molecular components, their known dimensions and
their their electron densities, it is possible to construct a model (Figure 15).

Its outer shell consists of a monolayer of protein, phospholipids and free
cholesterol. The core is mainly formed by cholesteryl esters. The latter
undergoes a phase transition from a highly ordered state at low temperatures
to a disordered one at physiological temperatures, which is perfectly reversible between 4 and 40° C. This transition is believed to play a role in
connection with the development of atherosclerosis.

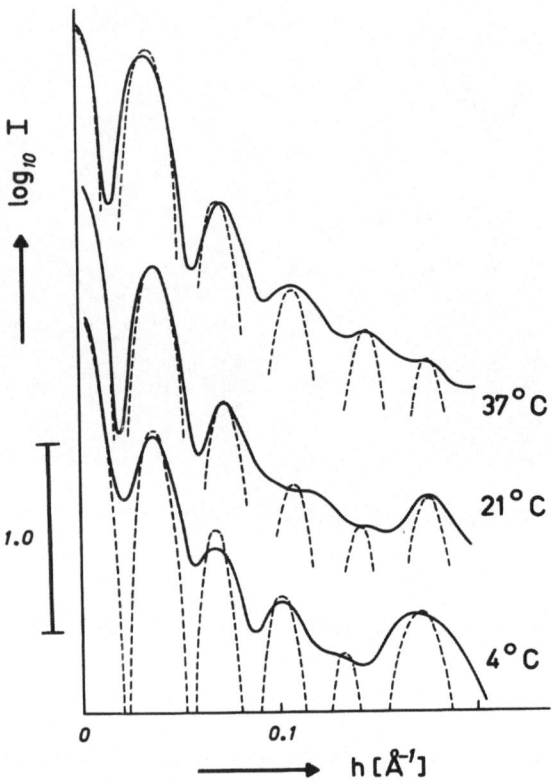

Figure 14: Scattering curves of human LpB.
(P. Laggner, G. Degovics, K.W. Müller, O. Glatter, O. Kratky, G. Kostner, and A. Holasek, Hoppe-Seyler's Z. Physiol. Chem. 358, 771 (1977))

With the help of a few typical examples, I have tried to demonstrate how several parameters and information on conformational properties of biological macromolecules can be derived. The small-angle technique is now well enough developed to be used by non-specialists, as the experimental equipment is easy to manipulate and the theory is rather simple. In fact, most of the essential results of the small-angle theory are contained in a number of computer programs.

For a particular application, it may be sufficient to use the method for the determination of only one or a few parameters, (like R, V, M, R_c, L, or an axial ratio, etc.), without performing a complete analysis.

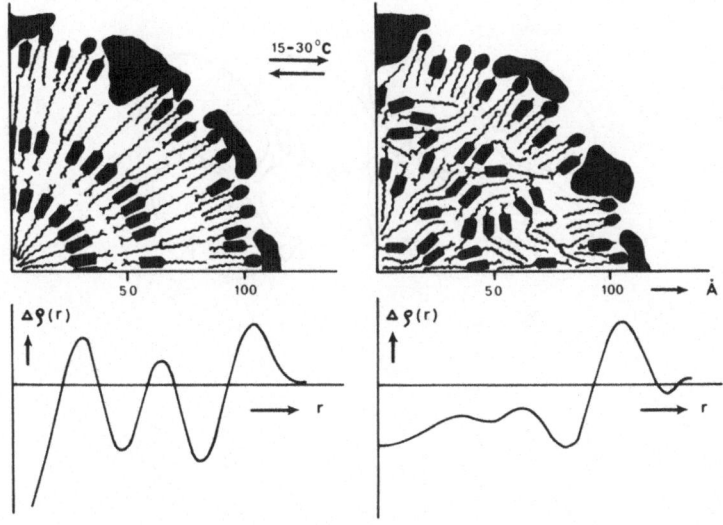

Figure 15: Idealized cross-sectional view of LpB as derived from the radial
electron density distribution below and above the transition
temperature.
(P. Laggner, G. Degovics, K.W. Müller, O. Glatter, O. Kratky,
G. Kostner and A. Holasek, Hoppe-Seyler's Z. Physiol. Chem. 358,
771 (1977))

2. Crystals [11]

The situation is quite different for crystals of biological macromolecules.
Here, it is neither reasonable nor feasible to determine only part of the
crystal structure. It is impossible in the present context to guide along
the way of such an analysis. So I shall mention only a few requirements
which have to be fulfilled.

A typical protein structure determination requires several man years of hard
work, to be done by excellent experts, it needs expensive equipment (more
than 1 Million D-Mark) and very much computer time.

The most limiting factor however is the availability of suitable crystals:
They have to show a low temperature factor, i.e. they should have a high
degree of ordering, they have to be big enough, and they have to form iso-
morphous derivatives, i.e. they have to specifically bind heavy atoms at
a specific site without change of structure. Figure 16 demonstrates schema-

tically a specific binding of heavy atoms. By calculating differences between
the loaded and the native protein, the structure of the latter can be elucidat

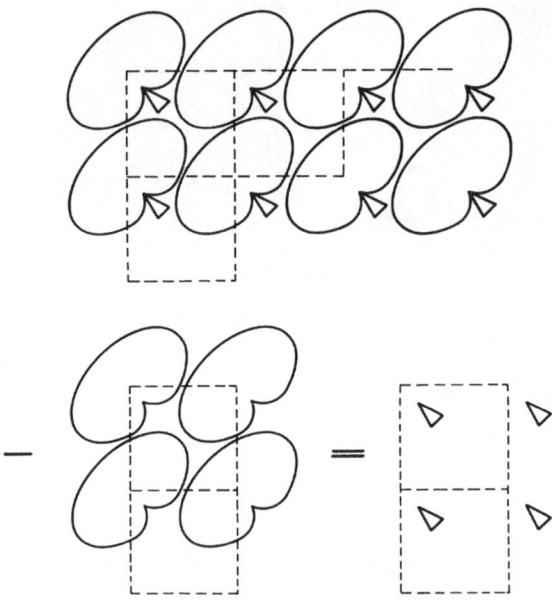

Figure 16: Loaded and native proteins.
(W. Hoppe, in: Biophysik, Springer-Verlag, Berlin-Heidelberg-
New York, pp 51-67 (1977))

The reward of this procedure is indeed considerable: with sufficiently good
crystals, one may get the whole of the protein molecule at essentially atomic
resolution. The pioneers in the field, M. Perutz and J. Kendrew, have performed the first complete structures of haemoglobin [12] and myoglobin [13]
(Figure 17) respectively. There is no other technique which yields such
a wealth of information on such a complicated system. About 200 protein
crystal structures were determined so far, and they have permitted considerable insight in the principles governing the folding of protein, in
enzyme mechanisms, etc.

It is true, the crystalline state is far removed from the state of the
proteins in vivo. Above all, the preparation of crystals involves precipitation from a solution by addition of large amounts of electrolytes or
organic solvents. Protein crystallographers, however, argue by pointing
out that their crystals still contain between 50 and 70 % water, that,
in other words, the crystalline state can be regarded as something like

an "ordered solution".

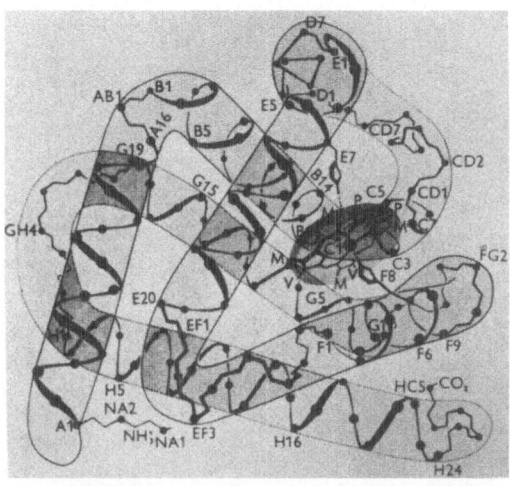

Figure 17: Secondary and tertiary structure of myoglobin, according to
J.C. Kendrew [12.13]

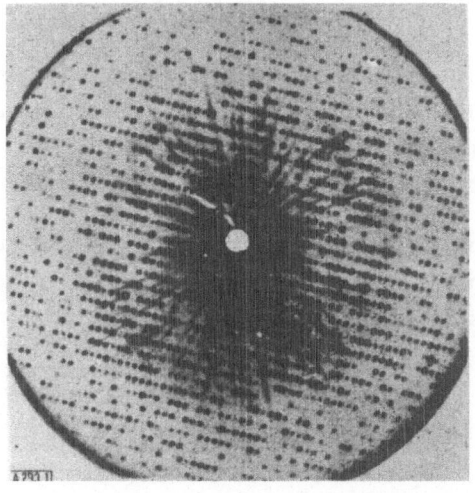

Figure 18: X-ray diagram of a myoglobincrystal according to J.C. Kendrew [13]

Ordering can be excellent indeed as is shown by the X-ray diagram of a myoglobincristal (Figure 18) and the electromigrograph of a Virus crystal (Figure 19).

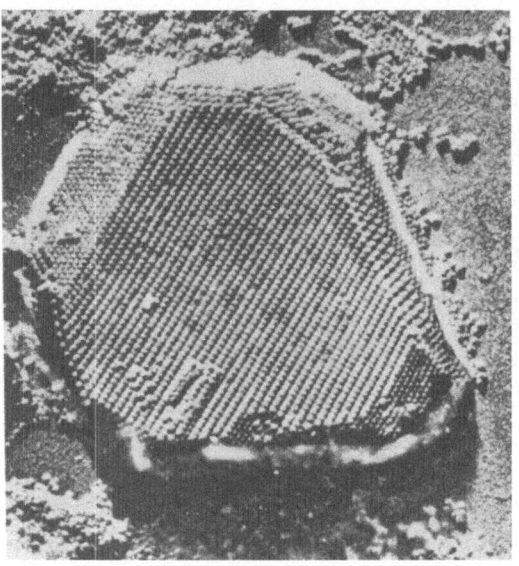

Figure 19: Electromicrograph of a crystal of southern bean virus.
(L.W. Lataw and R.W.G. Wyckoff, Arch. Biochem. Biophys. 67, 225 (1957)   )

It has to be mentioned, that there are several examples, which show a significant difference between the crystalline and dissolved state. One of these, the Myelomaglobulin, has been demonstrated. Much further work will be necessary to clear up problems of this sort.

The third group of objects we have to discuss are Biopolymers, which are more or less insoluble in water. They have a pronounced regular structure in one or two dimensions. We will confine ourselves to the one-dimensional ordered ones, the

3. Fibers [14, 15]

Globular proteins consist of polypeptide chains folded up in a well defined manner. In contrast, biological fiber molecules consist of parallel, chain-like molecules. The chains can be fully extended, as it is the case in cellulose, or they can have a more complex shape, for example a helix.

The basis for the X-ray structure determination is the fiber diagram, which is the scattering diagram one observes when the X-ray beam hits the fiber in a direction perpendicular to its axis. Figure 20 shows the fiber diagram of Desoxyribonucleicacid, Figure 21 that of cellulose.

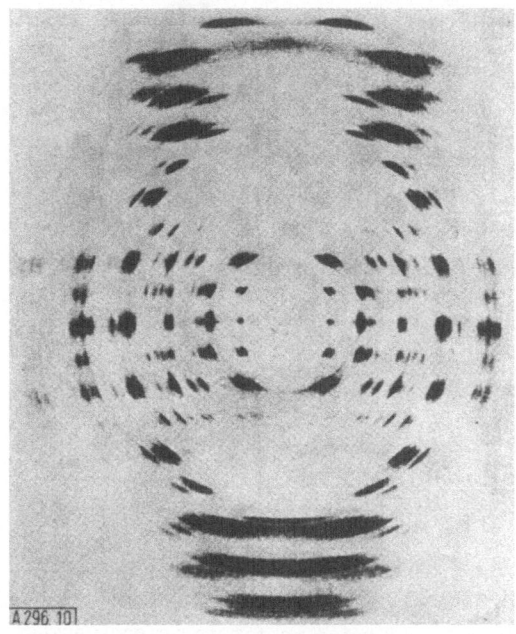

Figure 20: Fiber diagram of Desoxyribonucleicacid.
(M.H.F. Wilkins, Angew. Chem. 75, 429 (1963))

The reflections are arranged along lines of hyperbolic shape, which we call layer lines. Their distance from the middle line, the so-called equator, can be used to compute the fiber period, i.e. the distance between equivalent points along the fiber axis. Fiber period and layer-line distance are inversely proportional.

The fiber period contains two glucose residues in the case of cellulose,two amino acid residues in the case of silk fibroin. The conformation of the silk fibroin chain, which is fully extended, is called the β-form. In the case of keratin, which can, under appropriate conditions, also assume the β-conformation, the native state, however, consists of a contracted confor- mation, the so-called α-form. For a long time, the molecular conformation of the α-form was the subject of intense speculations. Since the individual

amino acids cannot shrink, a model had to be invented which could account
for the 5,4 Å translation period by arranging the amino acids in a some-
how folded-up chain.

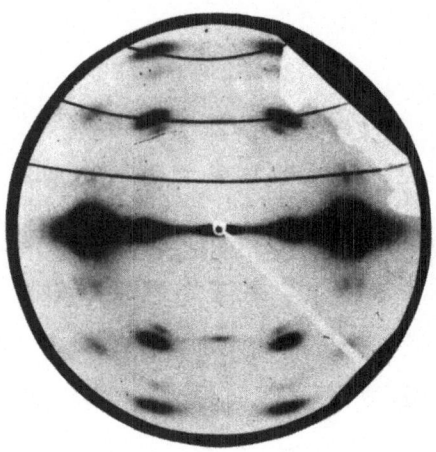

Figure 21: Fiber diagram of cellulose

It required the unorthodox concept of Linus Pauling [16] to achieve this goal.
The concept was so ingenious, that it was even rewarded with the Nobel Prize.
Before Pauling, everybody had tried to fit an integer number of amino acids
into one fiber period, which turned out to be impossible. Pauling showed,
that the 5,4 Å period corresponds to one pitch of a helical chain, which
contained a noninteger number of amino acids, on the average 3,6. Thus,
the observed 5,4 Å fiber period is not a translation period in the strict,
mathematical sense. This very unorthodox concept of the α-helix could
satisfactorily account for all experimental observations.

Time forces me to confine myself to this one spectacular example from fiber
crystallography; the more so, since the one example, which had the greatest
impact on the whole of biology, the double helix by Watson and Crick [17],
can safely be assumed to be familiar to all of you.

References (Books and review papers)

1. A. Guinier and G. Fournet; Small Angle Scattering of X-Rays, J. Wiley
    Sons Inc., New York and Chapman & Hall, London (1955).

2. W. Beeman, P. Kaesberg, J.W. Anderegg and M.B. Webb, Size of Particles
    and Lattice Defects, in: Handbuch der Physik, Springer-Verlag,
    Vol. 32, (1957).

3. O. Kratky, X-Ray Small Angle Scattering with Substances of Biological
    Interest in Diluted Solution, in: Progress in Biophysics, Pergamon
    Press, Oxford-London-New York-Paris, Vol. 13, 105-173 (1963).

4. H. Brumberger, Proceedings of the Conference held at Syracuse University,
    June 1965, on "Small-Angle X-Ray Scattering", Gordon and Breach,
    Science Publ. New York-London-Paris (1967).

5. O. Kratky and I. Pilz, Recent Advances and Applications of Diffuse X-Ray
    Small-Angle Scattering on Biopolymers in Dilute Solutions, Quarterly
    Reviews of Biophysics 5, 481-537 (1972).

6. I. Pilz, Small-Angle X-Ray Scattering, in: Physical Principles and
    Techniques of Protein Chemistry, Part C, Academic Press, New York-
    London, pp 141-243 (1973).

7. H. Pessen, T.F. Kumosinsky and S.N. Timasheff, Small-Angle X-Ray Scattering,
    in: Methods of Enzymology, Academic Press, New York-London, Vol. 27,
    151-209 (1973).

8. O. Kratky and I. Pilz, A. Comparison of X-Ray Small-Angle Scattering
    Results to Crystal Structure Analysis and other Physical Techniques
    in the Field of Biological Macromolecules, Quarterly Reviews of
    Biophysics 2, 39-70 (1978).

9. I. Pilz, O. Glatter and O. Kratky, Small-Angle X-Ray Scattering, in:
    Methods of Enzymology, Academic Press, New York-San Francisco-London,
    Vol. 61, 148-249 (1979).

10. P. Laggner and K.W. Müller, The structure of Serum Lipoproteins as Analysed
    by X-Ray Small-Angle Scattering. Quarterly Rev. of Biophysics 11,3,
    371-425 (1978).

11. T.L. Bundell, and L.N. Johnson, Protein Crystallography, Academic Press,
    London-New York-San Francisco (1976).

12. M.F. Perutz, Röntgenanalyse der Hämoglobine, Angew. Chem. 75, 589-595
    (1963).

13. J.C. Kendrew, Myoglobin und die Struktur der Proteine, Angew. Chem. 75, 595-603 (1963).

14. W. Hoppe, Strukturanalyse mit Röntgenstrahlen, in: Biophysik, Springer-Verlag, Berlin - Heidelberg - New York, pp. 51-67 (1977).

15. L.E. Alexander, X-Ray Diffraction Methods in Polymer Science, Wiley-Interscience, John Wiley & Sons, New York - London - Sydney - Toronto, pp. 280-356 (1969).

16. L. Pauling, The Nature of the Chemical Bond and the Structure of Molecules and Crystals, Cornell University Press (1960).

17. J.D. Watson and F.H.C. Crick, The Structure of DNA, Cold Spring Harbor Symposia on Quantitative Biology 18, 123-131 (1953).

INTERDISCIPLINARY APPLICATION OF MODERN TRACE ANALYSIS TO BIOLOGICALLY ACTIVE SUBSTANCES

HANS GEORG LEEMANN, FRITZ ERNI and BERNHARD SCHREIBER
Analytical Research and Development,
Pharmaceutical Division,
SANDOZ Ltd., CH-4002 Basle, Switzerland

Abstract. The power of detection and selectivity of modern analytical instruments are at present so high that trace analysis is used on a large scale in all areas of natural science.
By means of a few selected subjects, the possibilities and limitations of trace analysis characterization of biologically active substances are presented. Particular emphasis is laid on the preparation of samples, such as separation from the matrix, resolution of trace components and enrichment techniques. Selectivity and detection sensitivity are discussed, taking the determination of individual heavy metal trace elements in tissue as an example.
A closer look is taken at HPLC in the column switching mode combined with step gradient elution as one of the possibilities of separating trace organic compounds, in this case cyclosporin A, from an excess of other bio-organic material.
When physical chemical methods fail, the RIA test can be the method of choice as demonstrated in a methodological optimization process for a single dose migraine study with ergotamine.
What the trace analyst finds particularly interesting and attractive are the combination of methods. By means of an example, the yes-no presence of LSD in human blood, a multiple combination of RIA, HPLC and fluorescence spectroscopy shows how difficult analytical problems can be solved in an elegant manner by skillful, sequential combination of separation and measurement steps.
Also a review of the most recent technical literature is provided.

1. General aspects of trace analysis

Trace analysis in the life science has revealed more and more how extremely small amounts of chemical compounds or elements can influence biological processes.

Biological processes are useful only if properly balanced and controlled. Organisms have developed a sophisticated chemical system for the measurement of relevant concentrations, transfer and processing of information followed by a balancing action to restore the dynamic equilibrium of life. Enzymes, hormones, vitamins and metal ions are examples of trace substances acting as chemical tools in the living organism.

Mankind has added a vast variety of chemical compounds interacting with biological processes in trace amounts.

The different fields of application of trace analytical methods in life science are given in figure 1.

Figure 1: Application of trace analysis in life science

Medical research and diagnosis (tissue, blood, urine)
Food & drug research development and control
Microbiological research and control
Packaging materials research and control
Environmental analysis
Forensic chemistry
Engergy conversion

Trace analysis deals with very small concentrations of chemical species. Determination methods have to cope with the following problems:

a) Sufficient sensitivity and selectivity of the detector

b) Control of the physical and chemical properties influencing small amounts of material such as:
   - loss by adsorption, coprecipitation or due to volatility
   - omnipresence of trace species in the environment leading to contamination
   - differing chemical behaviour due to the mass acting principle
   - inhomogeneous distribution of the trace constituent in the matrix.

Consequently, procedures should adhere to the following principes:

- assure homogeneity of trace constituent when drawing the sample
- minimize the number of steps, number of vessels, surface contact and
  temperature during sample preparation or assay
- use reagents with certified purity and known blank values
- validate the method by
  - trace species spiked with radioactive atoms
  - analysis of certified standard materials
  - comparison with a second, independent method

Three selected fields from our activities in trace analysis have been chosen
to stand for the innumerable possibilities of trace analysis application.

a) Determination of trace elements in biological tissue by plasma emission
   or X-ray fluorescence spectroscopy,

b) Application of HPLC to the trace analysis of organic compounds,

c) Radioimmunoassay for the determination of trace amounts of drug substances
   or their metabolites.

2. Determination of trace elements in brain tissue by OES or XRF

The action of trace elements in biological systems has drawn increasing
attention during the past decade. The topic is described in detail in the
series of Prof. Helmut Sigel of the University of Basle [1].

In figure 2 and 3 the elemental composition of the human body is given in
grams/body weight.

Figure 2: Elements in the human body (gram per 70 kilograms)

Oxygen	45500
Carbon	12600
Hydrogen	7000
Nitrogen	2100
Calcium	1050
Phosphorus	700
Sulfur	175
Potassium	140
Chlorine	105
Sodium	105

	Figure 2: continued
Magnesium	35
Iron	4,2
Zinc	2,3
Silicon	1,4
Rubidium	1,1
Fluorine	0,8

Figure 3: Elements in the human body (gram per 70 kilograms)

Zirkonium	0,3
Copper	0,1
Strontium	0,1
Barium	0,1
Aluminium	0,1
Niobium	0,1
Lead	0,08
Antimony	0,07
Tin	0,03
Iodine	0,03
Cadmium	0,03
Vanadium	0,02
Manganese	0,02
Selenium	0,02
Arsenic	0,01
Boron	0,01
Nickel	0,01
Titanium	0,01
Molybdenum	0,005
Cobalt	0,003

H,C,N,O,P,S and Cl together with Na, K, Ca and Mg make up 99.988 % of the total weight. The rest, 8 grams, is distributed into about 30 trace elements [2-4] influencing the functions of the organism to an extent not believable some years ago [5,6].

At this time, elements given in figure 4 are known to be indispensable for human life. In clinical laboratories, Na, K, Ca, Fe and Cl are determined on a routine base.

ELEMENTS KNOWN TO BE ESSENTIAL FOR LIFE

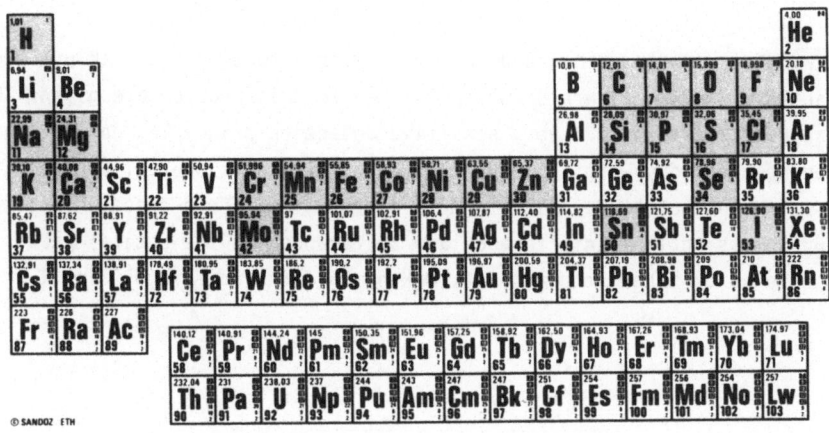

Figure 4: Elements known to be essential for life

F, Cu, I, Mg, Mn and Zn are becoming increasingly important, promoted by the availability of interference free methods such as atomic absorption or ion selective electrodes. The list will increase in the near future when plasma emission spectroscopy will be available to routine laboratories.

Deficiency diseases are known for Fe, F, I, Zn, Cu, Se and probably also for As, Co and Cr, whereas endemic intoxication by a high level of elements is known for As, F, Fe and Se. Artificial intoxication can be caused by nearly all elements. Industrial vaste, medical treatment or pesticides are the main sources.

The pharmaceutical industry is interested in metal trace determinations in two main fields:

a) research into the mechanism of drug action

b) purity and quality control of the products.

I should like to expand on the first point:

The content of aluminium in different parts of the brain and the walls of arteries is supposed to increase with age [7]. It might therefore be an indicator for the action of geriatric drugs.

In another study, copper and manganese were found to alter their concentration in the brain after longterm treatment with dopamine agonists against parkinsonism [8]. To prove these hypotheses a suitable assay technique to measure changes of trace amounts of Cu, Al and Mn in a biological matrix had to be developed.

Sample weights of some 100 mg and concentrations between 0.1 and 10 ppm had to be anticipated. In figure 5 two approaches to this problem are given. The first approach works with a very sensitive detection principle: optical emission spectroscopy with an inductively coupled plasma source. Good sensitivity of the detection minimizes sample preparation to one dissolution step only. But emission line interference and differences in the optical background have to be taken into account, making calibration more difficult.

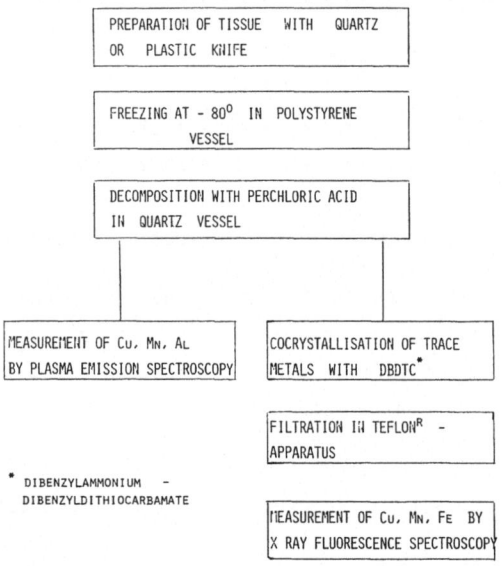

Figure 5: Determination of trace elements in brain tissue

The use of X-ray fluorescence spectroscopy avoids this problem. However, lower sensitivity forces us to include an enrichment step to the sample preparation procedure by using coprecipitation of the trace metals with dibenzylammonium-dibenzyldithiocarbamate [9]. Which procedure gives optimum results has to be proved for each problem.

Analytical data for the two determinations are given in the following
table:

	Inductively Coupled Plasma-Optical Emission Spectroscopy ICP - OES				X-Ray Fluorescence Spectroscopy XRF / enrichment		
	Al	Fe	Mn	Cu	Fe	Mn	Cu
Detection limit 1	0.1	0.06	0.06	0.1 ug	0.8	0.6	0.5 ug
Detection limit 2	0.5	0.3	0.3	0.5 ppm	0.8	0.6	0.5 ppm
Dynamic Range	0-200	0-100	0-100	0-200 ppm	0-50	0-20	0-50 ppm

Definitions:

Detection limit 1   minimum amount of element in micrograms, which can be
                     measured in such a way that a signal significantly
                     higher than the blank signal is obtained

Detection limit 2   minimum concentration of the element in the sample
                     matrix, if 1 g of the material is decomposed to give
                     10 ml of digest

Dynamic Range   concentration range of an element in the sample covered
                     by the method of determination.

The comparison shows, that a combination of an enrichment method with a non-
destructive determination (XRF) can achieve the same or even better analytical
performance than an intrisical trace analytical method like plasma emission
spectroscopy.

The determining factors are
- sample amount available
- number of elements to be determined
- form of the sample accepted by the spectrometer
- single/multi channel instrument
- destructive/non- destructive determination.

This example clearly demonstrates that trace analysis is not only a matter of
detection limits which a method can achieve under certain conditions, but much
more a result of the analytical concept used in solving the problem.

3. Trace analysis of organic compounds

Compared to trace element determinations with highly selective and sensitive detection systems like atomic absorption and X-ray fluorescence spectroscopy there is a general lack of detection selectivity and sensitivity [10] for measuring traces of organic material in biological matrices like body fluids and tissues [11].

This is the main reason for the important role of separation science in the determination of most organic traces in biological materials. Microphase extraction techniques could be one answer to the problem of obtaining sufficiently concentrated samples of organic trace compounds [12, 13]. In this connection mention should be made to the little remembered "applejack" concentration procedure successfully used with organic compounds sensitive to the effects of heat and light in an aqueous system by solidly freezing and then collecting slowly thawing volumes [14].

Beside the traditional separation techniques, such as thin-layer chromato-graphy or gaschromatography the combination of GC (especially glass capillary GC [15,16] and mass spectroscopy is probably the most powerful analytical instrument at present [17]. It combines an excellent separation technique with the sensitive and selective detection principle of mass spectroscopy.

For compounds with low volatility or for such which are thermally labile, factors which both apply to many organic compounds, high performance liquid chromatography (HPLC) today plays a key role in the determination of such trace amounts.

The combination of LC and MS may bring new possibilities for the near future [18-20].

The following example shows some strong points of a new HPLC technique used in the case of the determination of the cyclic peptide cyclosporin A (OL 27-400) in body fluids [21].

The sensitivity for this molecule (figure 6) is insufficient with either of the normal HPLC detection principles (UV-VIS, fluorescence, RI)

The only possibility for chemical derivatisation [22], here in the case of the cyclosporin A molecule is to make use of its α-amino acid moiety, and, indeed, we were successful in producing a derivative with a newly synthesi-zed reagent, naphtyl selenyl chloride, showing a welcome bathochromic shift and at the same time a remarkable increase in sensitivity [23].

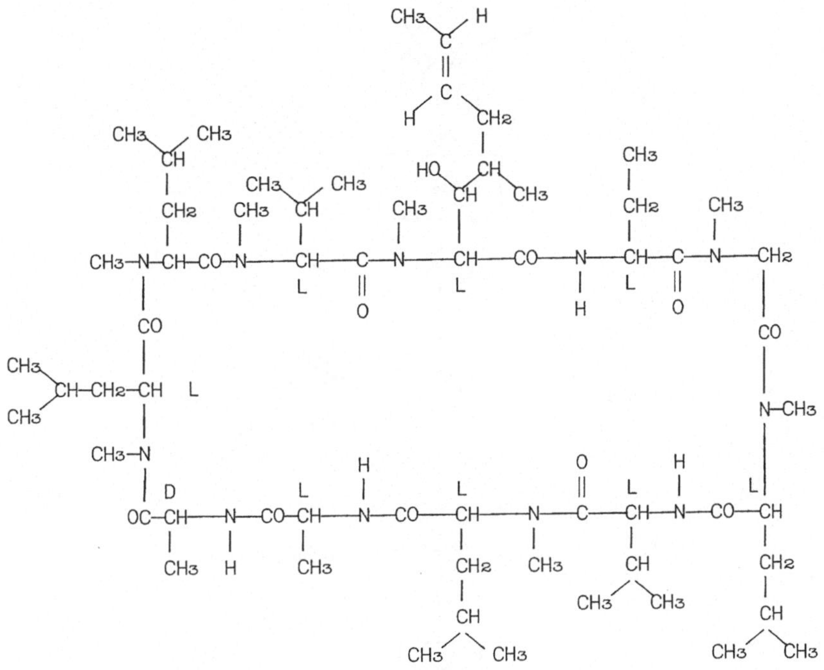

Figure 6: Cyclosporin A

However due to its slow reaction kinetics and lack of specificity the reagent proved to be of little use for the determination of cyclosporin A in biological media.

Thus, only UV-detection at a short wavelength (about 210 nm) yields an adequate HPLC signal but with a high background from interference by the biological matrix [24].

However, with a sophisticated HPLC column switching technique [25] it became possible to determine cyclosporin A in the low ng/ml range in blood samples, a range which has not been attained even by RIA. The column switching technique used is described in figure 7.

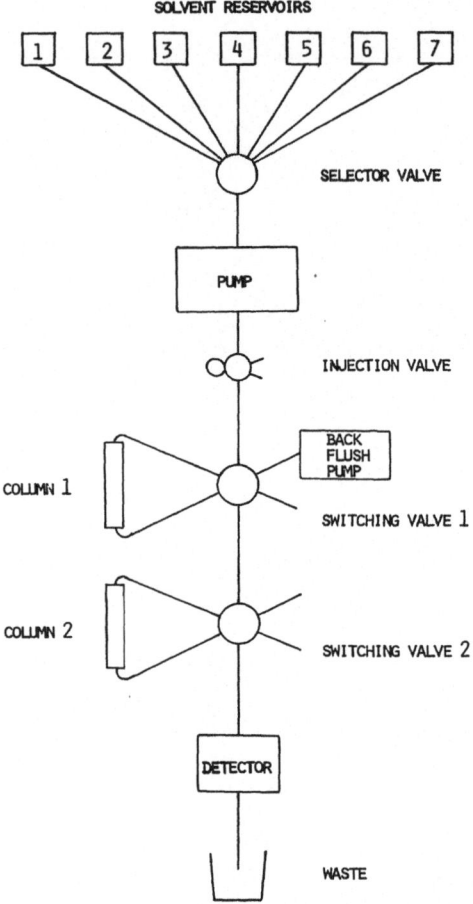

Figure 7: Schema of HPLC column switching system

For the analysis 2 columns are used in series with the possibility to
by-pass each of the columns by using 2 normal sixport injection valves
as switching valves (figure 8).

The first column is mainly for sample clean-up. First the sample is concentrated on top of column 1 by using well known sample concentration effects
on reversed phase columns [26].

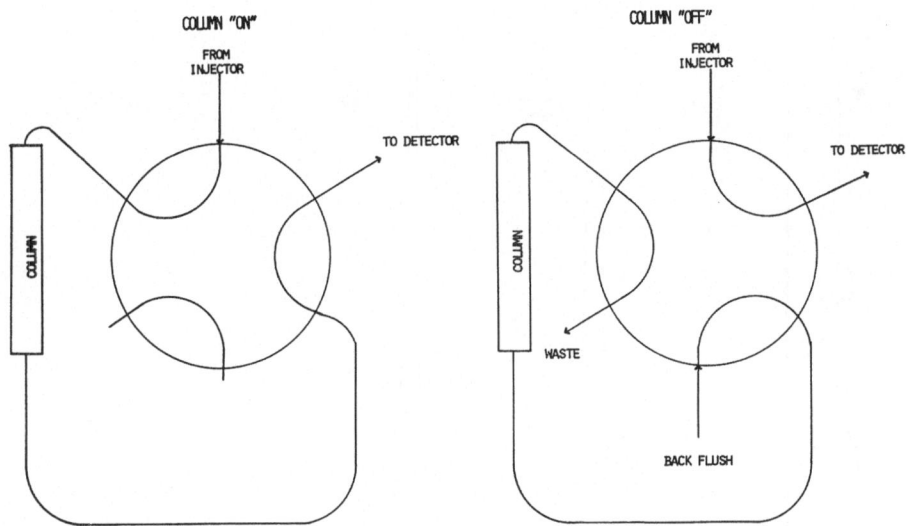

Figure 8: The 6-port valve as switching valve

Up to some milliliters of biological sample can be injected and concentrated on the first column without any loss of separation efficiency. Most of the very polar compounds of the matrix elute during the sample concentration procedure and the following online sample cleanup. Only a very small part of the eluate from column 1 is "cut" to column 2. In this case the cut is called a heart-cut (figure 9).

The strongly retained background peaks can then be eluted from column 1 using backflushing for cleaning the column. Figure 10 shows a typical HPLC chromatogram using column switching to determine cyclosporin A in plasma. At the bottom of the chromatogram the column functions are indicated. Figure 11 shows a chromatogram of cyclosporin A with an approximative detection limit of 2ng/ml in plasma.

The use of different types of columns in the switching system gives many possibilities to increase the selectivity of a system [28]. Using as a first column a stationary phase with generally smaller capacity factors (k'), a second concentration on top of column 2 is achieved. This leads to narrower peaks (meaning higher efficiency) of the total separation system. The two different columns also have different separation selectivities which is especially important for trace analysis. Only a small part of eluate from the first column is separated on the second column, meaning there are less peaks to interfere.

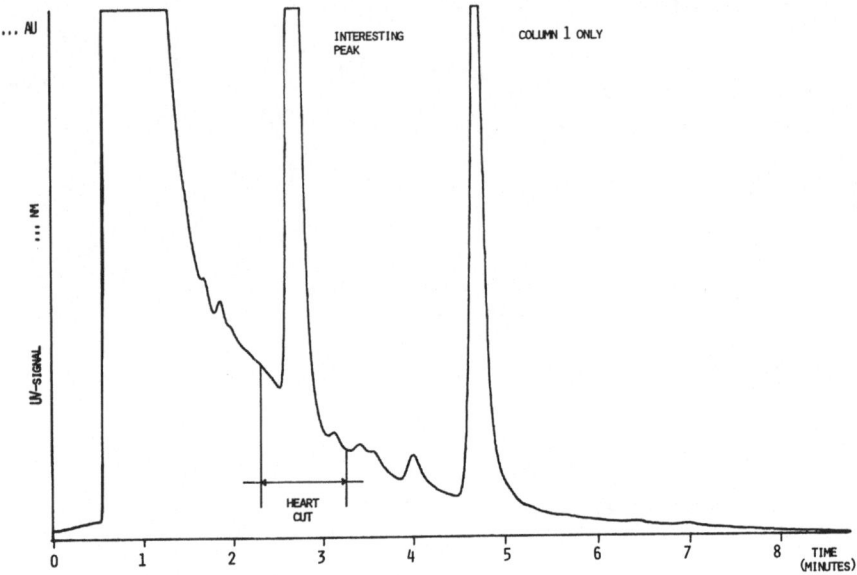

Figure 9: HPLC chromatogram of cyclosporin A in serum with "heart cut" segment

It can be shown from the information theory [26] that with this twodimensional separation system the information produced by column switching increases as the differences in the selectivities of the two columns increase. In addition the combination of column switching with step gradient elution gives even more flexibility in using not only different stationary phases but also different mobile phases for sample cleanup, on the first column and final separation on the second column, respectively.

Routine application of these column switching techniques showed that the columns can be used for several hundred biological samples indicating that the first column has been effectively flushed and regenerated, and that the second column is sufficiently protected from the well known effect of column plugging or destruction when complex samples are injected.

HPLC column switching is suitable for automation and can easily be assembled together from commercially available components.

It may be the method of choice when the compounds to be analysed are chemically labile and where the risk of causing artefacts with classical cleanup techniques is high [25].

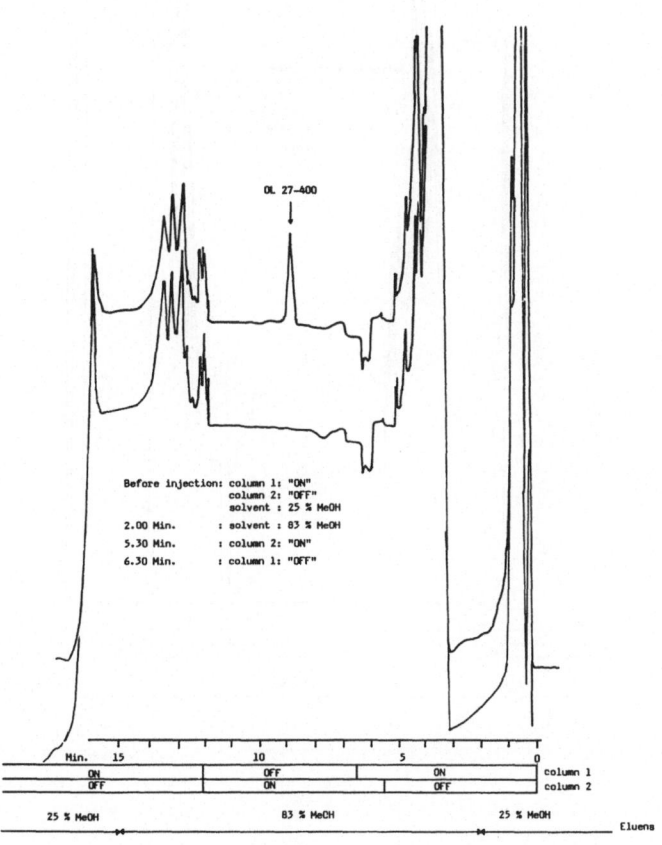

Figure 10: HPLC chromatogram of cyclosporin A in serum with column functions

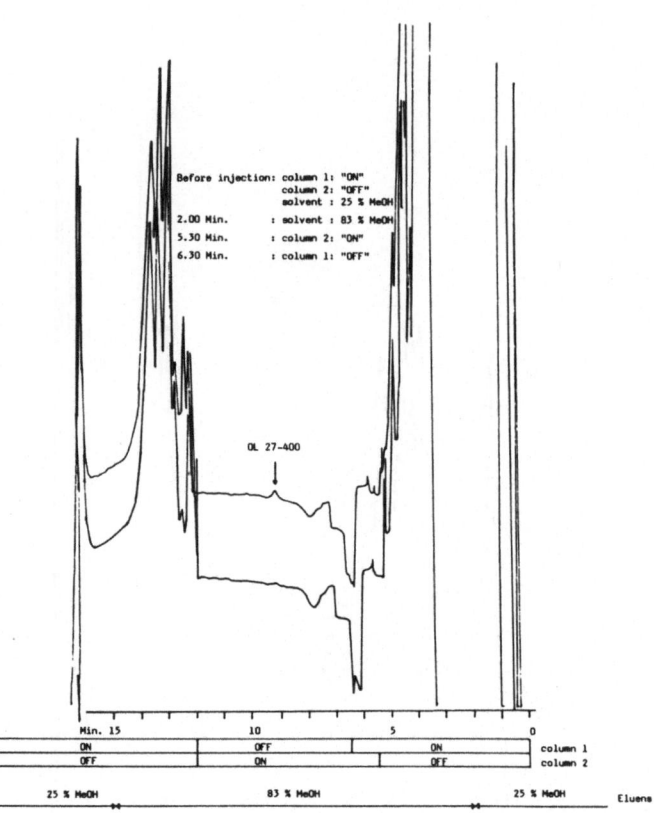

Figure 11: HPLC chromatogram of cyclosporin A in serum with detection limit of 2 ng/ml

4. Radioimmunoassay

Ligand binding assays are characterized by their high specificity and sensitivity. We also find these properties in the radioimmunoassay (RIA) where they are supplemented by high precision and by simplicity of measurement.

Since the classical immunoassay principle was first described by Berson and Yalow [28], the RIA test has in the past few years become one of the most important measuring techniques in biopharmaceutical and clinical trace analysis [29-37].

Active ingredients of pharmaceuticals usually have a molecular weight of < 1000 and are therefore themselves not capable of inducing the desired antigen- antibody reaction.

They will become immunogenic by binding the active ingredient covalently onto a suitable protein by means of a chemical reaction. The immune reaction which is induced by this conjugated protein produces an antiserum in a vertebrate species.

The use of radiolabelled haptens (antigens) permits simple, rapid and precise quantitative determination of pg quantities of active ingredient in a biological medium (figure 12).

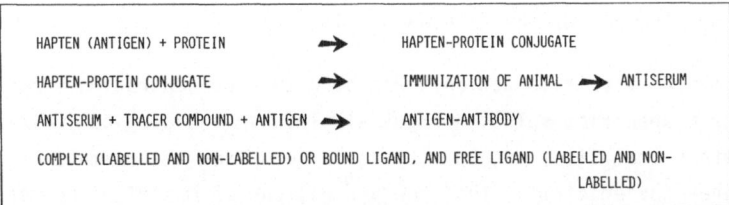

Figure 12: Principles of immunization with low molecular weight compounds and radioimmunoassay

Today, RIA is successfully used in biopharmacy as a method of organic trace analysis for the study of absorption, distribution, metabolism and elimination. Figure 13 shows the optimisation experiments which were necessary in order to be able to obtain a quantitative analysis of a singledose pharmacokinetic study with ergotamine on a healthy subject during an acute attack of migraine [38].

RIA-TEST FOR ERGOTAMINE

SANDOZ LTD, BASLE

HAPTEN-PROTEIN CONJUGATE	9,10 -DIHYDROERGOTAMINE + BOVINE SERUM ALBUMIN	1-HYDROXYLMETHYLERGOT - AMINE + BOVINE SERUM ALBUMIN	6 -NOR-6-CARBOXYMETHYL- 9,10 -DIHYDROERGOTAMINE + BOVINE SERUM ALBUMIN
SPECIES OF ANIMAL	RABBIT	SHEEP	RABBIT
ANTISERUM	WHOLE SERUM	WHOLE SERUM	γ- IMMUNOGLOBULIN
TRACER	$\left[9,10 - {}^3H \right]$ ERGOTAMINE	$\left[2- {}^3H \right]$ ERGOTAMINE	$\left[3 - {}^3H \right]$ DIHYDROERGOTAMINE
BUFFER pH 6	PHOSPHATE SUCCINATE	PHOSPHATE SUCCINATE	CITRATE
SEPARATING AGENT	CHARCOAL	DEXTRAN COATED CHARCOAL	PLASMA COATED CHARCOAL
DECTECTION LIMIT	0.47 NG/ML	0.35 NG/ML	0.03 NG/ML

Figure 13: RIA-test for ergotamine, Sandoz LTD., Basle

A first RIA test was developed using tritium labelled ergotamine; the sensitivity of the method however proved to be too low. The second test with
labelling at position 2 did not lead to success either. Also the chemical
instability of ergotamine in polar solutions presented difficulties so that
in a third test, the much more stable 9,10-dihydroergotamine was taken as the
tracer compound still carrying the same peptide part as ergotamine. By iso-
lating the γ-globulin fraction of the anti-serum instead of whole serum, and
by modifying the separating reagent, the limit of detection of the test could
be lowered by a factor of ten to 0.03 ng/ml.

It would prove difficult to get anywhere near to a concentration range of
30 pg/ml for a specific, selected ergot alkaloid in a biological matrix
using any other analytical method. It can be seen from the example of the
RIA test given for ergotamine that the process can be optimised by modifying
the variables, provided the following prerequisites, which are valid for the
RIA test in general, are fulfilled:

- The compound (hapten) must be immunologically active or one must be able
  to make it so;
- One must be able to produce the compound in a very pure form;
- One must be able to radiolabel the compound without changing the specific
  immune activity on incorporation of the radioactive isotope.

A very pure form is necessary because

- On the test, the compound is to be used as a standard of known concentration,
- It is to be radiolabelled and is a part of the RIA and thus has an
  influence on the accuracy,
- It is used to form the specific antibody in the organism of the vertebrate.

It also should be noted that the RIA represents a splendid yes-no analysis
for incorporation in a complicated enrichment isolation and determination
process, which apart from anything else, should stimulate the imagination
for meaningful combinations of methods.

The scheme of analysis which is shown in figure 14 exploits the particular
advantages of each method, HPLC, fluorescence spectroscopy and RIA, to the
greatest effect in the case of the analysis of LSD. Thus, although RIA is
not sufficiently specific to be used alone, the rapid initial screening by
RIA allows negative samples to be rejected. The presence and quantity of
LSD in RIA positive samples can then be confirmed by HPLC with fluorometric
detection, followed by fluorescence spectroscopy, HPLC/RIA and HPLC on silica
depending upon the degree of certainty of identification required. Levels
down to 0.5 ng of LSD per ml can be detected [39].

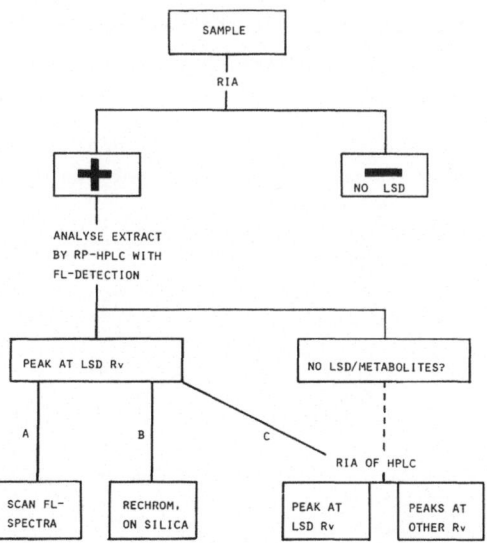

Figure 14: Analysis of LSD in blood - P.J.Twitchett et al., J.Chromatogr.150,
           73 (1978)

In future - this is my personal feeling - the RIA will be used mainly in clinical laboratories due to its handiness in application. HPLC column switching technique combined with step gradient elution, however, will be the technique of choice in an early drug development phase, when flexibility and prompt access to test results are required.

The three examples cited:
- the determination of trace metals in a biological medium with its extremely high sensitivity and specificity but also with its problems of contamination by the environment;
- the determination of trace organic compounds where separation science, such as column switching technique combined with step gradient elution comes to its full bearing to compensate for the lacking of selectivity and sensitivity of the detection systems available;
- and, ultimately, the elegance of the radioimmunoassay particularly powerful when used in combination with physico-chemical separation and determination techniques, provided a radiolabeled compound is available;
have been meant to stand for the impressive array of methods available to the trace analyst of today.

However, the rich selection of methods offered does not only require the analyst to be familiar with the methods and their potential applications, but rather to possess a sound judgement in deciding which method or combination of methods can supply the best information on an analytical problem in the field of trace substances. To that end, one must not only have knowledge, but also imagination, pleasure in treading unbeaten tracks and courage to tempt the unusual.

References

1. H. Sigel, Metal Ions in Biological Systems, Marcel Dekker Inc. New York, Basel, (1978)

2. H. Aebi, Z. Praeventivmed. 1, 137-149 (1956)

3. Wissenschaftliche Tabellen GEIGY, 7. Auflage, 228-241 (1965)

4. E. Baumgartner, Mtt. Geb. Lebensm. Hyg. 70, 34-42, (1979)

5. H.A. Schroeder, Trace Elements and Nutrition, Faber and Faber, London (1976)

6. F. Kieffer, Sandoz Bulletin 51, 52, 53 (1980), Sandoz AG, Basel

7. J.R. McDermott, A.I. Smith, K. Iqbal and H.M. Wisniewski, Lancet Oct.1, 710 (1977)

8. W.J. Weiner, P.A. Nausieda and H.L. Klawans, Neurology 28, 734 (1978)

9. H.R. Linder, H.D. Seltner and B. Schreiber, Anal. Chem. 50, 896 (1978)

10. D.M. Hercules, ed., Detectors for Trace Organic Analysis by Liquid Chromatography: Principles and Applications in: Contemporary Topics in Analytical and Clinical Chemistry, Vol.2, Plenum Publishing Corp., 1978, New York.

11. Symposium on Drug Analysis in Biological Material-Principle Aspects, Stockholm, Nov. 1-3, 1978.

12. W.J. Serfontein, D. Botha and L. de Villiers, J. Pharm. Pharmac. 27, 937, 973 (1975)

13. W. Dünges, Prä-chromatographische Mikromethoden R.E. Kaiser ed., Dr.Alfred Hüthig Verlag, Heidelberg, Basel, New York, 1979

14. I.S. Forrest, A.G. Bolt and M.T. Serra, Life Science 5, 473 (1966)

15. S.R. Lipsky, W.J. McMurray, M. Hermandez, J.E. Purcell and K.A. Billeb, J. Chromatogr. Sci. 18, 1 (1980)

16. M.L. Lee and B.W. Wright, J. Chromatogr. 184, 235 (1980)

17. W.H. McFadden, J. Chromatogr. Sci. 17, 1 (1979)

18. F. Erni and A. Melera, J. Chromatogr. in press.

19. A. Melera and F. Erni, J. Chromatogr. in press.

20. W.H. McFadden, J. Chromatogr. Sci. 18, 97 (1980)

21. A. Rüegger, M. Kuhn, H. Lichti, H.R. Loosli, R. Huguenin, C. Quiquerez and A. von Wartburg, Helv. Chim. Acta 59, 1075 (1976).

22. J.F. Lawrence, J. Chromatogr. Sci. 17, 113 (1979)

23. J.C. Gfeller, A.K. Beck and D. Seebach, Helv. Chim. Acta 63, 728 (1980)

24. W. Niederberger, P. Schaub and T. Beveridge, J. Chromatogr. 182, 454 (1980), Biomed. Applic.

25. F. Erni, H.P. Keller, C. Morin and M. Schmitt, J. Chromatogr. in press.

26. F. Erni and R.W. Frei, J. Chromatogr. 130, 169 (1977)

27. F. Erni and R.W. Frei, J. Chromatogr. 149, 561 (1978)

28. S.A. Berson and R.S. Yalow, Clin. Chim. Acta 22, 51 (1968)

29. S.A. Berson and R.S. Yalow, Radioimmunoassay: A Status Report (1971) in: Immunobiology, Sinnauer Associates Inc., Stanford, Con. 06905.

30. S. Spector, Ann. Rev. Pharmacol., 13, 359 (1973)

31. B.A. Peskar and B.M. Peskar, Eur. J. Drug Metab. Pharmacokinet., 4, 163 (1977)

32. G.P. Mould, G.W. Aherne, B.A. Monis, J.D. Teale and V. Marks, Eur. J. Drug Metab. Pharmacokinet., 4, 171 (1977)

33. W.M. Hunter, Handbook of Experimental Immunology, Vol. 1 (Immuno-chemistry), Blackwell Scientific Publications, Oxford, (1978)

34. R. Augstin, Handbook of Experimental Immunology, Vol. 3 (Application of Immunological Methods) Blackwell Scientific Publications, Oxford, (1978)

35. B.F. Erlanger, Pharmacol. Rev. 25, 271 (1973)

36. V.P. Butler, J. Immunol. Methods, 7,1 (1975)

37. K. Lübke and B. Nieuwboer, Immunologische Teste für niedermolekulare Wirkstoffe, Georg Thieme Verlag, Stuttgart (1978)

38. H. Eckert, J.R. Kiechel, J. Rosenthaler, R. Schmidt, E. Schreier in:Ergot Alkaloids and Related Compounds, B. Berde, H.O. Schild Eds., Springer-Verlag, Berlin (1978), p. 767

39. P.I. Twitchett, S.M. Fletcher, A.I. Sullivan and A.C. Moffat, J. Chromatogr. 150, 73 (1978).

IV. MICROCHEMISTRY IN ENVIRONMENTAL SCIENCES

FORMATION AND CONVERSION OF INORGANIC FINE PARTICLES IN THE ATMOSPHERE

CYRILL BROSSET

Swedish Water and Air Pollution Research Institute,
Box 5207, S-402 24 Gothenburg, Sweden

Abstract. The interaction of water and ammonia in the gas phase with sulphuric acid droplets has been described in terms of a phase diagram. The behaviour of nitric and hydrochloric acid in this system has briefly been mentioned.
The apparent non equilibrium state between the composition of graphite rich particles and precipitation at ground level has been discussed.
The difficulty of obtaining reliable data concerning coarse particles has been pointed out. An example of their reactivity as a function of particle size has been given.

1. Introduction

Most investigations of micro-components in the atmosphere have until now been focused on separate phases. However, it is likely that better understanding of atmospherical processes and observed concentrations would be obtained if the respective studies emphasized whole systems.

This paper is following such a line. It is tried to evaluate some interactions between the gas phase and condensed phases (aerosol particles) and in a few cases to establish potential equilibrium conditions.

2. Systems discussed

In the following three types of inorganic aerosols will be discussed

a) Particles derived from sulphuric acid droplets

b) Particles containing a graphite phase

c) Particles mainly constituted from inorganic non-soluble compounds

According to Whitby's particle classification (Figure 1) [1] it is seen that particles of the categories 1 and 2 are belonging to the accumulation mode and those of 3 are the coarse particles.

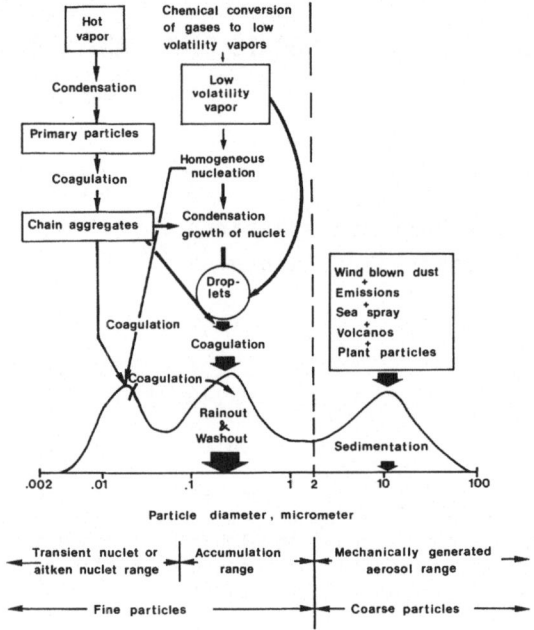

Figure 1: Particle classification and conversion scheme according to K.Whitby

If we now try to define these three types of particles as chemical systems we have to start by chosing the appropriate components.
It seems to be reasonable to start with a very limited number.

The following have been chosen

H_2O, O_2

$SO_2 (\rightarrow H_2SO_4)$

$NO (\rightarrow NO_2 \rightarrow HNO_3)$

NH_3

graphite

metal oxides

The oxidation reactions of SO_2, NO and NO_2 will not be dealt with here. Hence, any oxidized compound of sulphur and nitrogen can be chosen as compo- nents depending upon what is most convenient in different cases. This is in- dicated above by the formulas in brackets.

By means of these components the three cases may now be described in terms of the following chemical systems:

a) $H_2SO_4(1)- H_2O(g) - NH_3(g) - HNO_3(g) - \ldots$

b) graphite(s) $- H_2O(g) - NO(g) - SO_2(g) - O_2(g)$

———————— $H_2SO_4(1)- H_2O(g) - NH_3(g) - NHO_3(g) \ldots$

c) MeOH(s) $- H_2O(g) - SO_2(g) - NO_2(g) - O_2(g) - NH_3(g) - HNO_3(g)$

The systems are here described in a way that indicates the types of reactions which lead to their formation. For that reason in the case of system b) it has been indicated which components interact with the graphite particles to produce a sorbed phase and thereafter (the long dash) which of them that participate in the production of an outer layer similar in composition to the system a).

In the following the system a) will be described in more detail as this system is now rather well understood. For the systems b) and c) the problems which mainly have to be solved for their better understanding will be outlined.

3. The system $H_2SO_4(1)- H_2O(g) - NH_3(g) \ldots$

The phase diagram

Many investigations have shown that a large part of particles in the accumulation mode mainly consist of more or less acid ammonium sulphate with slight concentrations of nitrate and chloride.

We also know that their main precursor is sulphuric acid produced by the oxidation of SO_2.

The sulphuric acid drops have then primarily reacted with the humidity and ammonia in the air and usually to a lesser extent with gasous nitric and hydrochloric acids. The final composition of such a particle will be determined by the equilibrium conditions between the condensed phase(s) (solution, sometimes also crystalline phases) and the gas phase.

These equilibrium conditions of course may be summarized in terms of a phase diagram.

Such a diagram for the system H_2SO_4 - $(NH_4)_2SO_4$ - H_2O which represents the acid part of the system H_2SO_4 - H_2O - NH_3 is seen in Figure 2.

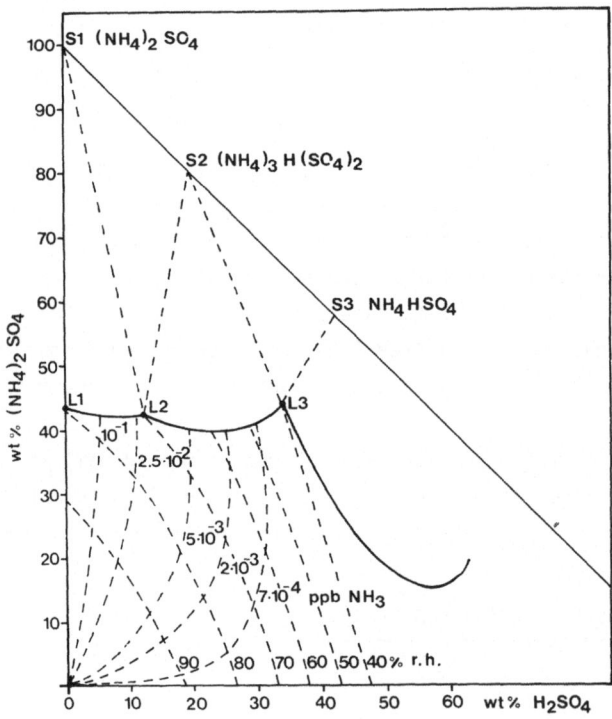

Figure 2: Phase diagram for the system $(NH_4)_2SO_4$ - H_2SO_4 - H_2O at $25^{\circ}C$

In this graph the solubility curve L1 - L2 - L3 is based on the 60-year-old data obtained by d'Ans [2] and applies at $25^{\circ}C$. The tielines for relative humidity (r.h.) have been determined by Tang et al. for $30^{\circ}C$ [3]. However, according to them, r.h. does not alter very much within a narrow temperature range. Their tielines should, therefore, with good approximation apply at $25^{\circ}C$ as well.

The tielines for the partial pressure of ammonia in the gas phase (pNH_3) have been obtained by interpolation based on calculated values for some 30 points within the area concerned (see Table 1) [4].

Interpretation of the diagram

To facilitate understanding in the discussion to follow, the designations used in the diagram and the meaning of its different parts will be explained.

The line S1 - S2 - S3 -

Point S1 represents the pure phase $(NH_4)_2SO_4$, point S2 the phase $(NH_4)_3H(SO_4)_2$ and point S3 the phase NH_4HSO_4. Thus, the line parts S1-S2 and S2-S3 correspond to mixtures of phases S1 and S2, and S2 and S3, respectively.

The line L1 - L2 - L3 -

This is the solubility curve determined by d'Ans. The part L1-L2 represents solutions in equilibrium with S1. In the same way, the part L2-L3 represents solutions in equilibrium with S2. The solution at point L2 is in equilibrium with both S1 and S2; the solution at point L3 in equilibrium with both S2 and S3.

At points L2 and L3 there are, in this three-component system, consequently four phases, which means one degree of freedom. At chosen constant temperature, both r.h. and pNH_3 are thus given.

At the curve parts denoted L1-L2 and L2-L3 there are three phases and thus two degrees of freedom. Consequently, at chosen constant temperature, there is a further choice of e.g. either r.h. or pNH_3.

The area below the solubility curve

In this area, no solid phases are present, which means that the number of phases is two and, thus, the number of degrees of freedom is three. At chosen constant temperature there is a further choice of e.g. both r.h. and pNH_3. A chosen such pair of values determines the composition of the solution.

Relevant special cases

The use of the phase diagram for determining the composition which a sulphuric acid aerosol may attain in air is illustrated in Figure 3.

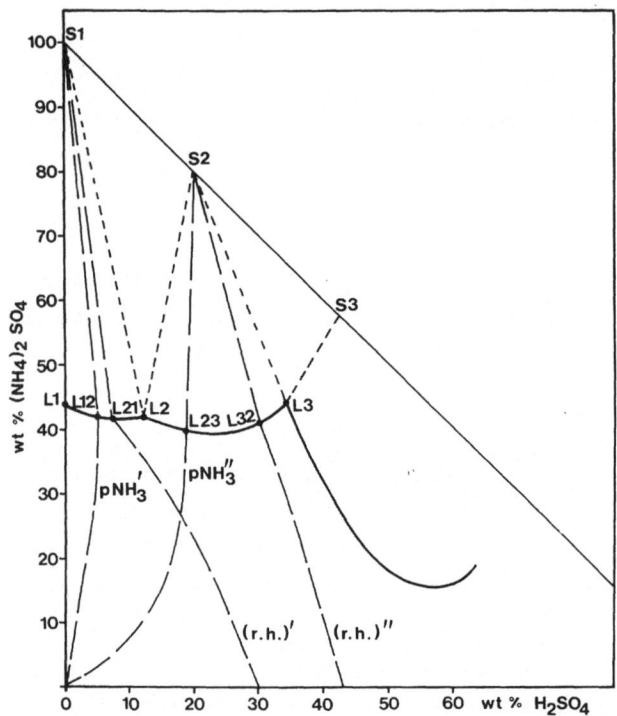

Figure 3: Phase diagram describing four special cases. See text

In this figure, the phase diagram from Figure 2 has been limited to four special cases, which are combinations between two tielines for pNH_3 (pNH_3' ≃ 0.1 ppb and pNH_3'' ≅ 5 · 10^{-3} ppb) as well as two tielines for r.h. ((r.h.)' - 75% and (r.h.)" - 50%). As will be seen from Figure 3, all points of intersection between these tielines and the solubility curve (L12, L21, L23 and L32) are located to the left of L3. The reason for this choice is i.a. that the area in the phase diagram to the right of L3 is less well investigated.

In the atmospheres corresponding to the four combinations mentioned, a sulphuric acid droplet may be assumed to react in the following manner.

Case 1: The combination $(r.h.)'$, pNH_3''

This is the simplest of the four cases. The curves intersect in the liquid range. The coordinates of the intersection point give the equilibrium composition of the liquid phase which in the present case will be about 27 wt% $(NH_4)_2SO_4$, 18 wt% H_2SO_4 and thus 55 wt% H_2O.

The course followed by the sulphuric acid droplet to reach this point depends on whether water or ammonia uptake is the fastest. Since the water vapour concentration in gas phase here is much higher than the ammonia concentration, it is possible that the water uptake, at least initially, is more rapid. This means that, in the first step, the sulphuric acid droplet will attain a composition corresponding to some point on the lower part of the curve $(r.h.)'$. After this, ammonia is taken up until the point $(r.h.)'$, pNH_3'' is reached.

Case 2: The combination $(r.h.)'$, pNH_3'

As will be seen from the diagram, there is no liquid phase composition that can be in equilibrium with this combination in the gas phase. The sulphuric acid droplet introduced into this atmosphere is likely to take up water and ammonia until some point on the curve $(r.h.)'$ is reached. This curve is then followed to point L21. Now, however, the liquid phase has to disappear. In this process, the system will follow some course within the area L21-S1-L12, while S1, i.e. $(NH_4)_2SO_4$, is being precipitated.

Case 3: The combination $(r.h.)''$, pNH_3''

This case is analogous to case 2. Phase S2 will be the final form, i.e. $(NH_4)_3H(SO_4)_2$.

Case 4: The combination $(r.h.)''$, pNH_3'

In this case, the sulphuric acid droplet is likely to take up water and ammonia until some point on the curve $(r.h.)''$ is reached. The curve is then followed to point L32. The system is still very far from equilibrium and the liquid phase will therefore disappear. In this process, the system takes a course between lines L32-S2 and L12-S1, which may mean precipitation at the same time of phases S2 and S1. Eventually, some point on the line S1-S2 is reached. The location of the point is determined by kinetic conditions.

It is important to point out that under atmospheric conditions supersaturation is likely to occur frequently.

A useful alternative form of the diagram

In certain cases, when determining the phase composition of the system in equilibrium, it may be helpful to plot log pNH_3 versus r.h. . An example of such a r.h.-log pNH_3 diagram is given in Figure 4. It corresponds to the conditions presented in Figure 2.

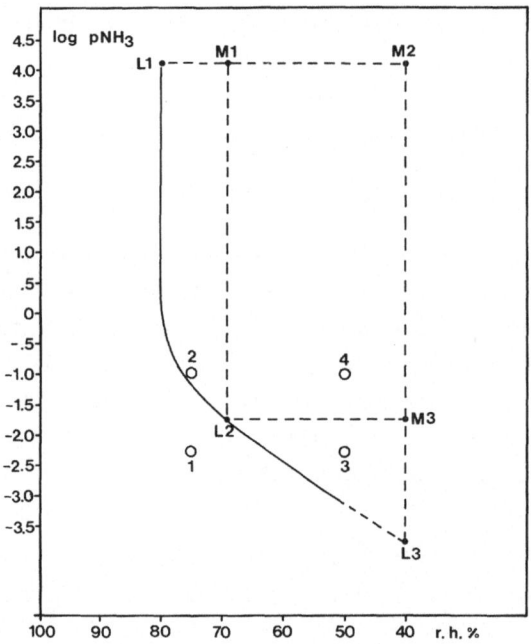

Figure 4: r.h.-log pNH_3 diagram at 25°C

The curve in Figure 4 represents the (r.h., pNH_3) combinations located on the curve L1-L2-L3 in Figure 2. The points L1, L2 and L3 in Figure 4 thus have the same physical meaning as the corresponding points in Figure 2.

The meaning of the areas introduced in Figure 4 is easily seen:

- A point located within the area below the curve L1-L2-L3 represents pure liquid phase. Thus, point 1 corresponds to case 1 above.

- A point within the area L2-L1-M1 represents the pure phase S1. This is illustrated by point 2 corresponding to case 2 above.

- A point within the area L2-L3-M3 represents the pure phase S2. Thus, point 3 corresponds to case 3 above.

- A point within the area L2-M1-M2-M3 represents a mixture of the phases S1 and S2. This is illustrated by point 4 corresponding to case 4 above.

This diagram form is very useful for the study of how the phase composition of the system changes as a result of changes in r.h. and pNH_3. Such an application will be illustrated in the following section.

Application of the diagram to actual atmospheric conditions

In the foregoing, the phase composition in the system $(NH_4)_2SO_4 - H_2SO_4 - H_2O$ was studied as a function of the variables t, r.h. and pNH_3. In the real situation (referring here to the conditions in central Sweden) all these parameters are likely to show a more or less pronounced daily and seasonal variation. That this is the case for temperature and relative humidity is a well-known fact.

Using a method developed by Ferm [5] 24 h means of ammonia concentration are being measured as of spring 1977 in Sweden. A clear seasonal (Figure 6) and a slight daily variation of the concentration following the air temperature have been established.

Let us now assume that equilibrium adjustment in the system in question is rapid relative to the existing 24 h variation in t, r.h. and pNH_3. In such a case it would be possible to obtain a rather detailed picture of the variation with time of the phase composition. The prerequisite for this is access to phase diagrams within relevant temperature ranges and data for r.h. and pNH_3 in the form of sufficiently short-term mean values.

Today, some of these prerequisites are not at hand.
Owing to the low concentration (approx. 0.001-10 ppb) of ammonia in the atmosphere, it is at present often necessary to use a sampling time (24 h) which is somewhat too long. Furthermore, phase diagrams for other tempera-tures than $25^{\circ}C$ are not available. In central Sweden, the relevant tempera-

ture ranges are, however, primarily from -5° to $+5^{\circ}C$, from $+5^{\circ}$ to $15^{\circ}C$ and from $+15^{\circ}$ to $+25^{\circ}C$.

In spite of these shortcomings, an attempt has been made to show how it is possible to follow the phase composition by means of a r.h.-log pNH_3 diagram. This attempt is illustrated in Figure 5. The data introduced in this figure were obtained through measurements at Rörvik, a clean air area about 40 km south of Gothenburg, during three days in October 1978 and three days in March 1979.

On the first occasion, the temperature was mostly between 8° and $12^{\circ}C$ where-as on the later occasion, it ranged between -6° and $+3^{\circ}C$.

The values plotted in this figure are 6 h means, the ammonia concentration being estimated from 24 h means on assumption that the 24 h variation (seemingly small here) largely followed the temperature.

There are, as mentioned, no phase diagrams available for the temperature ranges concerned here. The solubility curve drawn into Figure 5 was obtained by introducing solubility data valid for $10^{\circ}C$. (Locuty and Laffitte, 1934) in Figure 2 and assuming that the respective tielines do not shift their position too much at a temperature drop from $25^{\circ}C$ ($30^{\circ}C$ for r.h.) to $10^{\circ}C$.

The solubility curve thus obtained can probably be used for the values from October but only as a rough approximation for the values from March. However, nothing else is at hand at present.

In spite of this, Figure 5 at least gives a qualitative picture of what could have happened with the phase composition of the particles in question during the three-day periods involved. It was asumed that equilibrium was attained which, however, might not have been the case.

Assume that sulphuric acid aerosol was introduced into the air around the sampling site on the 16 October 1978 and remained there for the three follow-ing days. The diagram in Figure 5 shows that, initially, this could result in the formation of a mixture of phases S1 and S2. As reaction in solid state is likely to be slow here, nothing should happen until about 27 h later, when point a1 has been passed and the solid particles have been transformed into liquid drops. About 12 h later, point b1 is passed which should lead to the crystallisation of phase S1. The particles remain in this state until point c1 has been passed, at which time they return to liquid state. Later, when point d1 has been reached, phase S1 can recrystallise.

133

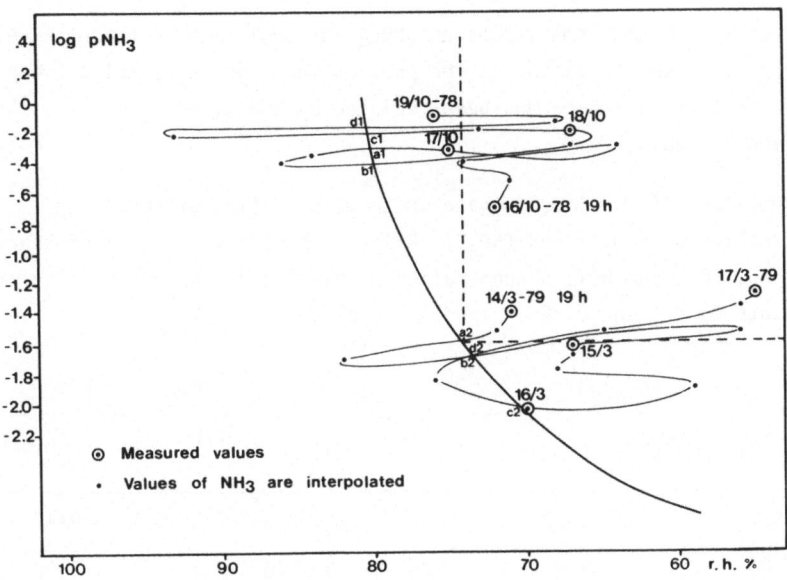

Figure 5: Values of r.h. and pNH$_3$ at Rörvik during two three-day periods

In the same manner, a sulphuric acid aerosol is assumed to have been introduced into the air on the 14 March 1979. Here, too, phases S1 and S2 may
crystallise initially, then transforming to liquid phase about 8 h later
when point a2 has been passed. After another 6 h, point b2 will be passed
and phase S2 can crystallise. This state remains for about 32 h, after
which point c2 has been passed and the particles transform to liquid phase.
After another few hours, point d2 will be passed and phase S2 can recrystallise.

Thus, the diagram predicts that the particles from the October period should
have a stoichiometric composition approximately corresponding to $(NH_4)_2SO_4$
and those from March corresponding to $(NH_4)_3H(SO_4)_2$. Analysis of daily
particle samples taken on these days has given, for the ratio of NH_4^+/H^+,
very high values (22 and higher) for October samples and values from
2.8 to 1.7 for the March samples, which can be considered to be a fair agreement.

Including HNO_3 and HCl in the diagram

Finally a few words on the reaction of $HNO_3(g)$ and $HCl(g)$ with the ammonium sulphate particles.

In the same way as for NH_3 some values of $pNHO_3$ and pHCl have been tentatively calculated for a system consisting of the previous one, where 10 mol % (resp. 5 mole %) of the total sulphate has been replaced by $NaNO_3$, resp. NaCl [4]. The result is given in Table 1.

Table 1: Composition of the liquid phase in terms of ratios of total concentration of (NH_4^+) and (SO_4^-). Relative humidity, partial pressures
of NH_3, HNO_3 and HCl. Concentration of nitrate and chloride in solution
amounts to 0.1 and 0.05, respectively of total sulphate.

No.	Solid phase	$\dfrac{2C_{(NH_4)_2SO_4}}{C_{H_2SO_4} + C_{(NH_4)_2SO_4}}$	R.H. %	P_{NH_3}	P_{HNO_3} ppb	P_{HCl}
1	$(NH_4)_2SO_4$	2.00	80	$1.3 \cdot 10^4$	$6.9 \cdot 10^{-5}$	$3.0 \cdot 10^{-4}$
2	$(NH_4)_2SO_4$	1.86	79	$3.9 \cdot 10^{-1}$	2.7	$1.1 \cdot 10^1$
3	$(NH_4)_2SO_4$	1.74	77	$1.4 \cdot 10^{-1}$	8.1	$3.9 \cdot 10^1$
4	$(NH_4)_2SO_4$	1.53	72	$3.2 \cdot 10^{-2}$	$4.0 \cdot 10^1$	$2.0 \cdot 10^2$
5	$(NH_4)_2SO_4$ + $+(NH_4)_3H(SO_4)_2$	1.42	69	$2.0 \cdot 10^{-2}$	$9.7 \cdot 10^1$	$3.7 \cdot 10^2$
6	$(NH_4)_3H(SO_4)_2$	1.05	55	$1.7 \cdot 10^{-3}$	$2.5 \cdot 10^3$	$2.2 \cdot 10^4$
7 +)	$(NH_4)_3H(SO_4)_2$ + $+ NH_4HSO_4$	0.98	40	$1.3 \cdot 10^{-4}$	$1.3 \cdot 10^4$	$6.8 \cdot 10^5$
1'	none	2.00	88	$8.0 \cdot 10^3$	$9.0 \cdot 10^{-3}$	$2.4 \cdot 10^{-4}$
5'	none	1.42	81	$2.8 \cdot 10^{-2}$	$3.5 \cdot 10^{-1}$	$1.2 \cdot 10^2$
7'	none	0.98	60	$1.8 \cdot 10^{-3}$	$2.2 \cdot 10^3$	$1.3 \cdot 10^4$
7''	none	0.98	78	$5.5 \cdot 10^{-3}$	$1.9 \cdot 10^2$	$4.6 \cdot 10^2$
+) uncertain values						

As is seen stoichiometrically neutral $(NH_4)_2SO_4$ solution will correspond to very high pNH_3 and low $pHNO_3$ but already a solution with a NH_4^+/SO_4^{2-} ratio of 1.05 will correspond to a $pHNO_3$ value of $2,5 \cdot 10^3$ ppb, i.e. much higher than is observed in the ground layer of the atmosphere. This explains the very low NO_3^- concentration observed in acid particles as illustrated in Table 2.

Table 2: Comparison of particle composition during white and black episodes. (Swedish westcoast, 1975)

episode type	N equiv./M^3						NH_4^+/H^+	soot UG/M^3
	SO_4^{2-}	NO_3^-	$SO_4^{2-} + NO_3^-$	NH_4^+	H^+	$NH_4^+ + H^+$		
black	308	33	341	342	7	349	51	23
white	317	<3	$\approx$ 320	233	75	309	3.1	2.3

Non equilibrium conditions ·

The discussion above was based on the assumption that equilibrium was prevailing. In fact the observation of the particle composition indicated that this sometimes could be the case.

There are, however, other cases when non equilibrium condition seems to be at hand. Two cases seem to be important.

During the last years we have followed the ammonia concentration in air at several places in Sweden and at the same time established the concentration of NH_4^+ and H^+ ions in rainwater.

From the concentration ratio NH_4^+/H^+ the correspondent equilibrium concentration of NH_3 in the air can be calculated. The calculated value is finally compared with the observed one. The result is given in Table 3. As is seen the observed values are 1-2 orders of magnitude higher. This means that the concentration observed in rainwater probably represent equilibrium condition at the cloud base. This seems to be true even for other components.

Another case of deviation from equilibrium is discussed in the next section.

Table 3: Ammonia concentration observed in the air at ground layer compared to concentrations calculated from the composition of precipitations. (Swedish westcoast)

1979 Date	μekv/l NH_4^+	H^+	NH_4^+/H^+	NH_3 calc. $nmol/m^3$	NH_3 obs. $nmol/m^3$	NH_3 obs $/NH_3$ ca
July						
16	8.3	22	0.38	0.16	38.5	241
17	11.7	45	0.26	0.11	31.8	289
18	15.0	53	0.28	0.11	27.1	246
19	11.1	16	0.69	0.28	17.8	64
20	27.8	23	1.21	0.49	78.4	160
Sept.						
2	121	212	0.57	0.23	7.5	33
8	67	529	0.13	0.05	8.5	170
9	101	264	0.38	0.16	55.2	345
11	37	47	0.79	0.32	1.5	5
12	493	≈ 0	stort	högt	38.1	-
13	128	91	1.41	0.57	20.5	36
14	29	40	0.73	0.30	18.9	63
16	43	156	0.28	0.11	20.3	185
17	27	38	0.71	0.29	27.4	94
18	19	56	0.34	0.14	41.6	297
20	36	43	0.84	0.34	4.6	14
21	22	6	3.7	1.5	22.0	15
23	27	36	0.75	0.31	32.7	105
25	58	145	0.40	0.16	25.9	162
26	38	62	0.61	0.25	12.5	50

4. The system containing graphite

In Table 2 some data concerning two types of fine particles are given. They are called white and black particles according to their colour. Since Novakov has shown that this colour depends on the presence of graphite it would be better to call them "graphite poor" and "graphite rich" particles. These particle types are observed episodically, the black especially during winter time. It is seen, however, that this graphite rich particles have a high NH_4^+/H^+ ratio corresponding to high pNH_3 value.

But according to recent measurements (Figure 6) NH_3-concentration is very low during winter time.

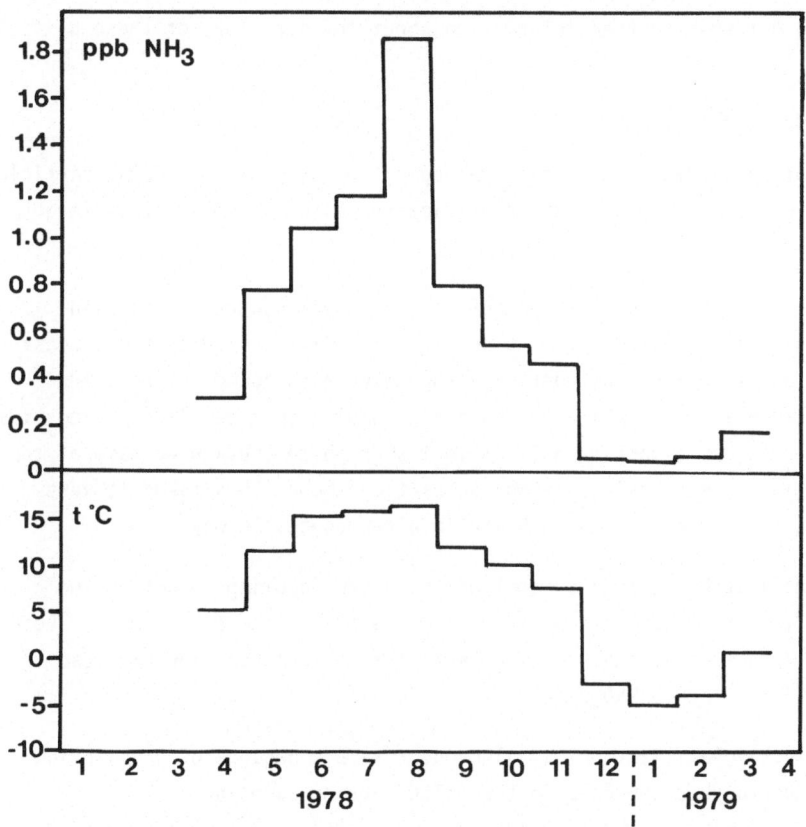

Figure 6: Monthly means of ammonia concentration and temperature (Swedish westcoast)

Earlier we had found that the composition of the graphite poor particles was in good agreement with equilibrium conditions. This seems not to be the case for the graphite rich ones. The reason for this has been indicated by Gundel et al.[6].

It seems as if the graphite phase which is produced by combustion processes already in the stack at relatively high temperature reacts with NO to produce NH_3-derivatives bound to organic functional groups on the graphite layer.

In this way there may be built up an alkaline sorbed layer. The outer layer of the same particles may, of course, attain equilibrium with the atmosphere. However, when these particles are leached in water or acid for subsequent analysis at least a part of the sorbed could be soaked out in form of the NH_3 increasing the NH_4^+-content of the solution. It would be very interesting to get some further information about the structure of these particles.

5. Coarse particles

Because of the variety of species belonging to the group of coarse particles interaction of gases with these particles represent a number of different problems.

A general feature of the coarse particles is their neutral or alkaline character. This means that acid gases will readily react with them. Consequently, as is well known, most airborne nitrate is found in the coarse particle mode and its tail entering the accumulation mode. At the same time there is probably almost no nitrate in the graphite containing tail of the accumulation mode entering the coarse particle mode. This makes it very difficult to sample airborne nitrate in a reproduceable way.

These complications demand a sampling procedure enabling to get in the sample all particle sizes containing nitrate and at the same time to avoid reaction between acid and alkaline particles as such reaction may lead to the volatilization of HNO_3.

Further such volatilization may also occur simply because of equilibrium changes due to pressure drop in the filter during sampling.

There are until now not very many investigations of direct interactions between coarse particles and gases.
However, some important cases of reaction between SO_2 and fly ash have been studied. A few results obtained in our laboratory (not yet published) might be mentioned.

In order to get some information of the possibility of catalytic sulphate formation in a stack plume the following measurements were performed. Parts of different fly ash samples were analyzed for water soluble sulphur compounds (mostly sulphate). Other parts were exposed for SO_2 under controlled conditions using a gas containing 1 ppm SO_2, having an r.h. of 70 % and a temperature of $25^\circ C$. After a fixed time of exposure the sample was extracted with water and the obtained solution analysed for sulphur compounds. It

turned out that different fly ash species gave very different results sometimes showing a very important, sometimes poor rate of formation of sulphite
and sulphate.

For a certain fly ash these measurements were performed on different size
fractions of the ash. A plot of the sulphur sorbed (converted) by surface
unit towards the logarithm of the diameter characterizing a size fraction
of the particles was made. It is given in Figure 7. As is seen here during
the exposure time only small amounts of SO_2 have been converted by the small
particles probably because of the acid reaction. However, when the particle
diameter exceeds $\sim$ 3 µm. The conversion rate increases drastically. Most of
the sulphur sorbed by the particles in this region has been found to consist
of sulphite.

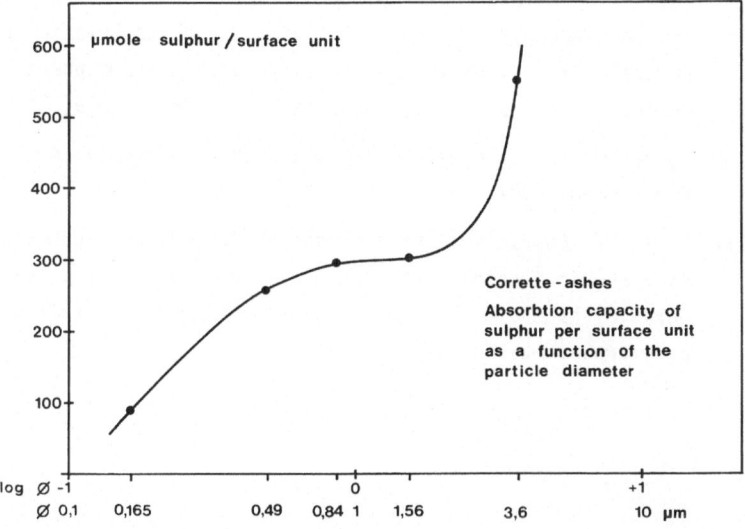

Figure 7: Sulphur absorption on fly ash

The above mentioned is a special case. Other coarse particle samples will
give different and probably today unpredictable results. We are thus still
very far from a complete understanding of these systems.

References

1. K.T. Whitby and B. Cantrell,   Atmospheric aerosols - characteristics
        and measurement. In Proceedings ICESA International Conference
        on Environmental Sensing and Assessment, Las Vegas, Nevada,
        September 14-19, 1975. Institute of Electrical and Electronics
        Engineers, Inc., New York, N.Y. IEEE Catalog 2 (29-1):1

2. J. D'Ans, Z. anorg. Chem. 65 (1909) 228

3. J.N. Tang, et al., Aerosol growth studies - IV. Phase transformation of
        mixed salt aerosols in moist atmosphere. J. Aerosol Sci.
        9:505-511, 1978

4. Ying-Hua Lee, C. Brosset, Interaction of gases with sulphuric acid aero-
        sol in the atmosphere. Paper presented at the WMO symposium on
        the Long Range Transport of Pollutants and its Relation to
        General Circulation Including Stratospheric/Tropospheric Ex-
        change Processes. Sofia, Bulgaria, 1-5 October 1979

5. Martin Ferm, Method for determination of atmospheric ammonia. Atmospheric
        Environment, Vol. 13, pp 1385-1393, 1979

6. L.A. Gundel, et al., Characterization of particulate amines. LBL-8696.
        Atmospheric Aerosol Research. Annual Report 1977-78, Lawrence
        Berkeley Laboratory, University of California, Berkeley,
        California.

MICROCHEMICAL CHARACTERIZATION OF AEROSOLS*

TIHOMIR NOVAKOV
Lawrence Berkeley Laboratory,
University of California,
Berkeley, California 94720

*This work was supported by the Biomedical and Environmental Research Division
of the U.S. Department of Energy under contract No. W-7405-ENG-48, and by the
National Science Foundation.

Abstract. This paper outlines a methodology to quantitatively differentiate
classes of ambient particulate carbon. The analysis of results provides an
assessment of the amounts of primary and secondary carbonaceous material, as
well as the amounts of total carbon, organic carbon, and black or graphitic
carbon. Two experimental approaches have been used in this study. The first
involves a systematic comparison of average ambient black carbon and total
carbon ratios with those of sources. The second approach relies on the evolved
gas (CO_2) thermal analysis method as a means for "fingerprinting" the organic
and black carbon components of source and ambient particles.

1. Introduction

The main purpose of chemical characterization of aerosol particles is to provide
information about their chemical composition. This information should ultimately
help in identifying their sources, formation mechanisms, and fate in the atmosphere.

Carbon-, sulfur-, and nitrogen-containing particles account for most of the anthro-
pogenically generated particulate burden in urban areas. Considerable attention
has been given to understanding the origin and speciation of the sulfur and nitro-

gen components, but until recently relatively little effort has been directed toward the carbonaceous aerosol, which is often the single most important contributor to the submicron aerosol mass. The objective of this paper is to outline a methodology developed in our laboratory to quantitate the amounts of different classes of carbonaceous particulates collected at various urban locations in the United States. The analysis of the results provides an assessment of the relative amounts of primary and secondary particulate carbon at these locations.

Carbonaceous particles in the atmosphere consist of two major components - graphitic or black carbon (sometimes referred to as elemental or free carbon) and organic material. The latter can be either directly emitted from sources (primary organics) or produced by atmospheric reactions from gaseous precursors (secondary organics). For the sake of clarity, we define soot as the total primary carbonaceous material, i.e., the sum of graphitic carbon and primary organics.

Black carbon can be produced only in a combustion process and is therefore definitely primary. Because of this, black carbon can be used as a tracer for primary carbonaceous particles. The problem of differentiating the primary and secondary components would be simple if black carbon were the only primary component. However, because many sources besides black carbon produce substantial amounts of primary organic material, the differentiation of these two components can be achieved only by a systematic study of large numbers of samples collected directly from sources, source-dominated environments, and well-aged ambient air. The ambient samples should also be collected in areas with widely different atmospheric chemical characteristics (e.g., degree of photochemical activity).

Two approaches have been used in our studies. The first essentially involves a systematic study of 24-hr average black carbon to total carbon ratios, since measurements of this ratio from a number of source samples give insights into the relative black to total carbon ratio of primary emissions and the source variabilities. Secondary material will not contain the black component, but it will increase the total mass of carbon and will therefore reduce the black to total carbon fraction.

For example, photochemical gas to particle conversion reactions should be most pronounced in the summer in the Los Angeles air basin, while in the winter these reactions should play a much smaller role and the primary component should be much more important. These different photochemical conditions should manifest themselves in the ratio of the black carbon to total carbon

of these particles. That is, under high photochemical conditions one would
expect this ratio to be significantly smaller than under conditions obviously
heavily influenced by sources.

This approach to the identification and quantitation of primary and secondary
carbonaceous aerosols involves a systematic comparison of particulates collected
from a wide range of ambient sites as well as combustion sources. Ambient
particulates are sampled at sites that differ significantly in meteorology,
photochemical activity, and source composition. Source samples have been obtained at a tunnel and a parking garage, and from direct source samplings.

The second approach in our studies relies on the evolved gas (CO_2) thermal
analysis method as a means of "fingerprinting" the organic and black carbon
components of source emissions, source-enriched samples, and ambient particles.
The differences between these should correspond to the secondary component.

2. Methods

In our field experiments the samples are collected in parallel on prefired
quartz fiber and Millipore filter membranes. The Millipore filter is used for
X-ray fluorescence (XRF) elemental analysis and for the LBL laser transmission
technique [1]. The latter technique gives a measurement that is proportional to
the amount of light-absorbing (black) carbon present on the filter. The quartz
filter is used for total carbon determination by a combustion method similar
to that described by Mueller et al. [2].

A schematic representation of the LBL laser transmission (optical attenuation)
apparatus is shown in Fig. 1.

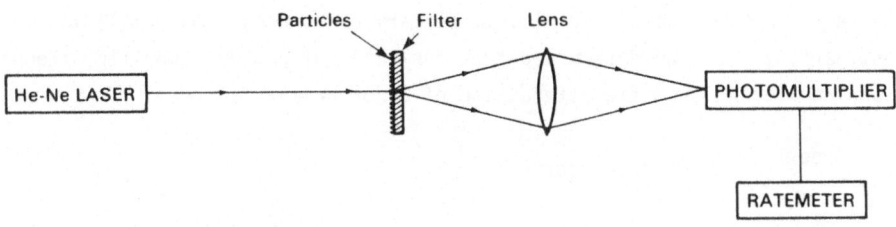

Figure 1: Schematic representation of the optical attenuation (laser transmission) apparatus (from Ref. 1)

This apparatus compares the transmission of a 633-nm He-Ne laser beam through a loaded filter relative to that of a blank filter. The loaded filters are placed in the beam with the loaded side towards the laser; after multiple scattering through the filter substrate, the light is collected by an f/1 lens and focused on a photomultiplier tube. The data presented in this paper were obtained from particles collected on Millipore filters. This technique is based on a principle similar to that of the opal glass method used by Weiss et al.[3] and measures the absorbing, rather than the scattering, properties of the aerosol. The relationship between the optical attenuation and the black carbon content can be written as:

$$[C_{black}] = (1/K) \times ATN, \tag{1}$$

where $ATN = -100 \ln(I/I_0)$. I and I_0 are the transmitted light intensities for the loaded filter and for the filter blank.

Besides the black carbon, particulate material also contains organic material which is not optically absorbing. The total amount of particulate carbon is then:

$$[C_{tot}] = [C_{black}] + [C_{org}]. \tag{2}$$

We define specific attenuation (α) as the attenuation per unit mass of total carbon:

$$\sigma \equiv \frac{ATN}{[C_{tot}]} = K \times [C_{black}] / [C_{tot}]. \tag{3}$$

The determination of specific attenuation therefore gives an estimate of black carbon as a fraction of total carbon.

The proportionality constant K, which is equal to the specific attenuation of black carbon alone, was recently shown to have an average value of 20[4]. In principle the percentage of soot (i.e., primary carbonaceous material) in ambient particles can be determined from the ratio of ambient specific attenuation and an average specific attenuation of major primary sources:

$$[Soot]/C = \sigma_{ambient}/\sigma_{source}. \tag{4}$$

The thermal analysis method used in our studies is a modified version of the apparatus originally developed by Malissa et al.[5]. Our version[6] enables measurement of optical attenuation simultaneously with the evolution of CO_2. Thermal analysis is used to obtain total carbon, black carbon, organic carbon, and carbonate carbon. A schematic representation of the thermal analysis apparatus used in our studies is shown in Fig. 2.

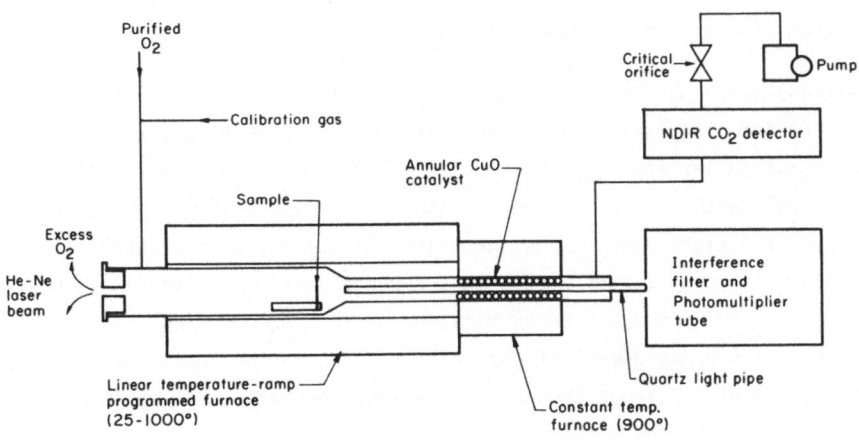

COMBUSTION OPTICO-THERMAL ANALYSIS (EGA-CO$_2$)

Figure 2: Schematic representation of the thermal analysis apparatus

The main components of this apparatus are a quartz tube and a temperature-
programmed furnace. The tube is mounted axially inside the furnace. The
particulate sample, collected on a prefired quartz filter, is placed in
the quartz tube so its surface is perpendicular to the tube axis. The tube
is constantly supplied with pure oxygen. The excess oxygen escapes through
an axial opening at the end of the tube, while the remainder of the oxygen
(and other gases evolved during analysis) passes through a nondispersive
infrared analyzer at a constant flow. In addition to the variable temperature
furnace, the apparatus also contains a constant temperature furance, usually
kept at about 850°C. The segment of quartz tube inside the constant temperature furnace is filled with a copper oxide catalyst to ensure that carbon-
containing gases evolved from the sample are completely converted to CO_2.
This is especially important at relatively low temperatures when complete
oxidation to CO_2 does not occur.

The actual measurement consists in monitoring the CO_2 concentration as a
function of the sample temperature. The result is a "thermogram" - a plot
of the CO_2 concentration vs. temperature. The area under the thermogram is
proportional to the carbon content of the sample. The carbon content is

quantitated by calibrating with a calibration gas (CO_2 in oxygen) and by measuring the flow rate through the system. This calibration is crosschecked by analyzing samples of known carbon content. The thermograms of ambient and source aerosol samples reveal distinct features in the form of peaks or groups of peaks that correspond to volatilization, pyrolysis, oxidation, and decomposition of the carbonaceous material.

To determine which of the thermogram peaks corresponds to black graphitic carbon, the intensity of the light beam produced by a He-Ne laser is monitored by a photomultiplier and displayed by the second pen of the chart recorder, simultaneously with the measurement of the CO_2 concentration [6]. In actual experiments the light penetrating the filter is collected by a quartz light guide and filtered by a narrow band interference filter to eliminate the effect of the glow of the furnaces. An examination of the CO_2 and light intensity traces enables the assignment of the peak or peaks in the thermograms corresponding to the black carbon because they appear concomitantly with the decrease in sample absorptivity.

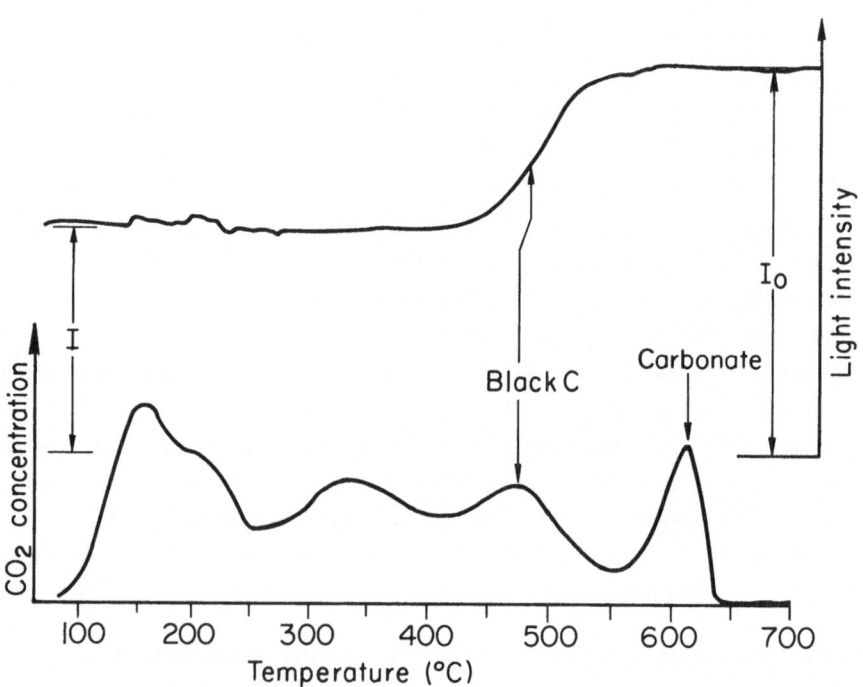

Figure 3: Thermogram of an ambient sample showing carbonate, black carbon, and several forms of organic material

In Fig. 3, a complete thermogram of an ambient sample is shown. The lower trace represents the CO_2 concentration vs. the sample temperature, while the upper curve corresponds to the light intensity of the laser light beam that reaches the detector during the temperature scan. Inspection of the thermogram shows that a sudden change in the light intensity occurs concomitantly with the evolution of the CO_2 peak at about 470°C. The light intensity I_0, after the 470°C peak has evolved, corresponds to that of a blank filter. This demonstrates that the light-absorbing species in the sample are combustible and carbonaceous. We refer to these species as black carbon. The carbonate peak evolves at about 600°C; and as carbonate is not light absorbing, it does not change the optical attenuation of the sample. In addition to black carbon and carbonate, the thermogram in Fig. 3 also shows several distinct groups of peaks at temperatures below ∿ 400°C that correspond to various organics.

The thermogram in Fig. 3 was obtained with a 1.46-cm-diameter disc cut out of a sample collected on prefired quartz fiber filters. The temperature ramp rate was 10°C/minute. The integrated area under the CO_2 trace is proportional to the total carbon concentration. For this sample the total carbon concentration, determined by thermal analysis, was 17.9 µg $[C]/cm^2$. The black carbon, determined from the thermogram, composes 14% of the total carbon. This value can be crosschecked by using the optical attenuation and total carbon data. The specific attenuation for this sample, determined in a separate measurement, is $\sigma \equiv ATN/C = 3.00$. The estimated percentage of black carbon (as a percent of total C), determined from measurement of optical attenuation and total carbon only, is $100 \times 3.0/20.0 = 15\%$. This value is in excellent agreement with the percentage of black carbon determined directly from the CO_2 thermogram.

3. Results and Discussion

The data presented in this paper consist of information obtained from analyses of 24-hr samples (collected weekdays) and multi-day samples (collected over weekends) [4]. Table 1 lists the routine sampling sites with the beginning date of sampling. In this section we present data on total average 24-hr concentrations of total carbon, specific attenuation, and estimated black carbon concentrations for ambient and source samples. By determining an average specific attenuation value for sources, the soot (total primary carbon) fraction can be estimated from Eq. 4.

Figure 4 shows the variations of 24-hr total carbon (weekends excluded) at the Fremont, California, site. These data cover the period from July 1977 to January 1980.

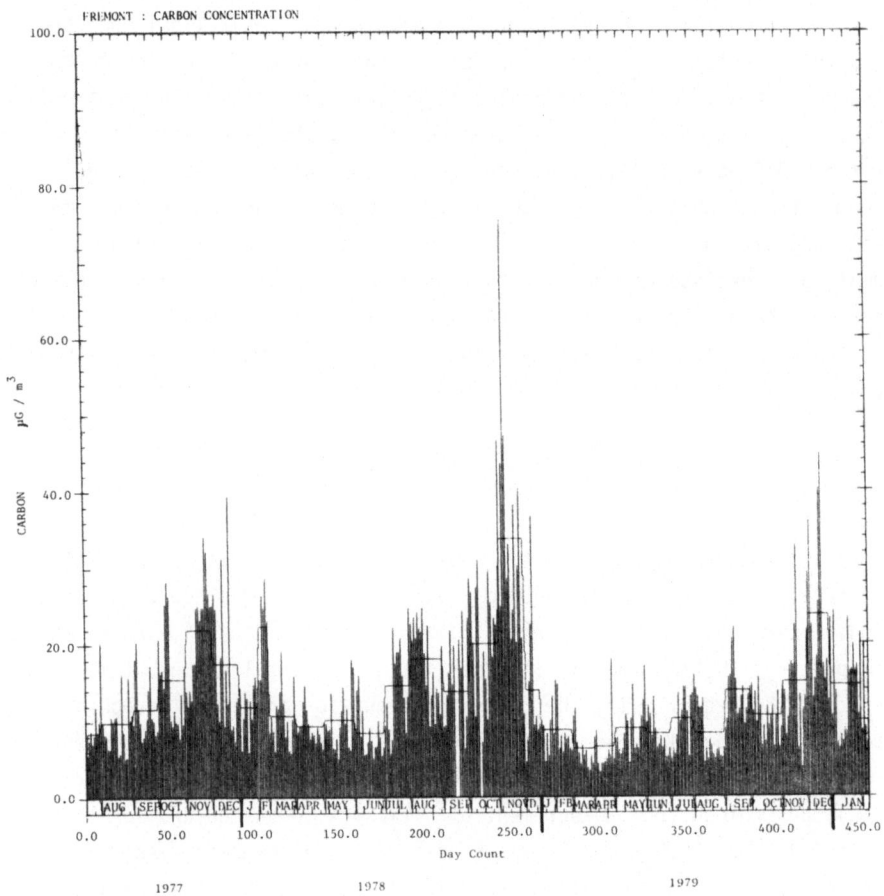

Figure 4: Variations in the daily total carbon concentration at the Fremont, California, site (from Ref. 4)

The 24-hr histogram superimposed on the bar diagram represents the monthly averages. It is evident that there are significant day-to-day variations in total carbon. The maximum and minimum daily concentrations differ by an order of magnitude. The monthly averages are at peak values during the November-December periods of each year. The variations in optical attenuation for the same samples are represented in Fig. 5.

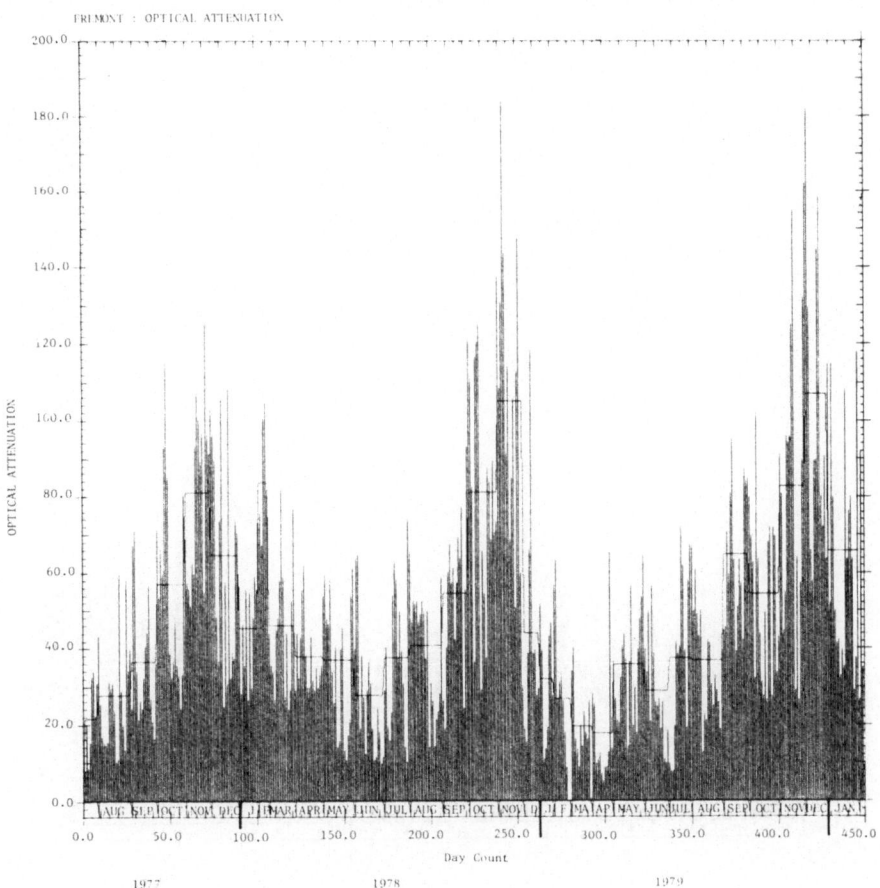

Figure 5: Variations in the optical attenuation at the Fremont, California, site (from Ref. 4)

The pattern of ATN values resembles that of total carbon and shows similar seasonal variations. The specific attenuation (ATN/C) variations represented in Fig. 6 are much less pronounced and show no clear seasonal variations. Similar features of total C, ATN, and ATN/C are also observed at other sites.

The data on total carbon, black carbon (from Eq. 3), and specific attenuation are presented in Tables 2 and 3. These results imply that there is a correlation between the black carbon and the total carbon content at every site studied.

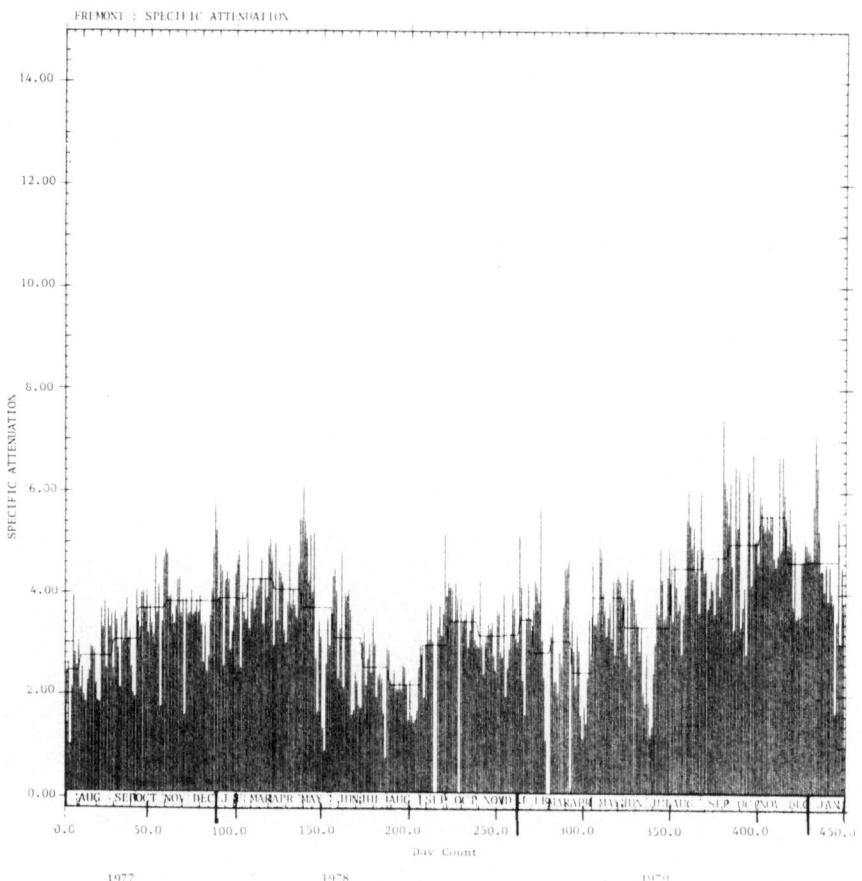

Figure 6: Variations in the specific attenuation at the Fremont, California, site (from Ref. 4)

Furthermore, a study of a number of source samples shows that there is also a strong correlation between optical attenuation and total carbon for these samples. The correlations between optical attenuation and total carbon for the three California sites, Argonne, and source samples are shown in Fig. 7 (a-e) [7].

Results obtained from ambient samples imply that the fraction of graphitic soot to total particulate carbon is approximately constant under the wide range of conditions occurring at a given site.

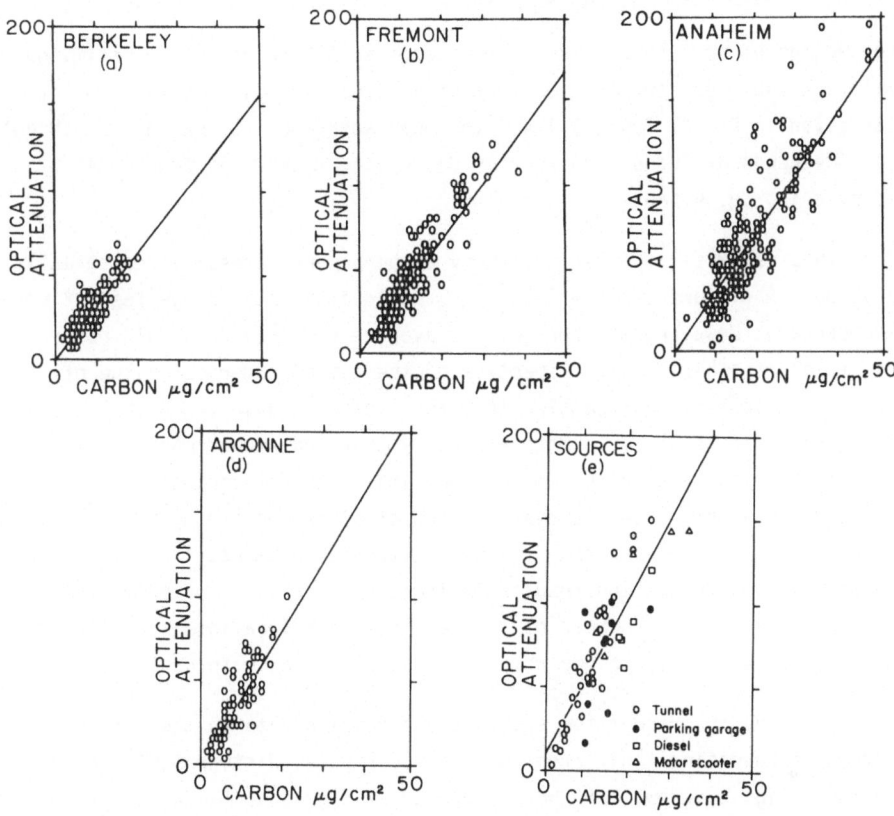

Figure 7: Plots of optical attenuation versus carbon loading in µg/cm² for particulate samples collected at Berkeley, Fremont, Anaheim, and Argonne, and from various combustion sources. The solid line represents the least squares fit of the data points (from Ref. 7)

On specific days, however, there can be large variations in the ratio, reflecting the variations in the relative amounts of organic and black carbon. The least squares fit of the data shows regional differences which are related to the fraction of black carbon due to primary emissions. These differences suggest an increase in the relative importance of the primary component for samples collected at Berkeley, Fremont, Anaheim, and Argonne.

Soot contains not only black carbon but also various organic material. Because the organic soot component does not absorb light, the specific attenuation of soot is much less than 20, the σ value of pure black carbon. Table 4 lists the average and extreme values of specific attenuation and the black carbon fraction of a number of source samples.

The percentage of soot in ambient carbonaceous particulates can be estimated by comparing the σ of sources with that of ambient samples. The fraction of soot is given in Eq. 4. Table 5 lists the mean specific attenuation of ambient samples (weekends excluded) in order of decreasing σ and soot fractions obtained by using Eq. 4 and σ_{source} = 5.85.

Based on this estimate, the New York City carbonaceous aerosol is essentially primary soot. A different value of σ_{source} would certainly change the estimated soot percentage. However, New York City's average soot content would nevertheless remain the highest, irrespective of the actual numerical value of σ_{source}. It is logical that samples from this location have the highest soot content because the site represents a heavily traveled street canyon. Fremont and Anaheim samples have on the average the smallest soot content, as may be expected, because both sites represent receptor sites. According to the above estimate, Denver has the smallest specific attenuation value. It is possible that high-altitude combustion results in increased emissions of primary organics; however, we note that the number of samples from this location is small compared to that from other sites, so the results should be taken with caution.

It is instructive to present the specific attenuation data in the form of histograms representing their frequency of occurrence. Histograms for New York and Fremont (Fig. 8) show that the occurrence of high specific attenuation samples is much greater for New York than for Fremont.

In Figure 9 the histogram of specific attenuations of a number of source samples is shown together with those for New York and Fremont. The distribution for sources looks similar to the distribution for New York. This supports the inference that the New York samples, on the average, consist almost entirely of primary carbonaceous material. Histograms for other sampling sites are shown in Figures 10 and 11.

```
HISTJGRAM OF ATN/  C FOR SITES NYC
FOR DATES 27 NOV 79 TO 14 APR 80
NPOINTS = 211    MEAN =  5.59    SDEV =  1.34    NO WKENDS

    18 *                              X
       .                              X.
       .                        X   XXX .X
       .                        XXXXXX XX
       .                       X XXXXXX XX
       .                      .XXXXXXXXX XXX
       .               XX   XXXXXXXXX.XXX X
       .              .XX.XXXXXXXXXXXXXXX XX
       .           . . .XXXXXXXXXXXXXXXXXXXXY.. .   .
       .         X   X X XXXXXXYXYXXXX XXXXXXXXXXX XX    X X
        .........*.........*.........*.........*.........*
       0.        2.00      4.00      6.00      8.00      10.00    ATN/  C
```

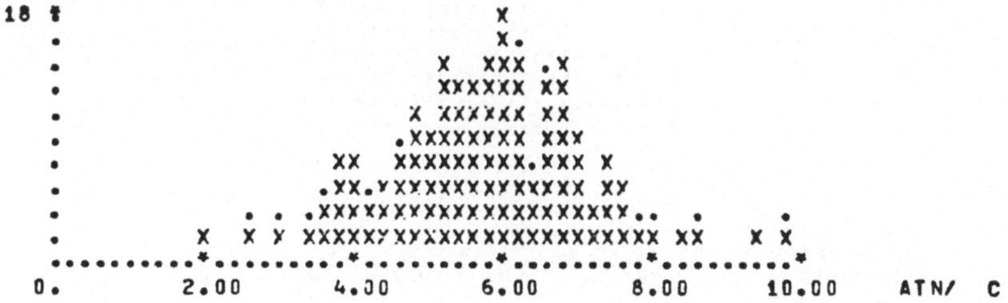

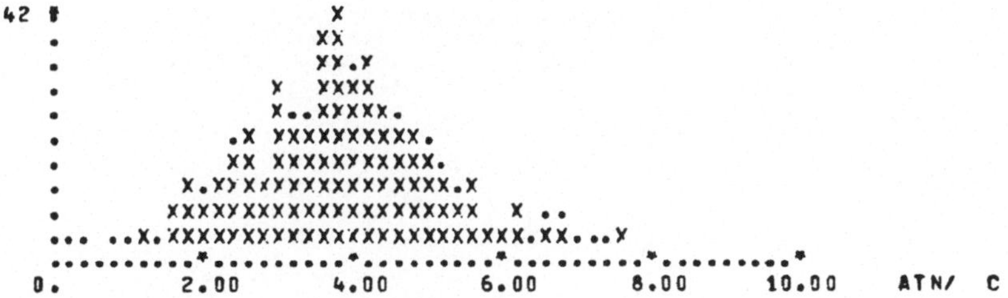

```
    42 *                   X
       .                   XX
       .                   XY.Y
       .               X   YXXX
       .               X..XYXXX.
       .             .X YXXXXXXXXX.
       .             XX XXXXXYXXXX.
       .          X.XYX XYXXXXXXYXXXX.X
       .          XXXXYXXXXXXXXXXXXXXX  X ..
        ... ..X.XXXXYXXXXXXXXXXXXXXXXXXX.XX...Y
        .........*.........*.........*.........*.........*
       0.        2.00      4.00      6.00      8.00      10.00    ATN/  C

HISTOGRAM OF ATN/  C FOR SITES FRE
FOR DATES 18 JUL 77 TO 12 MAR 80
NPOINTS = 461    MEAN =  3.74    SDEV =  1.25    NO WKENDS
```

Figure 8: Distribution of specific attenuation for the New York and Fremont
          sites (from Ref. 4)

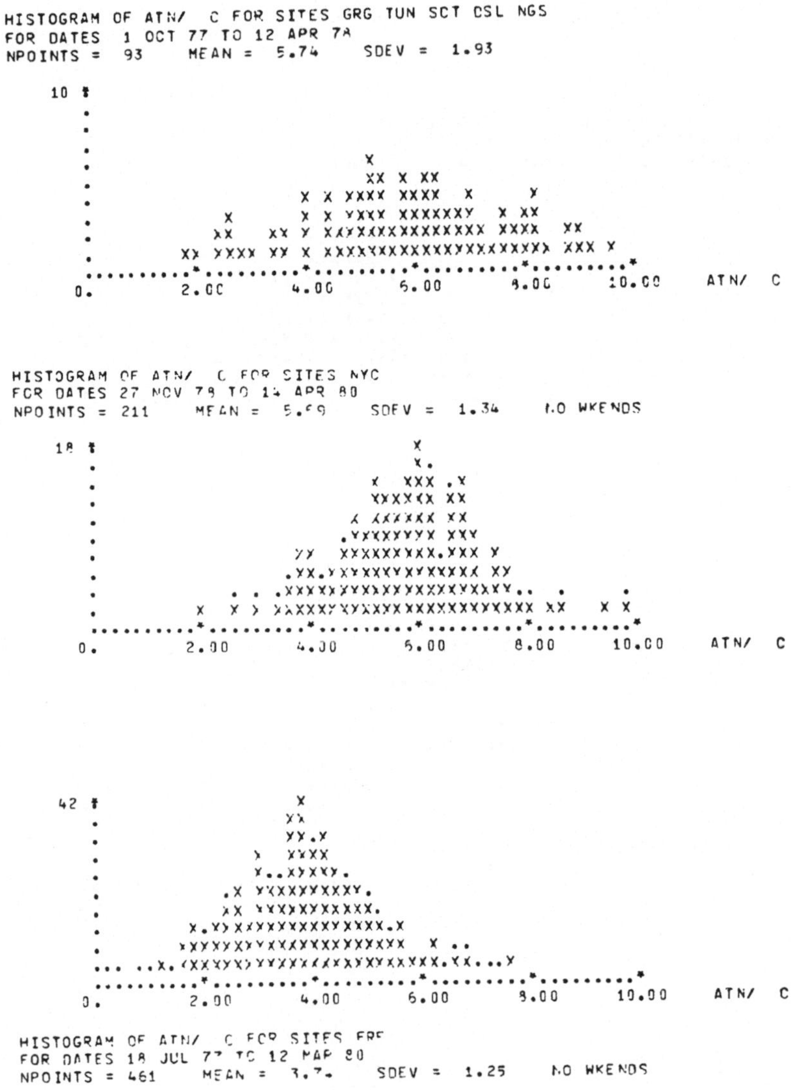

HISTOGRAM OF ATN/ C FOR SITES GRG TUN SCT CSL NGS
FOR DATES 1 OCT 77 TO 12 APR 7A
NPOINTS =   93   MEAN = 5.74   SDEV =   1.93

```
  10 *
     .
     .
     .
     .                            X
     .                         XX  X  XX
     .                       X X XXXX XXXX    X       Y
     .                   X   X X YXXX XXXXXXY   X XX
     .           X       XX X Y XX/XXXXXXXXXXXX XXXX   XX
     .          XX    XX XX Y X XXXXYXXXXXXXXXXXXXXXXX XXX X
     ...........*..........*..........*..........*..........*
     0.         2.0C       4.00       6.00       8.0C      10.C0    ATN/  C
```

HISTOGRAM OF ATN/ C FOR SITES NYC
FOR DATES 27 NOV 79 TO 14 APR 80
NPOINTS = 211   MEAN = 5.69   SDEV =   1.34   1.0 WKENDS

```
  18 *
     .                          X
     .                          X.
     .                    X   XXX .Y
     .                   XXXXXX XX
     .                   X XXXXXX XX
     .                  .YXXXYYYX XXY
     .            YY   XXXXXXXX.YXX Y
     .          .YX.YXYXXXYXYXXXX XY
     .      . .  .XXXXXXYXXXXXXXXXXXYXXXY.. .
     .      X   X >  XXXXXYYYXXXX XXXXXXXXXXX XX    X X
     ...........*..........*..........*..........*..........*
     0.         2.00       4.00       6.00       8.00      10.C0    ATN/  C
```

```
  42 *
     .                    X
     .                   X X
     .                   YY.Y
     .                  >  XXXX
     .                 Y..X>XXY.
     .                .X Y<XXYYXXXY.
     .                XX YXYXYXXXXX.
     .             X.Y>X XYYXXXXXYXXXX.X
     .            YYYYXY YXXXYXXXXXXXX  X ..
     ...  ..X. <XXXYX>YYYXX<XXXXYYXXXX.YX...Y
     ...........*..........*..........*..........*..........*
     0.         2.00       4.00       6.00       8.00      10.00    ATN/  C
```

HISTOGRAM OF ATN/ C FOR SITES FRE
FOR DATES 18 JUL 77 TO 12 MAR 80
NPOINTS = 461   MEAN = 3.74   SDEV =   1.25   NO WKENDS

Figure 9: Distribution of specific attenuation for source, New York, and
Fremont samples (from Ref. 4)

```
HISTOGRAM OF ATN/  C FOR SITES ANA
FOR DATES 22 AUG 77 TO 29 JAN 30
NPOINTS = 444     MEAN =  3.93     SDEV =  1.71     NO WKENDS

34 *                         X  .
   .                      .  X  XX
   .                   .X. < >.XX
   .                   X/X <.XXXX
   .                   X/'XXXXYYX   . .
   .                 X.XX> XXXXYXX>X<XX
   .                 XYXX>>XXXXXX> XVX>  .
   .                >>XXX>>X>XX>XXXVXXY.XX X
   .               XYYYVX'X>XXXXX>XXXXVVX X.
   .        XX XYXXXYXXX>X<VXYXX>,XXX>XXXYXXX.... YX. ..Y... .
    ........*.........*.........*.........*.........*
   0.      2.00      4.00      6.00      5.00     10.00     ATN/  C
```

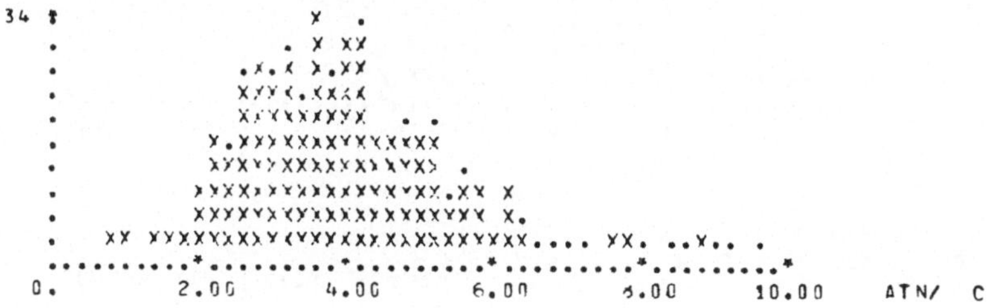

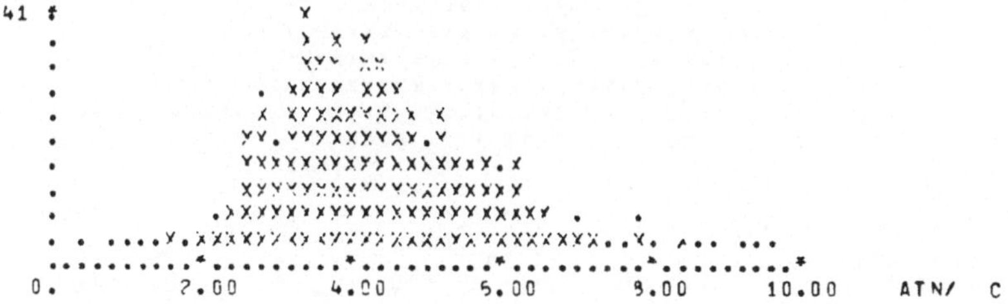

```
41 *                       Y
   .                      > X  Y
   .                      XYV  >.V
   .                   .  XYYY  XXY
   .                   X  XYXXXYX>X  X
   .                  YV.VYYXXXVXY.V
   .                  YYXXXXXYXYXXXXXXxX.X
   .                  XYYYYYXYVV VXXXYXXYX X
   .                .>XYYYY>YXYYXXYXXYYXXAXY  .   .
   . . ....Y.XYXXYY<>KYYYYXAXAYXYXXXXXXXX..X. A.. ...
    ........*.........*.........*.........*.........*
   0.      2.00      4.00      5.00      9.00     10.00     ATN/  C
```

```
HISTOGRAM OF ATN/  C FOR SITES BRK
FOR DATES  1 JUN 77 TO 17 APR 80
NPOINTS = 513    MEAN =  4.28    SDEV =  1.47     NO WKENDS
```

Figure 10: Distribution of specific attenuation for Anaheim, California, and Berkeley, California, sites (from Ref. 4)

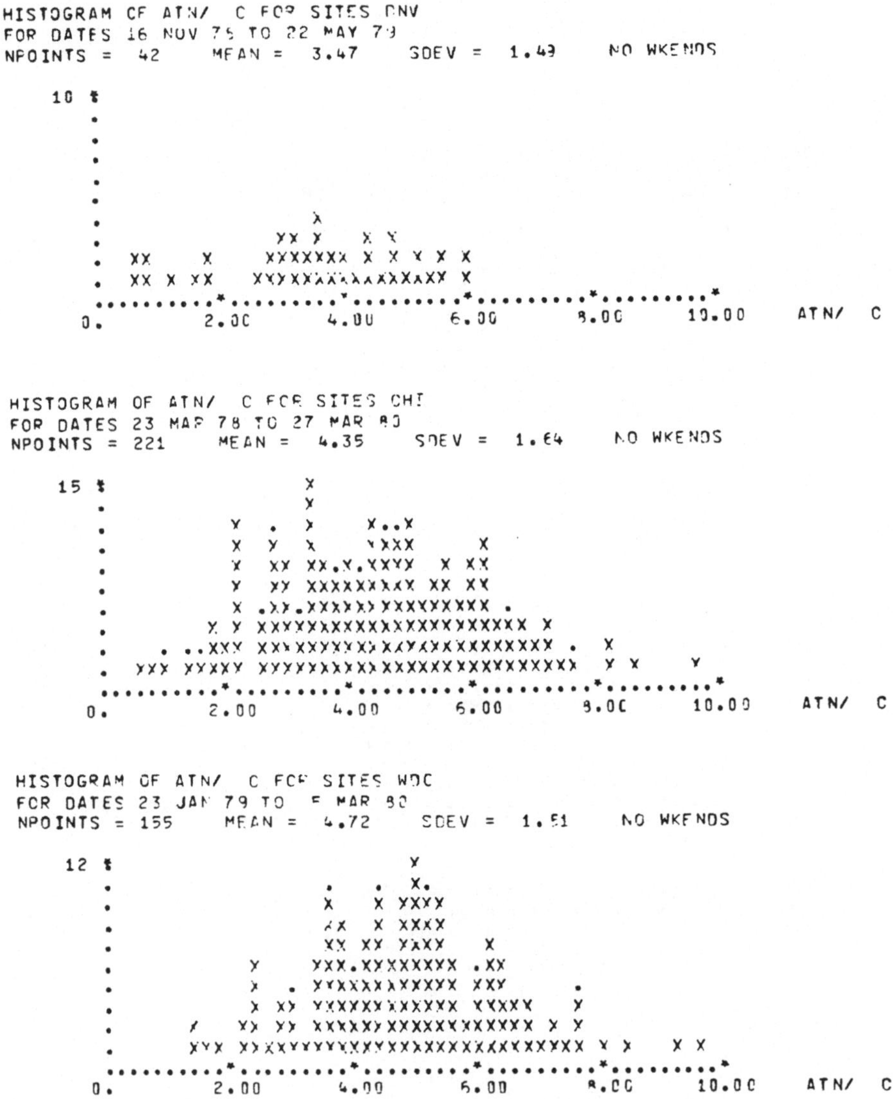

```
HISTOGRAM CF ATN/  C FOR SITES DNV
FOR DATES 16 NOV 75 TO 22 MAY 79
NPOINTS =   42     MEAN =   3.47     SDEV =   1.43       NO WKENDS

   10 *
      .
      .
      .
      .
      .
      .                      X
      .                    XX X     X X
      . XX     X         XXXXXXX X X X X X
      . XX X XX        XXYXXXXXXXAAAXXAXX X
      ...........*..........*..........*..........*..........*
      0.         2.00        4.00        6.00        8.00       10.00     ATN/  C


HISTOGRAM OF ATN/  C FCR SITES CHI
FOR DATES 23 MAR 78 TO 27 MAR 80
NPOINTS = 221     MEAN =   4.35     SDEV =   1.64       NO WKENDS

   15 *                    X
      .                    X
      .             Y  .  X    X..X
      .             X  Y  X    XXXX      X
      .             X  XX XX.X.XXYY  X XX
      .             Y  YY XXXXXXXXXX XX XX
      .             X .XX.XXXXXXX XXXXXXXXX .
      .             X X XXXXYXXXXXXXXXXXYXXXXX X
      . . ..XXX XXXXXXYYYXX XXXXXXXXXXXXXXX . X
      . YXX YYXXY XYYYXXXXXXXXXXXXXXXXXXXXX Y X   Y
      ...........*..........*..........*..........*..........*
      0.         2.00        4.00        6.00        8.00       10.00     ATN/  C


HISTOGRAM OF ATN/  C FCR SITES WDC
FOR DATES 23 JAN 79 TO  5 MAR 80
NPOINTS = 155     MEAN =   4.72     SDEV =   1.51       NO WKENDS

   12 *                    Y
      .                  .   X.
      .             X    X XXXX
      .             XX   X XXXX
      .             XX XX YAXY     X
      .        Y    YXX.XYXXXXXX .XX
      .        X  . YYXXXXXXXXXX XXY
      .        X XY YXXXXXXXXXXX YXXXX     Y
      .     /   YY YY XXXXXYXXXXXXXXXXXX X X
      .     XYX YYXXXYYYYYXXYYXXXXXXXXXAXXXXXX Y X    X X
      ...........*..........*..........*..........*..........*
      0.         2.00        4.00        6.00        8.00       10.00     ATN/  C
```

Figure 11: Distribution of specific attenuation for Denver, Colorado; Argonne, Illinois; and Gaithersburg, Maryland, sites (from Ref. 4)

It is clear from the results described so far that the ratio of black carbon to total carbon may vary on specific days. However, no large systematic differences are found as a function of the ozone concentration, which is viewed as an indicator of the photochemical activity [7]. This is graphically demonstrated in Figure 12, which shows the distribution of the ratios of the optical attenuation to total carbon content for ambient samples from all the California sites taken together, subdivided according to peak hour ozone concentration. Clearly there is no trend for high-ozone days to be characterized by aerosols which have a significantly reduced black carbon fraction. This places a low limit on the importance of secondary organic particulates formed in correlation with the ozone concentration.

Results in Table 5 suggest that the California sites have an organic component that occurs in excess of sources of source-dominated organics. This excess should be equal to the secondary organic material, which can be conveniently identified by the thermal analysis method.

We have already described how thermal analysis can be used to obtain the total carbon, black carbon, organic carbon, and carbonate carbon. The greatest strength of this method, however, is its ability to "fingerprint" source-produced carbonaceous particles and their contribution to the ambient aerosols. As an illustration, in Figures 13 and 14 we show the thermograms of a sample collected in Manhattan (high σ) and one collected in Berkeley (low σ). The common features of both thermograms are the black carbon peak, the peak at $\sim$ 340°C, and the peaks occurring below $\sim$ 250°C which correspond to the volatile organics. The Berkeley sample clearly shows the presence of a prominent peak at $\sim$ 380°C which is absent in the thermogram of the New York sample. This peak (or possibly group of peaks) may correspond to secondary organic species.

To check this hypothesis, we have performed solvent extractions on some of the ambient samples and obtained thermograms of the insoluble filter residues [8], since according to an operational definition of Appel et al. [9], "primary" organic carbon is cyclohexane-extracted carbon; "secondary" carbon is the difference between the total carbon extracted by the benzene, methanol-chloroform sequence and the cyclohexane-extracted carbon. The thermograms of sequentially extracted filters should thus show which peaks can be identified with primary and secondary organics.

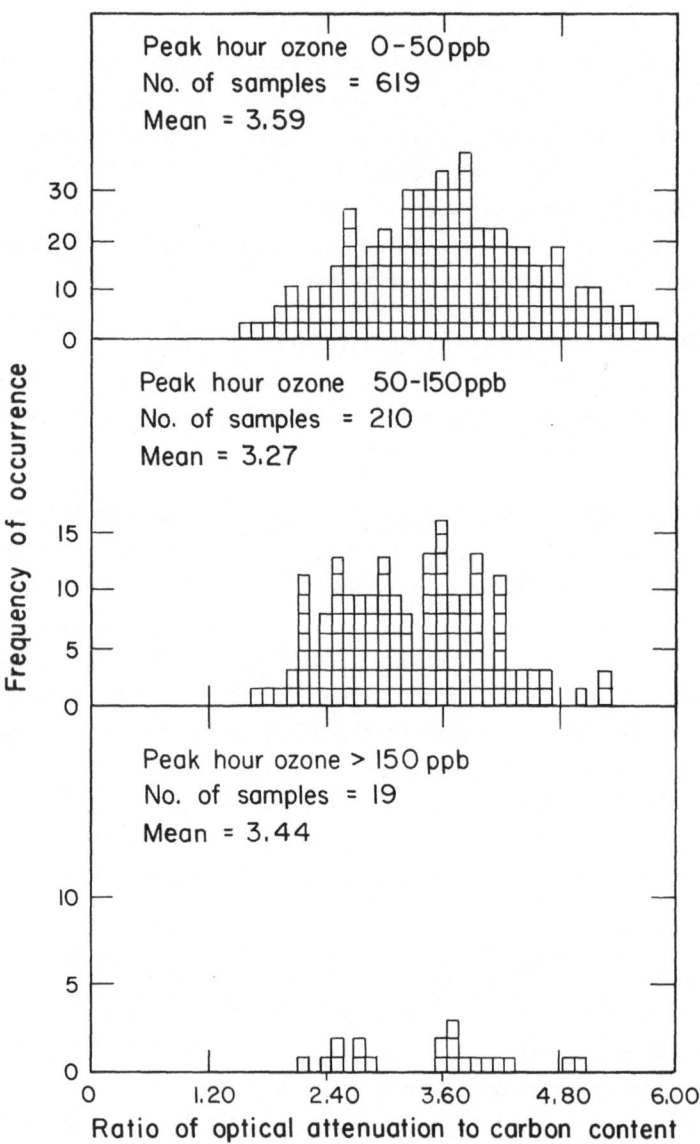

Figure 12: Distribution of the ratios of specific attenuation subdivided
according to the peak ozone concentration. Note that the means
of the distributions are only marginally smaller at larger ozone
concentrations, which puts a rather low limit on secondary
organics produced in correlation with ozone (from Ref. 7)

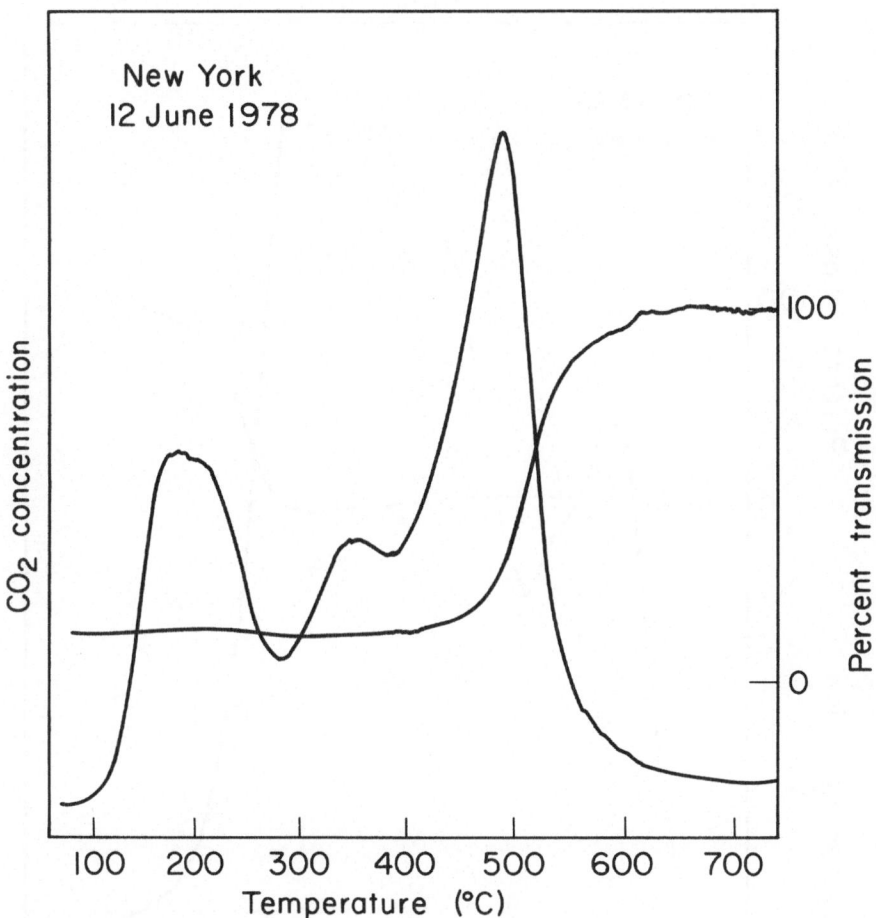

Figure 13: Thermogram of a New York sample with high specific attenuation

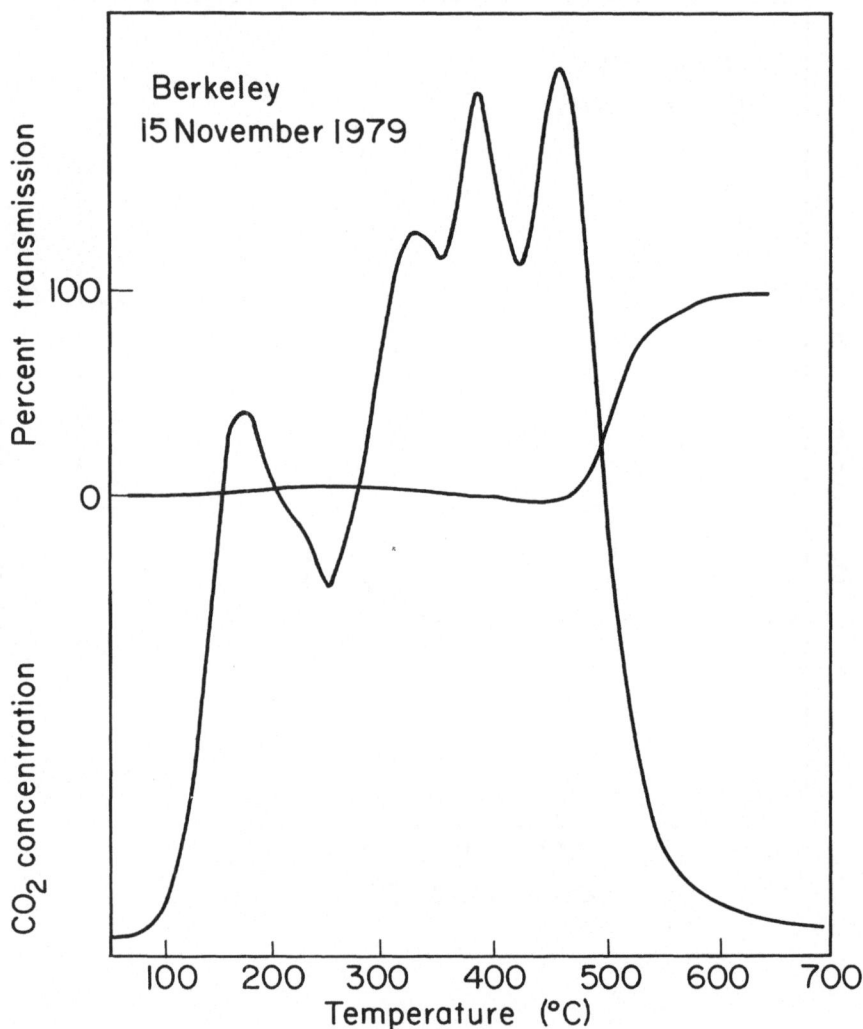

Figure 14: Thermogram of a Berkeley, California, sample with low specific
    attenuation

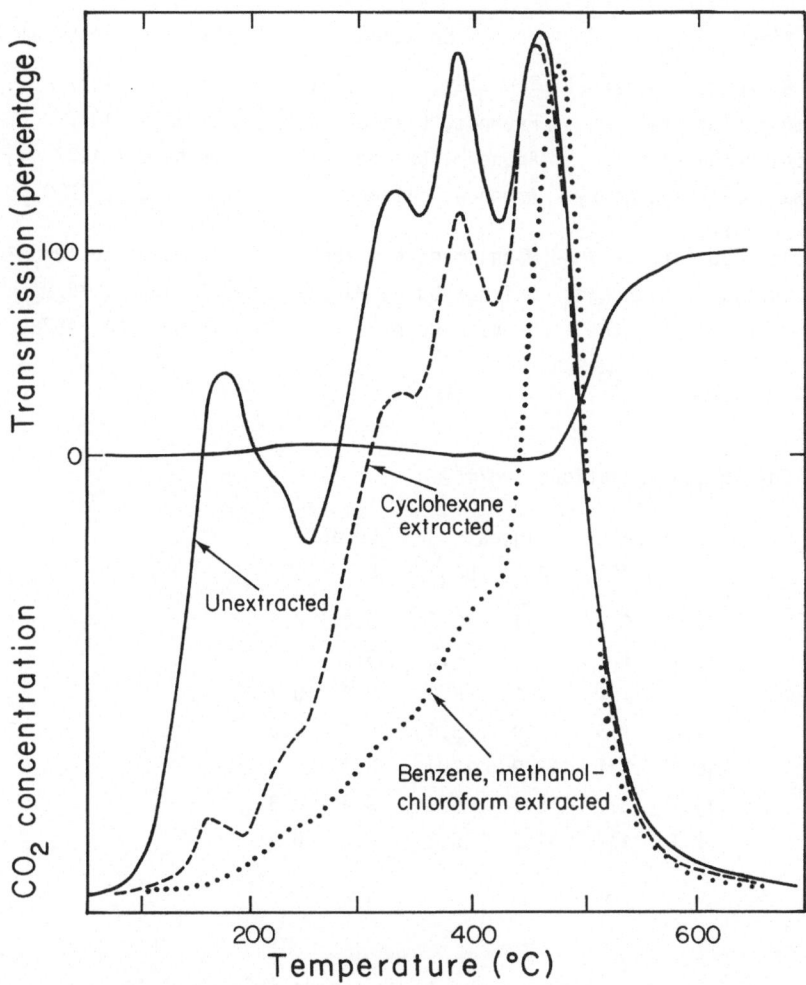

Figure 15: Thermograms of a sequentially extracted ambient (Berkeley) sample

The result of one such experiment with a sample collected in Berkeley is shown in Figure 15. Cyclohexane extraction has removed practically the entire volatile organics, which - according to the above definition - should be primary species. The peak at $\sim$ 380°C is removed only with the polar benzene-methanol-chloroform solvent. This is consistent with our preliminary assignment - that this peak is due to secondary species. The black carbon peak, as expected, was not removed by solvent extraction.

Table 1: LBL aerosol sampling sites.

Site	Location	Date of first sample
Lawrence Berkeley Laboratory	Berkeley, California	1 June 1977
BAAQMD monitoring station	Fremont, California	15 July 1977
SCAQMD monitoring station	Anaheim, California	19 August 1977
Argonne National Laboratory	Argonne, Illinois	22 January 1979
DOE Environmental Measurements Laboratory	Manhattan, New York	22 November 1978
National Bureau of Standards	Gaithersburg, Maryland	23 January 1979
Denver Research Institute	Denver, Colorado	15 November 1978

Table 2: Carbon concentrations ($\mu g/m^3$).

Site	Average		Highest		Lowest	
	C	BC	C	BC	C	BC
New York	15.2	4.2	53.1	12.6	3.4	0.6
Argonne	8.1	1.7	25.1	5.2	3.1	0.2
Gaithersburg	6.1	1.4	17.6	5.6	2.3	0.3
Denver	9.8	1.6	30.8	5.3	4.1	0.2
Anaheim	16.6	3.1	112.9	17.4	3.1	0.3
Fremont	12.0	2.1	75.6	9.2	3.4	0.3
Berkeley	6.7	1.3	31.7	5.2	3.0	0.3

Table 3: Specific attenuation (σ) and black carbon (BC) (% of total C) from ambient samples.

Site	Dates on file	# samples	Average		Highest		Lowest	
			σ	%BC	σ	%BC	σ	%BC
New York	Nov 78 - Apr 80	439	5.44	27%	11.1	56%	2.8	14%
Argonne	Jan 79 - Mar 80	438	4.30	22%	9.1	46%	1.1	6%
Gaithersburg	Jan 79 - Mar 80	381	4.33	22%	8.0	40%	1.8	9%
Denver	Nov 78 - May 79	141	3.23	16%	5.7	29%	1.4	7%
Anaheim	Aug 77 - Jan 80	852	3.70	19%	9.6	48%	0.8	4%
Fremont	Jul 77 - Mar 80	924	3.55	18%	8.3	42%	1.6	8%
Berkeley	Jun 77 - Apr 80	998	4.09	20%	9.2	46%	1.2	6%

Table 4: Specific attenuation (σ) and black carbon (BC) (% of total C)
of source samples.

Source	# samples	Average σ	Average % BC	Highest σ	Highest % BC	Lowest σ	Lowest % BC
Parking garage	12	5.4	27%	7.7	39%	2.25	11%
Diesel	6	5.6	28%	5.7	29%	3.5	18%
Scooter	9	5.1	26%	6.1	31%	4.2	21%
Tunnel	63	6.3	32%	12.5	63%	3.7	19%
Natural gas	6	2.6	13%	3.3	17%	1.9	10%
Garage and tunnel		5.85	29%				

Table 5: Mean specific attenuation of ambient samples.

Site	# samples	$\bar{\sigma}$	SDEV	Soot (%)
New York	211	5.69	1.34	97
Gaithersburg	155	4.72	1.51	81
Argonne	221	4.35	1.64	74
Berkeley	513	4.28	1.47	73
Anaheim	444	3.99	1.71	68
Fremont	461	3.74	1.25	64
Denver	42	3.47	1.49	59

References

1. a. H. Rosen, A.D.A. Hansen, L. Gundel, and T. Novakov, "Identification of the optically absorbing component of urban aerosols", Appl. Opt. 17, 3859 (1978).

   b. Z. Yasa, N. Amer, H. Rosen, A.D.A. Hansen, and T. Novakov, "Photo-acoustic investigation of urban aerosol particles", Appl. Opt. 18, 2528 (1978).

2. P.K. Mueller, R.W. Mosley, and L.B. Pierce, "Carbonate and noncarbonate carbon in atmospheric particulates", in Proceedings, Second International Clean Air Congress (New York, Academic, 1971).

3. a. R. Weiss et al., "Application of directly measured aerosol radiative properties to climate models", Proceedings, Symposium on Radiation in the Atmosphere, Garmisch-Partenkirchen, FRG, p. 469 (1976).

   b. R.E. Weiss, A.P. Waggoner, R. Charlson, D.L. Thorsell, J.S. Hall, and L.A. Riley, "Studies of the optical, physical, and chemical properties of light absorbing aerosols", Proceedings, Conference on Carbonaceous Particles in the Atmosphere, Lawrence Berkeley Laborator: Report LBL-9037, p. 257 (1979).

4. A.D.A. Hansen et al., Lawrence Berkeley Laboratory, unpublished results.

5. H. Malissa, H. Puxbaum, and E. Pell, "Zur simultanen relativkondukto-metrischen Kohlenstoff- und Schwefelbestimmung in Stäuben", Z. anal. Chem. 282, 109 (1976).

6. R.L. Dod, H. Rosen, and T. Novakov, "Optico-thermal analysis of the carbonaceous fraction of aerosol particles", Atmospheric Aerosol Researc Annual Report 1977-78, Lawrence Berkeley Laboratory Report LBL-8696, p. 2 (1979).

7. H. Rosen, A.D.A. Hansen, R.L. Dod, and T. Novakov, "Soot in urban atmospheres: Determination by an optical absorption technique", Science 208, 741 (1980).

8. L. Gundel et al., "Application of selective solvent extraction and thermal analysis to ambient and source-enriched aerosols", Atmospheric Aerosol Research Annual Report 1979, Lawrence Berkeley Laboratory Report LBL-10735, p. 8 (1980).

9. a. B.R. Appel, E.M. Hoffer, M. Haik, S.M. Wall, E.L. Kothny, R.L. Knights, and J.J. Wesolowski, Characterization of Organic Particulate Matter, Final Report to California Air Resources Board, Contract No. ARB 5-682 (1977).

   b. B.R. Appel, E.M. Hoffer, E.L. Kothny, S.M. Wall, M. Haik, and R.L.Knights, "Analysis of carbonaceous material in southern California atmospheric aerosols", Environ. Sci. Technol. 13, 98 (1979).

ANALYSIS AND TOXICOLOGICAL SAFETY EVALUATION OF ENVIRONMENTAL POLLUTANTS

DIETRICH HENSCHLER

Institute for Pharmacology and Toxicology,

University of Würzburg,

Versbacher Straße 9, D-8700 Würzburg,

Federal Republic of Germany

Abstract. Qualitative and quantitative changes in the exposure pattern of modern mankind to chemical, and the increasing awareness of new types of diseases caused by irreversible damages (mutagenesis, carcinogenesis, teratogenesis) have prompted a revision of safety testing strategies of drugs, and occupational and environmental chemicals. New elements of toxicological methodology are: pharmaco-kinetics, metabolism of xenobiotics, identification of toxic metabolites by examination of excretable and non-excretable reaction products with biomolecules, and short-term tests. Proper use of these elements in testing procedures decisively improves the prediction of toxic reactions in humans.

1. Introduction

I want to emphasize, in my presentation, the key role of modern analytical techniques, their outstanding sensitivity and specificity, in the evaluation of health effects in humans exposed to environmental chemicals, as well as in the predictive toxicological safety evaluation from laboratory experiments.

The significance of chemicals as health-impairing principles in modern societies is universally recognized. The lessons learned from the disasters after the intake of drugs (such as thalidomide, aminorex, chlormadinone, practolol), after exposure to occupational toxicants like vinyl chloride, asbestos, bis-(chloromethyl)-ether, 2-nitropropane and others, and negative experiences with pesticides, household-products, agricultural chemicals, etc., have initiated not only dramatic changes in the minds of politicians, of the public and of the scientific community but also many activities towards an improvement of preventive measures.

From a toxicologist's point of view, the justification for rapid and far-reaching actions is found in the following facts:

a) For the last 30 years, worldwide, there has been an exponentially increasing change in chemical conversion processes. This implicates an ever - increasing exposure of mankind to chemicals (figure 1).

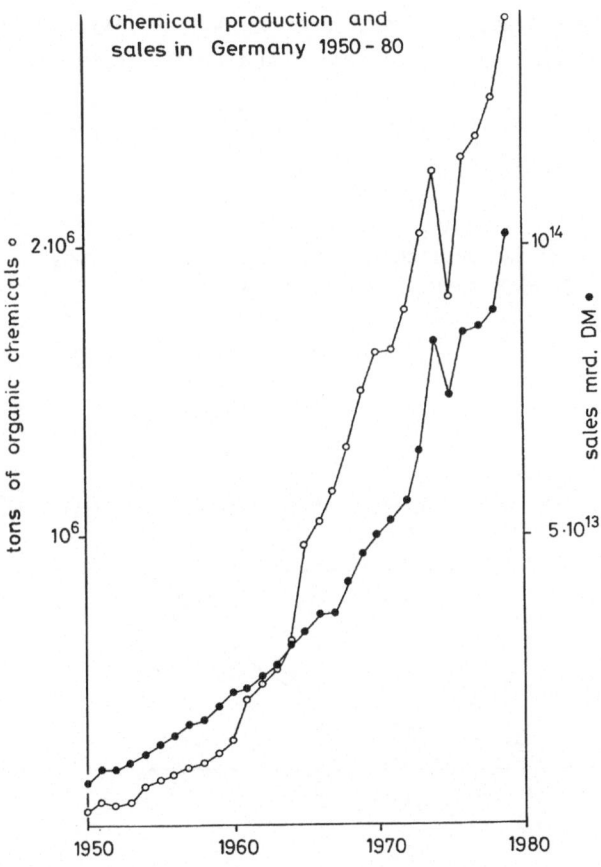

Figure 1: Increase in chemical production and sales in Germany 1950-1979 [1]

b) Chemists try to optimize the reactivities of their chemicals and the yields of production processes. Plastic monomers and laboratory chemicals are prominent examples of this phenomenon. This implies increasing toxic potentials of these chemicals.

c) The knowledge of chemically-induced diseases and the diagnostic tools for their detection are in the process of rapid development towards more sensitivity and precision. Not only are new methods of experimental toxicology introduced regularly; but there is also constantly improved sensitivity of test organisms for the detection of discrete changes of tissue structure and function.

d) The past decades have seen the manifestation of new types of diseases, and/or an increased awareness of their significance in public health. These are teratogenic, mutagenic and carcinogenic effects. The main features of these types of disease are: irreversibility, manifestation only after prolonged latency periods, and transfer to future generations.

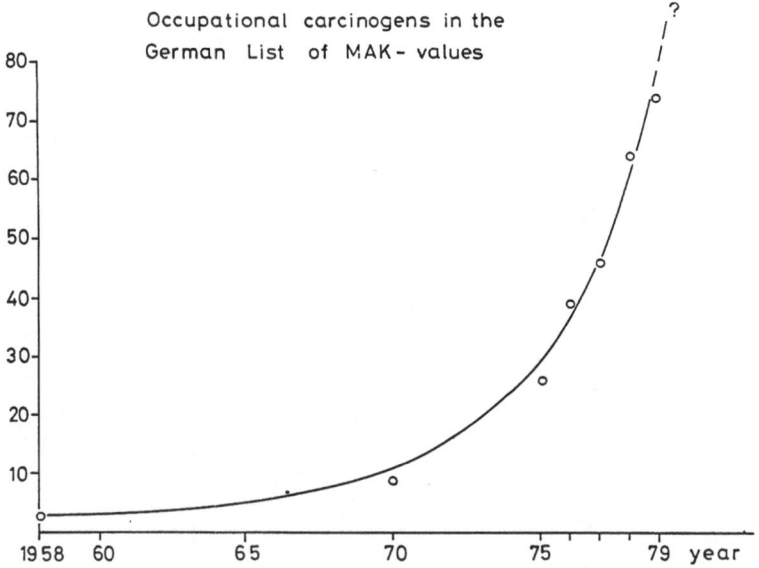

Figure 2: Increase in carcinogenic chemicals (proven or suspected) in the German List of MAK-values 1972-1980 [2]

To quote one example of this dramatic change: within just one decade the
number of carcinogenic occupational chemicals on the German List of MAK-
values has increased from 3 to more than 80, the tendency being exponential
(figure 2). The latency period is long; in the case of mesothelioma formation
from asbestos fibres it amounts to more than 3 decades at an average [3]. The
outstanding significance of chemically induced cancer is further documented
by the fact that medical treatment is very ineffective in most cases, that
carcinogenic stimuli are transferred transplacentally, the embryo being much
more susceptible than the adult organism, and that even germ cells may confer
an increased cancer risk on future generations.

2. Criticism of conventional toxicological safety evaluation

The traditional methods of toxicological testing with animal experimentation
are characterized by an ever - increasing number of procedures, of animal
species used, of animals per dose level, and by almost astronomical numbers
of computerized observations per test. Perfection in fulfilment of fixed
protocols, irrespective of rationale or necessity, rather than flexibility
and rationality govern the strategies of safety testing of new drugs and
chemicals at present. Two emotional elements further foster the tendency
towards expanded test procedures: the fear of toxicologists in industrial
laboratories of being incriminated with the omission of this or that test,
and the mistrust of authorities in the credibility of the laboratory records
which led to the establishment of "Good Laboratory Practice". The resulting
cemeteries of data can hardly be reviewed nor integrated for safety evaluation
by individual research workers.

The criticism with that unsatisfactory system may be summarized as follows:

a) The expenditure for routine testing of drugs and chemicals is dispro-
   portionally high as scaled against the economically feasible benefit.

b) The predictive value of the system is too limited.

c) The results of long-term toxicological tests are too delayed.

d) Fixed protocols inhibit scientific progress in the field of methodological innovation.

The latter point in particular renders routine testing rather unattractive
and frustrating to young scientists. They are no longer willing to follow
the principle of administering the test compound to the test organism and
then just recording the final results after a given period of time, instead
of searching for the fate of the compound on its course through living

systems and for the mechanisms of chemico-biological interactions.

3. New elements of toxicological methodology

The progress attained during the past decade or so is listed in table 1.
All these items aim at a careful study of the mechanism of action of a
chemical, making proper use of modern methods of microanalysis.

Table 1: New elements of toxicological methodology

```
1) Pharmacokinetics: Selection of most suitable species for extra-
                     polation to man

2) Metabolism:       Identification of toxic principles(s)

3) Excretable reaction products of toxic metabolites
                     (Urine, bile, feces)

4) Identification of reaction products with fixed tissue  constituents:
                     a) in target tissue(s) (e.g. DNA)
                     b) in blood as dose marker

5) Biological markers as short-term tests
                     a) in vitro
                     b) in vivo
```

The first new element is a complete elaboration of the pharmacokinetic
profile of a compound. Most important in this respect is the total balance
of the pathways through the compartments of an organism in order to have
at hand a quantitative basis for the evaluation of toxic principles. Comparisons of pharmacokinetic analyses in different species render a rational basis
for selecting - for the ensuing toxicity tests - that species which resembles
man as closely as possible.

Studies on the metabolic transformation of chemicals should provide knowledge
of the toxic chemical intermediates. A combination of pharmacokinetic and
metabolic information will then facilitate estimation of the relative rates
of bioactivation and deactivation reactions. These rates serve as a basis
for predicting the type and degree of toxic reactions, and possibly organ
specificity and species specificity as well. One such example has been worked
out recently with the potent carcinogen aflatoxin B_1 , where the high suscepti-
bility of rats as compared to the almost complete resistence of mice can be
explained by widely differing ratios of activating vs. detoxifying metabolic
processes [4].

Reactive metabolic intermediates cannot be isolated and identified as such
in most cases. However, their excretable reaction products with soluble
nucleophiles like glutathione or cysteine may easily be determined in the
excreta, and be used as dose-monitors in experimental animals and humans
as well. Mercapturic acid determination, an old and well-established analytical method, is therefore celebrating splendid revival for extended use and
benefit in the field of predictive toxicology.

A decisive step towards the identification and quantification of the biochemical lesion is seen in the analysis of reaction products with critical
target macromolecules, like DNA, for genotoxic effects. Many adducts of
electrophilic ultimate carcinogens with DNA-bases have been identified.
Radioactive labelling of test chemicals offers a means of studying the
persistence of such covalent binding to DNA, which in some cases correlates
well with the organ specificity of cancer [5]. Also, the ratio of binding to
critical vs. non-critical targets seems to allow for prediction of the
potency of carcinogenic effects [6].

Covalent binding may also be used as a dose marker in non-target organs.
Hemoglobin may act as a nucleophilic trap for direct or indirect electrophilic carcinogens. Adducts of alkylating mutagens and carcinogens like
vinyl chloride, ethylene or ethylene oxide, and most probably many others,
to the N_3 of histidine in hemoglobin can be determined by the most sensitive
techniques of mass fragmentography and reflect the exposure of humans to
that type of compound [7]. The amount of covalent binding to hemoglobin seems
to be proportionate to the tissue burden in target organs [8].

The final steps towards the quantification of genotoxic risks at the exact
site of effects are the tests for biological endpoints. These are, on the
in vitro-scale, the well-known short-term tests for mutagenicity with micro-
bia and cell cultures. Their limitation for risk assessment in humans derives
from the unsatisfactory specificity and quantification, which is due to a
variety of differences in the biology of test systems. A breakthrough is,
however, envisaged with the search for mutagenic effects in human cells in
vivo, such as in lymphocytes, bone-marrow cells and spermatocytes. Encouraging
results are already available with some cytostatic and antipsoriatic drugs.
Obviously, this area will rapidly develop into one of the most important
fields of modern predictive toxicology because it eliminates the many un-
certainties of interspecies variability.

4. New strategies for toxicity testing

How do we propose to incorporate these new elements of mechanistic studies into the regimen of safety testing?

Experts in the field of toxicology have long since unified their thinking, in that a reasonable and rational safety evaluation cannot be accomplished for chemicals as such, but only for certain types and amounts of exposure to chemicals. Therefore, the first step in any safety evaluation must be the identification of exposure. There are no problems in the case of pharmaceuticals which are handled according to well-defined dose-regimens. With occupational and environmental chemicals, however, analytical chemistry is called upon to specify and to quantify exposure outside and inside the human body.

Any toxicity testing starts with the study of acute effects; this is by no means identical with a LD_{50}- determination. In fact, the informative value of a LD_{50} is rather poor. Toxicologists are well aware of the limited reproducibility of these statistical figures, and efforts towards standardization are rather unsuccessful. Far more important is a careful investigation of the site and type of effects, the cause of death, reversibility or irreversibility of effects, late and secondary effects, success of therapeutic efforts, and - most important - the significance of the lesion in the hierarchy of structures and functions in the integrated organism.

The result of this acute toxicity investigation may lead to a rejection of the chemical because of an unacceptable risk. If not, the next steps in modern safety evaluation strategies consist of the elaboration of pharmacokinetics, metabolic conversion, and in vitro testing for genotoxic (mutagenic) potential. Each of these tests, or any combination of these, may again implicate rejection on account of unacceptable risks. The ensuing steps are repeated administration and subchronic tests, tests for reproductive performance including teratogenicity, and finally long-term (life-time) studies at different dose levels. Suitable decision-tree patterns facilitate the flexible and proper selection of necessary tests and the elimination of superfluous tests.

In essence, the use of modern analytical and microanalytical techniques in conjunction with mechanistic considerations allow for a more rational scheme of testing strategies, an incorporation of more economical short-term tests, and a higher efficiency in terms of prediction of untoward effects in humans.

And - last but not least - the science of toxicology will develop towards
a discipline which will be as attractive to the younger generations as e.g.,
chemistry, biochemistry, genetics, and real preventive medicine.

References

1. Verband der Chemischen Industrie e.V., Frankfurt/M., personal
    communication by Dr. W. Munte 5/12/1980

2. MAK-Werte-Listen, Deutsche Forschungsgemeinschaft, Bonn-Bad Godesberg,
    1972-1980

3. J.S.P. Jones, F.D. Pooley, P.G. Smith: I.N.S.E.R.M. Symp. Ser. Vol. 52,
    117, Paris 1976

4. G.H. Degen, H.-G. Neumann: 7th Congress IUPHAR, Abstr. No.1464, Paris 1978

5. P. Kleihues, G. Doerjer, J.A. Swenberg, E. Hanenstein, J. Bücheler,
    H.K. Cooper: Arch. Toxicol. Suppl. 2, 253 (1979)

6. W.K. Lutz: Mutat. Res. 65, 289 (1979)

7. L. Ehrenberg, S. Osterman-Golkar, D. Segerbäck, K. Swenson, C.J. Calleman:
    Mutat. Res. 45, 175 (1977)

8. H.-G. Neumann, H. Baur, R. Wirsing: Arch. Toxicol. Suppl. 3, 69 (1980)

TRACE ANALYSIS OF TOXIC ORGANIC SUBSTANCES IN THE ENVIRONMENT

FRANCIS W. KARASEK
Department of Chemistry, University of Waterloo
Waterloo, Ontario N2L 3G1, Canada

Abstract. The use of incinerators to destroy municipal garbage and create useful energy is a practice of long standing. The fly ash produced by this process contains trace levels of organic compounds. Microanalytical methods for these compounds involve Soxhlet extraction with benzene, condensation of the extract by a factor of 2000 and analysis by GC and GC/MS techniques.

Over 200 compounds have been identified which include toxic compounds such as polycyclic aromatic hydrocarbons, polychlorinated dibenzo-dioxins (PCDD), polychlorinated aromatics and phenols and the polychlorinated dibenzofurans. Specific analysis by GC/MS with SIM technique for all isomers of the poly-chlorinated dibenzodioxins permits survey studies of variations of these compounds generated at a given municipal incinerator. Correlation studies indicate the PCDD compounds are not directly related to formation of other organic compounds. Particle size fractionation studies of a single fly ash sample show concentrations of PCDD isomeric groups and total organic compounds which vary significantly with particle size.

1. Introduction

Solid particulate matter is present in the atmosphere of modern urban communities. These particles have trace quantities of organic compounds adsorbed on their surfaces in concentrations ranging from nanograms to picograms per gram. The organic compounds adsorbed are a complex mixture whose composition is related to the origin and history of the particles [1]. Although particulate matter originates from many different industrial and community sources, those of high current interest are the fly ash particles which are produced from combustion of municipal waste. Incineration of waste has been an accepted method of disposal since its inception at the municipal level in 1874. It is a worldwide practice whose use is in-

creasing. In the Netherlands the percentage of waste incinerated has increased steadily between 1960 and 1978 from 15 percent to 34 percent [2].
Many large cities throughout the world incinerate their waste in plants
which use the heat generated to produce energy. In France, for example,
the city of Paris derives most of its electrical and steam heating needs
from burning urban waste in three large, efficient incinerators.

Fly ash, composed of fine particles formed in the combustion zone, would
enter the atmosphere with the stack gases if they were not precipitated
electrostatically and collected in large hoppers. So efficient are the
precipitators that only 1 to 5 percent of the fly ash escapes with the
flue gases. An incinerator burning 500,000 tons of waste per year will
produce about 17,800 tons of fly ash. In Canada about 1.5 million tons
of waste per year are burned which produces an estimated 7500 tons of
fly ash escaping to the atmosphere along with 25,000 tons of vapours.
The urban location of municipal incinerators creates a ready mechanism
for distribution of fly ash and its adsorbed organic compounds into the
environment of densely populated areas.

Recent studies reveal that over 400 organic compounds are present as adsorbed species on fly ash particles from municipal incinerators. These
compounds include polychlorinated dibenzo-p-dioxins (PCDD), polycyclic
aromatic hydrocarbons (PAH), polychlorinated benzenes, chlorinated phenols
and polychlorinated dibenzofurans. [3,4,5,6] All these compounds have
varying degrees of toxic and carcinogenic properties, but it is generally
recognized that the PCDD compounds are the most toxic known.

In 1977 Hutzinger reported finding significant amounts of dioxins adsorbed
on precipitated fly ash particles from several incinerators in Netherlands [3]. Since then, fly ash samples from municipal incinerators in other
countries have been examined for the presence of dioxins. Evidence of the
presence of all the PCDD compounds has been found in each fly ash sample
analyzed, regardless of the type of garbage burned, the design of the incinerator or the detailed composition of the total organic mixture adsorbed on the fly ash particles. The diversity in composition of the garbage feed and conditions of combustion appear to form ideal conditions
for synthesis of the individual dioxins which are present in amounts as
high as 100 nanograms (10^{-9}g) per gram of fly ash.

Evidence for the toxic and carcinogenic effects of the dioxins lies in
the many toxicological studies made over the years with animals. The effect

on humans is based on extrapolating animal studies and analyzing individual
case histories of human exposure and its subsequent effects. If the picture
is unclear it must be remembered there are 75 possible compounds involved,
only a few of which have undergone extensive testing. A number of those
tested show fairly comparable toxicity [7]. Evaluation of test data must also
take into account the fact that only recently have we had microanalytical
methods available that can unambiguously separate and identify some of the
many closely related dioxin isomers.

Much of the toxicological testing of dioxins has been done with the
2,3,7,8-TCDD isomer [8]. The picture emerging reveals a very toxic substance
with oncogenic (creates tumors) and teratogenic (cause birth defects) properties. The effects differ depending on the test animal involved and dosage
given. The guinea pig has an acute toxicity LD_{50} level (that amount causing
death in 50 percent of the test population) of 600 nanograms per kg of body
weight, while the level for rats is 50 times greater and for monkeys 100
times greater. Rabbits and dogs are less sensitive. Some animals die of
lesions in the liver, others waste away with symptoms akin to starvation.
Oncogenic tests of rats found tumors of the liver, lung and mouth appearing
with a 2200 ppt dosage in food (equivalent to 100 nanograms/kg/day), while
22 ppt showed no effects for the lifetime of the animal.

Effects of the dioxins on humans is even less conclusive. Based on projections from guinea pig data, the chemical plant release of 2,3,7,8-TCDD
in Seveso, Italy in July, 1976 should have killed every human within miles.
Yet to date no deaths have occurred, only some chloracne and transient
liver toxicity. There has been no evidence of teratogenic effects. These
observations can be contrasted with many case histories. After a 1963
explosion in the Philips Duphar 2,4,5-Trichorophenol (a chemical which
has dioxin as an impurity) factory in Amsterdam 18 workers decontaminated
the plant using heavy protective clothing. Nine of the workers developed
chloracne, with three dying within two years. There is much of this indirect evidence of the danger of dioxins [9].

It is clear that much more study and work is necessary if the dangers to
human welfare of all 75 of these related dioxins are to be identified and
understood. Since all these compounds appear to be generated in the incinerator combustion, their total hazard taken as individual compounds of
the series must be considered. Ability to detect dioxins at the nanogram
level in the complex mixtures adsorbed on fly ash depends heavily on effective microanalytical methodology. An adaption of techniques developed for

analyzing trace organic compounds adsorbed on airborne particulate matter has given good results [1]. The techniques involve solvent extraction, gas chromatography and gas chromatography/mass spectrometry (GC/MS) with selected ion monitoring (SIM). Using this methodology studies have been made on a series of fly ash samples which indicate the dioxin concentration in relation to particle size distribution and concentration of other organic compounds present. Such data derived entirely from analytical results may lead to an understanding of the chemical pathways by which the dioxins are formed and produce information important for control of environmental contamination.

2. Experimental

2.1. Sample collection and storage

Once each week samples were collected in 1 to 3 kilogram quantities from the electrostatic precipitator of a municipal incinerator in an urban center in Ontario. Samples were taken from this single incinerator by the Ontario Ministry of the Environment personnel over a several month period. The samples were stored in closed glass jars at room temperature protected from light exposure. For the weekly variation study all samples were analyzed within 7 days of collection. After extraction the sample extracts were stored in a freezer at -15°C.

2.2. Analysis of samples

Each sample was analyzed in duplicate or triplicate using procedures described in detail in a previous study [4]. These procedures include: 16 hr Soxhlet extraction of 10g of sample with 200 mL of benzene, (Caledon Labs, Georgetown, Ont. - distilled in glass) condensation of the extract to 0.1 mL, and direct analysis using GC and GC/MS techniques. The concentrations of PCDD and PAH were determined using SIM in GC/MS analysis by comparison to response factors determined independently under identical SIM conditions for 1,2,3,4-TCDD and biphenyl, fluorene, anthracene, pyrene, and benzo(a)pyrene. Both packed and glass capillary columns were used in the GC/MS analysis. The packed column was a 2.6-m long, 2-mm i.d. glass column packed with high performance Aue packing containing 0.2 wt percent of carbowax 20M polymer. The glass capillary column was a 60-m long, 0.36 mm i.d. column coated with SE-52. Chromatographic conditions were: initial temperature, 90°C; program rate, 4°C/min; final temperature, 250°C then isothermal to end of analysis; injection port temperature, 250°C; helium carrier gas flow, 37 mL/min measured at 90°C.

A Hewlett-Packard 5992A GC/MS system equipped with a single floppy disk
and an X-Y plotter was used for standard GC/MS analyses and for SIM analysis. The HP Peakfinder software was modified by us to allow mass chromatograms and total ion current, in addition to individual mass spectra to be
stored on flexible desk. Spectra were scanned from 500 to 40 amu at
330 amu/second. Spectra taken at the lowest point of the GC peak valleys
between consecutive peaks were saved for background subtraction of spectra
taken at GC peak maxima. In the SIM mode 6 preselected ions were monitored
at dwell times of 166 milliseconds for each. The GC/MS instrument used a
membrane separator which operated in the usual manner with packed columns.
When using glass capillary columns it was necessary to build a special
adaptor for the column exit so that make-up gas could be added to sweep
the narrow GC peak rapidly by the membrane.

Chromatographic analyses were performed using both a Hewlett-Packard 5830A
and a 5880A gas chromatograph equipped with a flame ionization detector
(FID). The columns and GC run conditions used were the same as those for
the GC/MS analyses.

2.3. Fractionation study

A 324g sample of fly ash was separated into six size fractions using five
Tyler sieves (W.S. Tyler Co. of Canada, Ltd., St. Catherines, Ont.). The
brass screens had openings of 850, 250, 150, 106 and 63 µm. All sieves
were cleaned by ultrasonic agitation with an aqueous solution of Alcanox
detergent, followed by rinsing with water and methanol and air-drying.
After hand sieving was performed the separated fractions were stored in
solvent cleaned polypropylene containers sealed with screw caps.

Some particles were too fine for Soxhlet extraction. Therefore, all extractions were performed with ultrasonic agitation. Ten-gram samples were
added to individual flasks with 100 mL of benzene and subjected to ultrasonic agitation for one hour. The benzene was then decanted and the ultrasonic extraction was repeated two additional times with 100 mL of benzene
added each time. Benzene from the three extraction steps was condensed to
a final volume of 0.1 mL.

All glassware used in these studies was cleaned by ultrasonic agitation
in Alcanox solution, rinsed with deionized water and baked for one hour
at 300°C. A 300 mL portion of benzene solvent was carried through the
entire procedure as a blank.

3. Results and discussion

An FID-GC analysis of the fly ash extract using a glass capillary column gives over 400 peaks, which show those compounds extracted from the original fly ash when present at the nanogram to picogram/gram level (Figure 1).

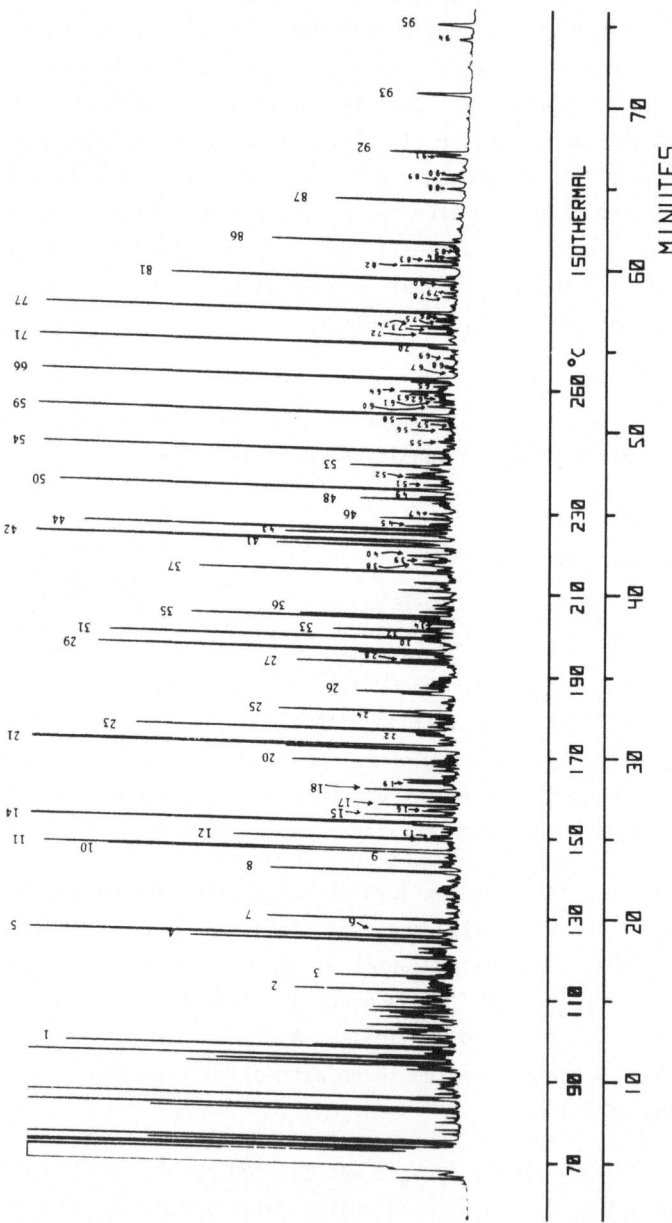

Figure 1: The GC-FID, glass capillary column chromatogram of organic compounds extracted from municipal incinerator fly ash. Numbered peaks have been identified by GC/MS (ref.5)

Because of the decreased sensitivity and resolution of the GC/MS system in the normal scanning mode not all of these compounds can be identified in a GC/MS analysis. Table 1 lists those identified, categorized as total numbers of a given type.

Table 1: Organic compounds adsorbed on municipal incinerator fly ash
identified by GC/MS

compound Type	numbers identified
n-Alkanes	33
Polycyclic Aromatic Hydrocarbons	24
Chlorinated Benzenes	12
Brominated Benzenes	4
Chlorinated Naphthalenes	10
Chlorinated Phenols	4
Phthalates	7
Chlorinated Benzofurans	6
Polychlorinated Dibenzodioxins	41
Polychlorinated Dibenzofurans	12
Other Types	80
	233

Although the relative distribution of these compounds varies from inciner-ator to incinerator, the general types of compounds found were present in all incinerators studied [4].

Chlorinated compounds appear to be a predominate species. From an examina-tion of the mass spectra associated with each GC peak, spectra charac-teristic of the PCDD compounds were easily recognized. Figure 2 shows spectra from each group of the dioxin isomers from the tetra-CDD to the Octa-CDD. Figure 3 shows GC/MS data in which the mass chromatograms of the molecular ions of each isomeric group were reconstructed along with the chromatogram to indicate the presence and numbers of all the PCDD separated. In this figure a large amount of hexachlorobenzene is re-vealed because of the coincidence of its fragment ion at m/z 285.9 with the parent ion of the tri-CDD.

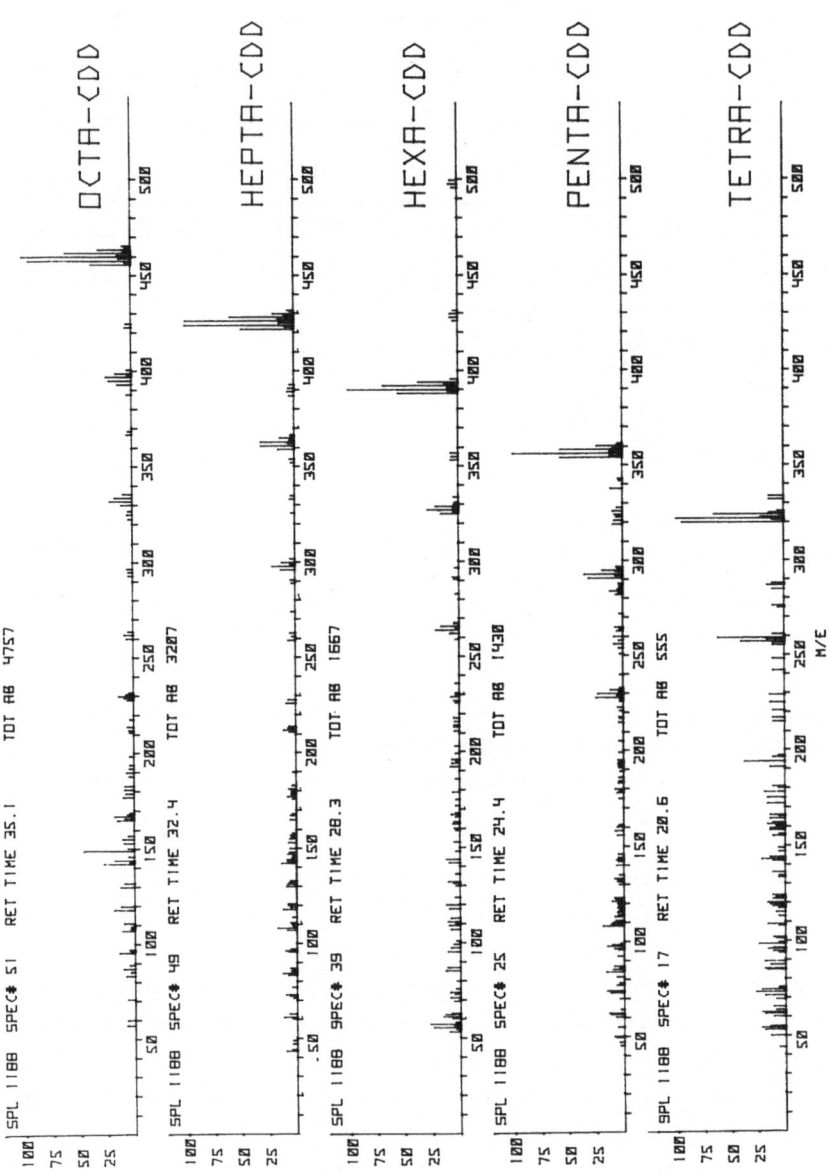

Figure 2: Mass spectra of polychlorinated dibenzodioxin isomers obtained from a GC/MS analysis of fly ash extract

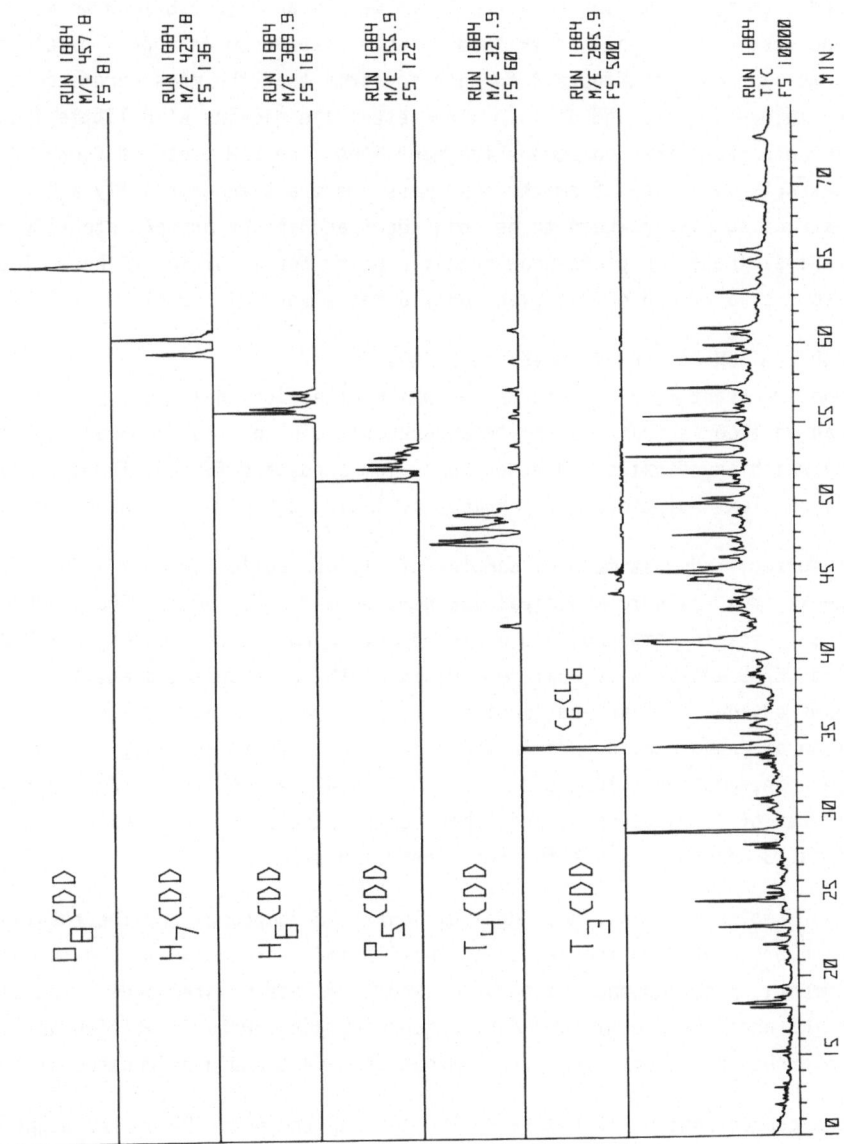

Figure 3: GC/MS data of reconstructed chromatogram and mass chromatogram
from analysis of fly ash extract

Quantitation and identity of specific isomers can be accomplished by
comparing SIM plots of the sample with those of the TCDD reference standards. In these studies only the 1,2,3,4-TCDD and 2,3,7,8-TCDD were used
and responses for all other isomers were assumed to be approximately the
same. The data in figure 4 further confirms that the mass chromatograms
of molecular ions and SIM plots represent the dioxins with little interference from other compounds fragment ions. The SIM plots of four
characteristic ions from the TCDD mass spectra taken for a fly ash
extract sample are seen to be coincident and in the proper intensity
ratios. These SIM plots show specific peaks for 15 of the 22 known TCDD
isomers; a packed column gives only 6 peaks for this sample.

CORRELATION DATA FROM SURVEY ANALYSES.
One important type of information which effective analytical methodology
can develop is that leading to an understanding of the chemical pathways
by which the dioxins and other toxic compounds are formed. The situation
is complex, but there are a number of indirect approaches one can use.

A survey study was made of samples of fly ash collected weekly for eight
weeks from a single municipal incinerator and analyzed for PCDD, PAH and
total organic compounds [10]. A packed column was used in both the FID-GC
and GC/MS analyses. Visual comparison of the FID-chromatograms shows
numerous qualitative and quantitative differences. The total concentration of organic compounds in the samples was estimated using the sum of
peak integration values from the FID-GC analyses and an average response
factor of 400 area counts/ng. These calculated concentrations varied from
1 to 23 μg/g of fly ash for the eight samples.

The samples were analyzed for individual PAH compounds and the totals of
each group of PCDD isomers using the SIM mode for the GC/MS and through
comparison to response factors for known amounts of standards. From the
GC/MS analysis the most abundant organic compunds were identified as
chlorobenzenes, chlorophenols, phthalate esters and n-hydrocarbons.

The concentrations of the tetra-CDD through the octa-CDD on the eight fly
ash samples are shown in Table 2. These data show not only the large
variations of concentrations of individual groups of isomers but also
some changes in the ratio of quantities of isomeric groups within a
sample

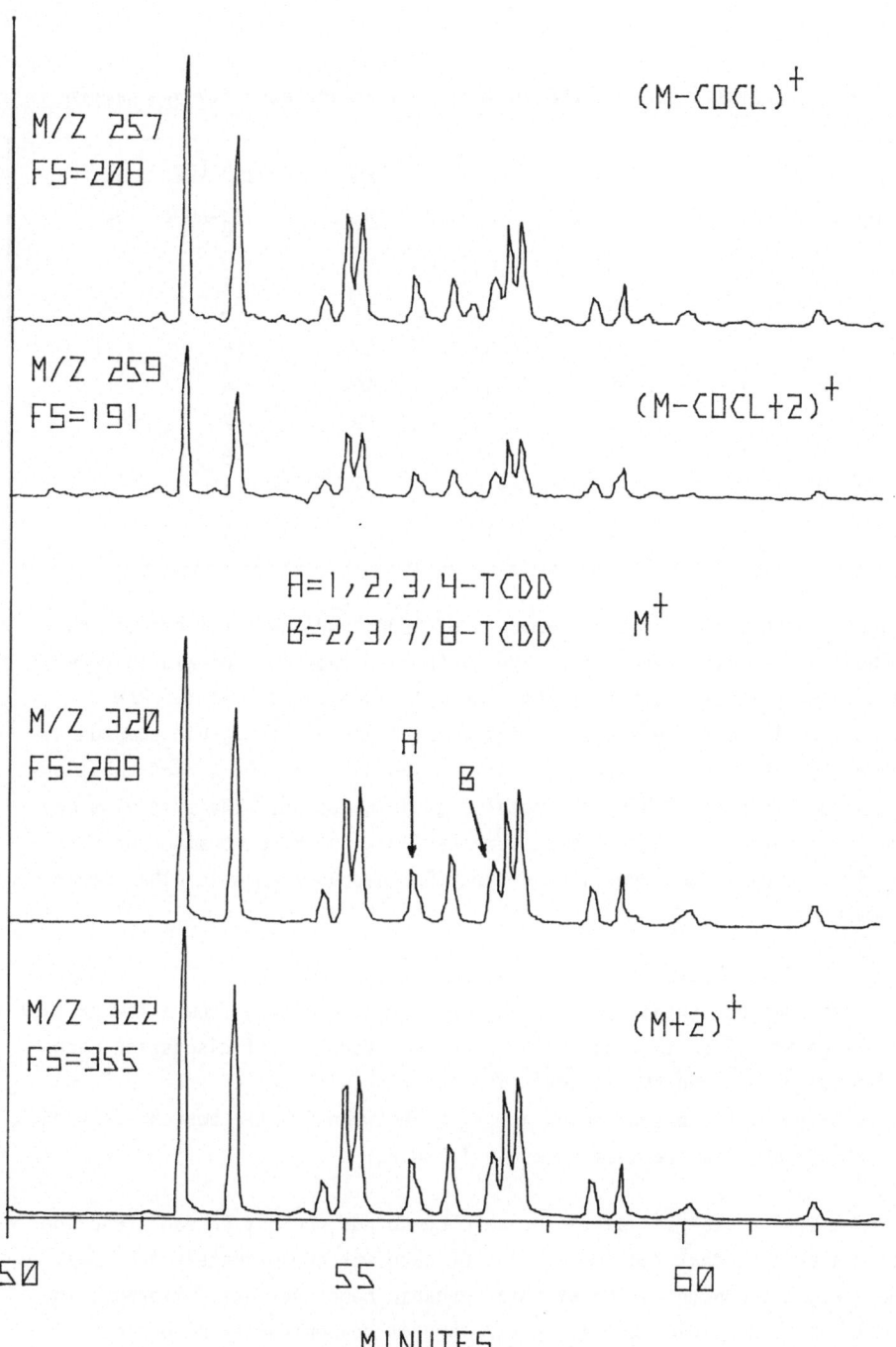

Figure 4: SIM plot of four characteristic ions for the tetra chlorodibenzodioxin isomers in GC/MS analysis of fly ash extract

Table 2: Concentration of PCDD in municipal incinerator fly ash samples

Sample	T_4-CDD	P_5-CDD	H_6-CDD	Concentration (ng/g)	
				H_7-CDD	O_8-CDD
1	12	19	23	8	2
2	10	18	16	11	5
3	15	34	53	43	26
4	27	45	39	16	4
5	19	33	34	19	7
6	10	18	26	17	5
7	3	3	2	1	1
8	12	15	13	5	1

Figure 5 shows how the total amounts of PCDD, PAH and total organic compounds varied with the sample. The plots also show the correlation between these three classes of compounds. It is evident that the PAH
concentration parallels the estimated total amount of organic compounds.
However, the PCDD concentrations do not exhibit a simple direct relationship to either the PAH or the total organic compounds. These results may
be interpreted to indicate that some variable in the combustion process
is more specific to formation of the PCDD compounds than to other organic
compounds.

SIZE FRACTION STUDY.
Fly ash samples contain a wide range of particle sizes, from a few particles
as large as 0.5 cm to those below 5 μm. For purposes of discussion each
fraction other than the largest size fraction will be referred to as having
an average particle size which is midway between its two boundary screen
sizes. These data are summarized in Table 3.

From a GC-FID analysis of the organic compounds present on each fraction
it was evident that the composition of each was approximately the same.
However, the concentration of total organic compounds was different. The
small 30 μm particles contained the largest concentration of organic
compounds by a factor of 2 to 4.

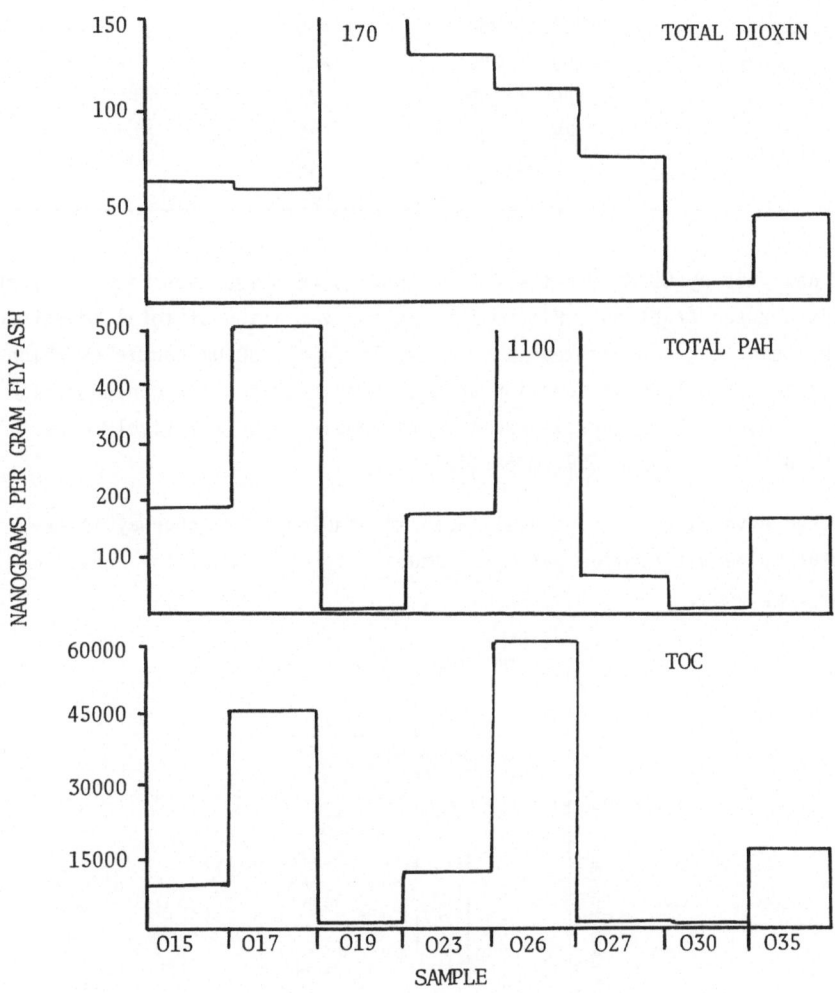

Figure 5: Plot of data obtained from GC and GC/MS analysis for organic compounds extracted from a series of weekly municipal incinerator fly ash samples

Table 3: Particle size distribution of separated municipal incinerator
fly ash fractions

Size fraction	Particle size range (μm)	Average particle size (μm)	Percent of total sample
1	850	–	3
2	250 to 850	550	14
3	150 to 250	200	8
4	106 to 150	175	10
5	63 to 106	80	27
6	63	30	38

SIM analyses by GC/MS for the PCDD compounds revealed significant differences between fractions. Figure 6 shows the variation of total amounts of
each isomeric group with particle size. The small 30 μm particles which
constitute the largest fraction show a relatively smaller concentration
of the tetra-CDD isomers. These isomers appear to be relatively more
abundant in the large 550 μm particles.

Further development of the analytical methodology now underway is expected to permit similar variation studies for all 75 of the individual
PCDD compounds.

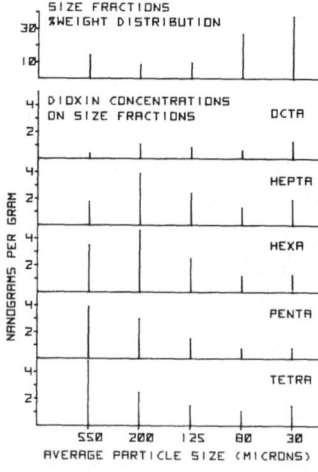

Figure 6: Plot of data obtained by GC/MS analysis of size fractions from
a single fly ash sample

References

1. F.W. Karasek, D.W. Denney, K.W. Chan and R.E. Clement, Anal. Chem., 50, 82 (1978).

2. J.W.A. Lustenhouwer, K. Olie and O. Hutzinger, "Chlorinated Dibenzo-p-dioxins and Related Compounds in Incinerator Effluents", Chemosphere (in press)

3. K. Olie, P.L. Vermeulen and O. Hutzinger, Chemosphere, 6, 455 (1977)

4. G.A. Eiceman, R.E. Clement and F.W. Karasek, Anal. Chem., 51, 2343 (1979)

5. K.W. Chan, "Trace Organic Analysis by Glass Capillary Columns and GC/MS/Computer Techniques", PhD thesis, University of Waterloo, Waterloo, Ontario (1980)

6. R.E. Clement, "Analytical Methodology for Trace Levels of Organic Compounds on Airborne Particulates and Municipal Incinerator Fly Ash", PhD thesis, University of Waterloo, Waterloo, Ontario (1980)

7. "Bioassay of Dibenzo-p-dioxin for Possible Carcinogenicity", Nat. Cancer Inst. Techn. Rep. Series No. 122, OHEW Publ. No. (NIH), 79 1377, US Dept. Health, Education & Welfare, Wash. D.C., USA, p. 138 (1979)

8. R.L. Rawls, Chem. & Eng. News, 57, Feb. 12, 23 (1979)

9. T. Whiteside, "The Pendulum and the Toxic Cloud", Yale University Press, New Haven, Conn. (1979)

10. G.A. Eiceman, R.E. Clement, and F.W. Karasek, "Variations in Concentrations of Organic Compounds Including PCDD and PAH in Fly Ash from a Municipal Incinerator", Anal. Chem. (in press)

V. MICROCHEMISTRY IN MATERIAL SCIENCES

PHYSICAL ASPECTS OF MICRO-CHEMISTRY

GERNOT EDER
Atomic Institute of the Austrian Universities
Schüttelstraße 115, A-1020 Wien, Austria

Abstract. Starting with diffraction and elastic scattering of photons by atoms and nuclei, strong, electromagnetic and weak interactions for various kinds of fundamental particles are discussed in the context of microchemistry.

1. Diffraction and elastic scattering

In order to investigate small areas it is reasonable to look directly at those areas. This means we use our eyes and microscopes, and photons of the visible part of the electromagnetic spectrum, where the wavelength λ is $\lambda = 400 \ldots 800$ nm. The energy of the photons is smaller than 3.1 eV. Thus the structure of the material to which we are looking is not changed. We expect that we can see sharp pictures of objects with linear dimensions $D > 1\mu m$ because we have learned that for $D \gg \lambda$ we have the limit of geometrical optics. But geometrical optics is never right, not even in this limit. On account of Heisenberg's uncertainty principle or equivalently on account of diffraction every shadow of an object becomes diffuse within a diffraction angle ϑ , where

$$\sin \vartheta = \lambda/D.$$

In a distance r behind the object the shadow is reduced to

$$D' = D - r \sin \vartheta = D - r \, \lambda/D$$

and it vanishes for $r = D^2/\lambda$. For larger distances we can see a specific angular distribution of the scattered photons only. This distribution is expressed by the differential cross section $d\sigma/d\Omega$ and the total elastic cross section

$$\sigma_{el} = \int \frac{d\sigma}{d\Omega} \, d\Omega = \int\limits_{o}^{\pi} \int\limits_{o}^{2\pi} \frac{d\sigma}{d\Omega} \, \sin\vartheta \, d\vartheta \, d\varphi$$

Also in the limit $D \gg \lambda$ we get $\sigma_{el} = 2\pi(D/2)^2$, which is twice the value of the geometrical cross section $\pi(D/2)^2$, because both the reflection of the object and the diffraction contribute the same ammount to the total scattering. An object with a diameter $D=1$ cm can give a shadow over 100 m. But for a diameter $D=10\mu m$ the shadow becomes diffuse behind a distance of $200\mu m$ already. Therefore in general microchemical objects are visible by their cross sections only, where the internal structure of the object modulates the cross section by a form factor $|F|^2$:

$$\frac{d\sigma}{d\Omega} \propto |F(q)|^2$$

Here $\hbar q = 2\hbar k \sin(\vartheta/2)$ is the transferred momentum in an elastic collision and $\hbar k = h/\lambda$ is the momentum of the incoming photon. The quantity F is essentially the Fourier transformed density distribution $\rho(r)$ of the object. For example, a Gaussian distribution

$$\rho(r) = (2/\pi R^2)^{3/2} \, Q \, \exp\{-2(r/R)^2\}$$

of a total charge Q yields a form factor

$$|F(q)|^2 = \exp(-q^2 R^2/4) = \exp\{-k^2 R^2 \sin^2(\vartheta/2)\}$$

In order to reduce the form factor from 1 in forward direction ($\vartheta = 0$) to the value $\frac{1}{2}$ for $\vartheta = 60°$ we need a wavelength

$$\lambda = \pi(\ln 2)^{-1/2} R = 3.77R.$$

For a wavelength $\lambda \gg 5R$ the form factor becomes constant and no structure can be seen at all. For $\lambda < 0.5R$ the form factor and the cross section decrease strongly with increasing scattering angle ϑ, but then the structure is hidden within the small angle scattering ($\vartheta < 8°$).

Coming from geometrical optics one would think that we could get ideal sharp picture if the wavelength is small enough. But we found that a microscopic structure of radius R should be detected by wavelengths

$$0.5R < \lambda < 5R$$

For larger wavelengths the structure is ignored by the incoming light and the structure is lost in the isotropic component of the scattered light. For smaller wavelengths the structure is sharp but too concentrated within the narrow cone of forward scattering. Thus each structure has an optimal spectral region of electromagnetic and other waves. If we are interested in linear

dimensions of interatomic distances (R=200 .... 500 pm) we have to use
X-rays of wavelengths

λ=100 pm ... 2.5 nm corresponding to quanta of energy
E=hc/λ =1240 eV nm/λ =1240 keV pm/λ =12.4 keV ... 0.5 keV.

For shorter wavelengths we can see the electronic structure of atoms and
ions. There the inner electrons contribute high charge densities. Let us
consider a neon-like ion with the second Bohr radius

R=211.7 pm/(Z-4.35)=24.5 pm(Al)... 2.7 pm(Pb).

Therefore we need X-rays with $\lambda \cong$ 10 pm or quanta of energy E $\cong$ 124 keV. But
the electronic structure of the atoms is diffused by the lattice vibrations
which have amplitudes in the same order of magnitude. These vibrations give
rise to an additional form factor

$$|F_{DW}(q)|^2 = \exp\{-\frac{1}{4} q^2 R_{DW}^2 \ (1-\frac{q^2}{4k^2}) \}$$

known as Debye-Waller factor. For an absolute temperature T smaller than the
Debye temperature θ the radius R_{DW} is given by

$$R_{DW} = \hbar \{\frac{6}{k_B \theta M} \ (1 + \frac{2\pi^2}{3} \ \frac{T^2}{\theta^2}) \}^{1/2}.$$

k_B is Boltzmann's constant and M is the mass of the atoms. Also for T $\rightarrow$ 0
we get a finite radius on account of the zero point vibrations of the atoms
around their equilibrium positions. For example, we find for aluminium

R_{DW}=16.6 pm (T=0) ... 36.1 pm (T=293 K).

This is in the same range as for the electronic charge distribution. This
feature is quite similar to the Doppler broadening of the natural width of
a spectral line.

Nuclear charge radii vary between 2.2 fm (He) and 6.6 fm (Pb). In order to
see nuclear structure we have to use gamma rays with $\lambda \cong$ 5 fm or E=1240 MeV$\times$
fm/ λ =250 MeV. The scattering cross section is determined primarily by the
proton distribution within the nucleus. The scattering is slightly modified
by the neutrons: Although their total charge vanishes, they have an internal
charge distribution in the radial region of about 0.4 fm. As we know the
nuclear form factur only for a momentum transfer $\hbar q < \hbar q_0 (q_0=3.5fm^{-1})$,we can
not detect a fine structure in the radial charge distribution in a range

$\Delta r < 2\pi/q_0$ = 1.8 fm.

A proton has a charge radius of R_p=0.92 fm and a neutron has a charge
structure with a characteristical radius R_n =0.4 fm. Therefore an investigation of the internal structure of strong interacting fundamental particles -

usually called "hadrons" - needs wavelengths $\lambda < 0.4$ fm or energies $E > 3.1$ GeV. In this region of high energy research one hopes to find the socalled "quarks" as constituents of fundamental particles. But the quarks, if they exist at all, seem to have odd properties: For short distances from their equilibrium positions they move asymptotically free like atoms in an ideal gas or protons and neutrons in the atomic nucleus. But if they are forced by an incoming high energy gamma particle to leave the confinement of a fundamental particle then such a high force is active that many fundamental particles are produced and they can be detected as jets of those particles. Thus the quarks have not been found so far as isolated particles like a dictator who is covered always by a cloud of secret police. We have seen that in micro-chemistry the direct geometrical picture of an object is substituted by the form factor of cross sections. But now the structure of an object is represented by jets of secondary particles which have not been involved originally. The situation is similar in the Auger effect, where we do not observe the characteristical X-rays but electrons which transport the same energy. In the same way internal conversion means that for heavy elements an excited nucleus does not emit a gamma quantum but an atomic electron is emitted by an inelastic collision with the nucleus and the electron accepts the nuclear excitation energy.

2. Types of interaction

So far we have discussed form factors of cross sections, but not the cross sections themselves. For a non-transparent macroscopic object the cross section is the geometrical one. But the transparency depends on the interaction between the incoming rays and the object. The interaction between photons and electric charges is determined by Sommerfeld's fine structure number

$$\alpha_f = 1/137.036.$$

This number is small. Therefore the cross section for the scattering of light on a mass M with a charge of Z protons is small too

$$\sigma = (8\pi/3) \; Z^2 \; \alpha_f^2 \; (\hbar/Mc)^2 = 2.002 \times 10^{-5} \; \text{fm}^2 \; Z^2 \; (M/\text{amu})^{-2}.$$

This formula of Thomson tells us that a gas of free non-relativistic charges is penetrable for light and that we see by X-rays primarily high electron concentrations or heavy elements. If we scatter electrons on neutral atoms of radius R and proton number Z then the Rutherford cross section

$$\frac{d\sigma}{d\Omega} = 4\ Z^2\ \alpha_f^2\ (\frac{1}{1+q^2R^2}\ \frac{R^2}{\hbar/M_e\,c})^2$$

looks similar to the photon cross section. But for small angles (qR << 1) we get roughly the geometrical cross section

$$\frac{d\sigma}{d\Omega}\ (\vartheta=0) = (2Z\ \alpha_f\ R^2\ M_e c/\hbar)^2 \cong 4R^2$$

because $R \equiv \hbar/Z\alpha_f M_e c$. Although the interaction is electromagnetic we get a large cross section in forward direction on account of the infinite range of the Coulomb force. For small electron energies (kR << 1) the total cross section is larger than the geometrical cross section

$$\sigma = 16\pi R^2\ (\frac{R}{\hbar/M_e c Z \alpha_f})^2\ \frac{1}{1+4k^2R^2} \rightarrow 16\ \pi\ R^2.$$

The electromagnetic character of the interaction is reflected by the strong decrease of the cross section with increasing angle ϑ and energy E_e of the electron

$$d\sigma/d\Omega(\vartheta=\pi)=(Z\alpha_f M_e c/2\hbar k^2)^2 = (Z\alpha_f \hbar c/4E_e)^2.$$

We see that electrons are strongly scattered. Thus they can not penetrate thick materials but they are effective for the investigation of surfaces. Positrons have the same scattering behaviour; they are only more expensive. But they have the advantage that the cross section for the annihilation process

$$e^+ + e^- \rightarrow 2\ x\ (\gamma + 511keV)$$

is highest for positrons at rest. Therefore energy and angle of the emitted gamma quanta are modified only by the state of the electron in the material. In this way the angular correlation of the two gammas can give information about the band structure in the neighbourhood of the trapped positron. The myon is the heavy sister of the electron ($M_\mu=206.378\ M_e$). After a mean life of 2.20 μs it decays into an electron and two neutrinos

$$\mu^- \rightarrow e^- + \bar{\nu}_e + \nu_\mu,\ \mu^+ \rightarrow e^+ + \nu_e + \bar{\nu}_\mu.$$

For materials science it does not seem very useful, because for a kinetic energy of E=100 eV only it has a wavelength of λ =8.54 pm and a mean range of 91 cm.

The picture changes if we use neutron optics. Thermal neutrons from a reactor can be made monochromatic by Bragg reflection. In an energy range of E=20.5 ... 3.3 meV we get neutron wavelengths

$$\lambda = 28.6015\text{pm } (E/eV)^{-1/2} = 200 \ldots 500\text{pm}$$

which are just in the region of interatomic distances. But in a material neutrons have other preferences than X-rays because neutrons interact by strong forces with atomic nuclei and on account of its magnetic spin moment

$$\mu_n = -1.91315\mu_N = -9.6629 \times 10^{-27} \text{ Joule/Tesla}$$

they interact with magnetic atoms too. The electron-neutron interaction can be neglected. Therefore the atomic form factor is reduced to the Debye-Waller factor. The elastic cross section due to nuclear forces is 8000 fm^2 for 'H and varies in the range $\sigma_s = 100 \ldots 3000 \text{ fm}^2$ for other nuclei. For heavy elements σ_s is restricted to the region $\sigma_s = (4 \ldots 8) \times \sigma(\text{geometrical})$. By X-rays we see primarily heavy elements, by neutrons we see primarily light hydrogen. Magnetic structures dominate by their influence on the polarization of the neutron spin.

A projectile similar to the neutron is the myon-hydrogen, an atom consisting of a proton and a negative myon with a binding energy of 2.524 keV, a mass of 1.111 neutron masses and with a first Bohr radius of 285 fm only. Therefore within a material this atom behaves like a fundamental particle of spin 0 or spin 1. In the spin-0-state the particle is non-magnetic. In the spin-1-state the myon-hydrogen has a magnetic moment

$$\mu = -6.09769\mu_N = -3.07982 \times 10^{-26} \text{ Joule/Tesla,}$$

roughly three times the magnetic moment of a neutron. The only problem is that it decays after a mean life of 2 µs. But the variation of the life-time depends on the structure of the target. If the myon decays then a high energy electron is emitted with a continuous spectrum. On the other hand, if the myon is captured by the proton then a monoenergetic neutron is emitted with an energy of $E_n = 5.1977$ MeV.

From high energy accelerators we have more short-life and strange particles which have not been used so far within material science. After a long period of low energy excitations nuclear physics is just using the tools of particle physics. A similar development can be seen in material science. Of course, there are difficulties: On the one hand, if the interaction is strong as for pions, protons and alpha particles, then we have a high yield for our detectors, but we have multiple scattering, various creation and decay processes, and the analysis of the whole phenomenon becomes very complicated. On the other hand, if the interaction between incoming particles and the

material is very weak as for neutrinos, then we have a high degree of pene-
tration, the processes are well defined and can easily be analysed, but the
number of events is smaller and thus the accuracy is low. The optimalization
of a method depends on the problem, which we want to solve:
In order to investigate short range correlations we have to use high energies
and if the target is very thick then we can get information about the internal
structure by weak interactions only.

The weakest interaction we know is gravitation. In a hydrogen atom the Coulomb
force exceeds the gravitational force by a factor of 2.2701×10^{39}. But on
account of the infinite range of the gravitational force we have huge effects
on planets and stars in our universe. (Also the Coulomb force is not strong.
But on account of its infinite range the Rutherford scattering diverges in
forward direction and the hydrogen atom has an infinite number of bound
states wheras the deuteron only has one single bound state.) It is possible
to estimate the density and the crude chemical composition of the earth's
crust by gravimetric measurements. Here we have a target with a thickness
of 30 to 60 km. The socalled "weak interactions" are stronger than gravitation but still weaker than electromagnetism. For example, the interaction
of a neutrino with an atomic nucleus has a cross section of about $2 \times 10^{-18} fm^2$.
This means that the whole earth has a cross section of about $7 \ km^2$, which
is 5×10^{-8} times the geometrical cross section of our planet. But this weak-
ness of interaction is necessary in order to investigate the chemical composition in the interior of the sun, a target with a diameter of $1.4 \times 10^6 km$.
From the fusion and decay processes in the centre of the sun only the
neutrinos can reach our detectors here on the earth and at present it is a
serious problem, why so few solar neutrinos can be detected compared to the
high energy output of the sun (4×10^{26} Watt).

3. Inelastic and other processes

So far we discussed elastic processes, where only the momentum but not the
energy of the incoming particle and of the target changed in the centre-of-
mass system. For microchemical analysis we need also other processes. For
optical spectroscopy we have to excite the valence electrons. This method
has high precision but is limited by the optical wavelengths, by the possibility of interpretation of complicated spectra and by the fact that the gas
phase is the best for atomic absorption. If we change to inner electron shells,
the interpretation becomes easier, but only more selected chemical effects
can be observed. For example, in ESCA-spectroscopy we use monochromatic

X-rays for the photo effect on inner shell electrons. From the energy spectrum
of the emitted electrons we can determine the binding energy of the electrons.
As this process is concentrated to the surface of a material, the surface preparation demands high precision. Otherwise we observe only secondary effects.
In X-ray resonance fluorescence we go deeper into the material and deeper into
the electron shells of the atom. There we use X-rays from a standard X-ray
generator. Also synchrotron radiation, which is produced in electron accelerators
has advantages with respect to the intensities and with respect to the wavelengths available. Also particle induced resonace fluorescence can be used;
for example by a proton beam. By the X-rays electrons are ejected from an atom.
The re-occupation of the bound states of these electrons gives rise to
characteristic X-rays of the corresponding element. Here again the interpretation is clear for an inner transition from the L- to the K-shell. Only
two lines $K_{\alpha 1}$ and $K_{\alpha 2}$ are emitted. Wavelength and intensity vary systematically with the proton number Z. Therefore the determination of trace elements
can be done with high precision, especially for heavy elements, because the
cross section increases strongly with the element number Z.

For the L- and M-lines we have more influence from the chemical structure; but
the lines increase in number and the interpretation becomes more complicated.

In the centre of an atom we have the influence of the chemical structure because of the hyperfine interaction between the atomic nucleus and the electromagnetic field produced by the electrons at the position of the nucleus.
Nuclei with odd mass number A=N+Z or odd proton number Z and odd neutron number
N have a finite spin and a finite magnetic moment in the order of few nuclear
magnetons μ_N. The splitting of the nuclear energy levels is proportional to
the magnetic field at the position of the nucleus. The magnetic field depends
on the electron distribution and in this way on the chemical structure of the
material. This fact is used in nuclear magnetic resonance spectroscopy and in
Mössbauer spectroscopy. Many nuclei have a nonspherical charge distribution.
For nuclear spin quantum number $I \geq 1$ this deformation can be determined by
the nuclear electric quadrupole moment. Nuclei in the region of the rare earths a
in the region of actinides have large electric quadrupole moments. Here the lev
splitting is proportional to the electric field gradient produced by the electro
at the position of the nucleus. This splitting can be observed in optical spectr
copy. Nuclei with an isomeric transition from an excited state to the ground sta
become smaller with respect to their radial charge distribution. The energy shif
is proportional to the corresponding change $\delta<r^2>$ and again to the electric

field gradient of the electron distribution. Those effects can be detected by Mössbauer spectroscopy.

Radio-isotopes are very useful tracers for the investigation of the dynamics of chemical reactions. Nuclear activities are also valuable for microchemical analysis:

There are chemical processes induced by radiation. Ionizing radiation can produce synthesis, modification and polymerisation of chemical products. By puls radiolysis one can study the kinetics of chemical reactions. For example, a puls of energetic electrons within 1 μs or shorter can induce free radicals like hydratized electrons and their reaction mechanisms can be studied within a small time scale reaching the range of manoseconds. Activation analysis is a very powerful tool of microchemical analysis: A target is bombarded with a beam of particles, usually by thermal neutrons of a nuclear reactor, because the absorption cross section increases with decreasing velocity. Neutron absorption leads to an excited state of an isotope of the same element. The excitation energy is emitted by gamma rays with energies specific for each isotope. In this way elements are detectable if their mass exceeds about 10^{-12}g. The method is non-destructive, very sensitive and applicable for light elements too. The only conditions are that the element has a natural isotope with a sufficient high neutron absorption cross section and that the produced isotope has a halflife not shorter than some milliseconds.

Nuclear structure calculations have shown that there could exist in nature super-heavy elements with proton number $Z \cong 114$ and mass number $A \cong 298$. They should be stable against spontaneous fission. As the element 114 and lead should have very similar chemical properties, this element could exist as a trace element in lead as a relic from a supernova explosion 5000 millions of years ago. Also in heavy ion collisions like $^{238}U + ^{90}Zr$ there could be superheavies within the variety of reaction products. The element could be identified by its characteristic K-lines with energies and wavelengths

$$E(\alpha 1) \cong 173 keV, \ E(\alpha 2) \cong 160 keV, \ \lambda(\alpha 1) \cong 7.2 pm, \lambda(\alpha 2) \cong 7.7 pm.$$

A project for the identification of superheavies by their L-lines remained without success so far.

Not only the hypothetical superheavies but also the transuranium elements with $Z > 98$ need special microchemical methods. Not many atoms of the new elements $Z=99...106$ have been produced. After a short life-time mainly in the range of seconds they decay either by spontaneous fission or by alpha decay. Just this instability allows an identification of an element by its decay products. A typical fission process would be

$$^{276}108 \rightarrow (^{132}Sn+130MeV)+(^{142}Ce+120MeV)+2n$$

with a total kinetic energy of about 250 MeV. Fission is accompanied by the emission of few neutrons wich are evaporated from the excited fission products. Also the alpha energies for a certain mass number A are very specific for an element. Let us take for example the mass number A=276. We can estimate the alpha energies for various elements

$$Z \qquad = 105 \quad 106 \quad 107 \quad 108$$
$$E_\alpha(\text{ MeV }) = 6.45 \quad 7.12 \quad 7.77 \quad 8.41$$

These are the maximum energies. As the daughter nuclei have lowlying rotational states there is also a fine structure of the alpha energies. For example we expect for Z=108 alpha energies

$$E_\alpha(\text{ MeV }) \quad = 8.41, 8.34, 8.18, \ldots.$$

with decreasing intensity. The energies are high and the life-times are short. Therefore a single atom or nucleus can be identified. Nuclear physics and particle physics have developed methods which can be used for detection of elements if only a few atoms are available. Especially for anti-chemistry this fact is very important. Anti-matter consists of anti-protons and anti-neutrons in the atomic nuclei, which are surrounded by positrons. In the atoms of anti-hydrogen a positron rotates around an anti-proton. Spectroscopically matter and anti-matter can not be distinguished. Only when hydrogen and anti-hydrogen come together can we observe annihilation processes of the type

$$e^{+}+e^{-} \rightarrow 2\gamma + 1.022 \text{ MeV}, \quad \bar{p} + p \rightarrow \pi^{+} + \pi^{\circ} + \pi^{-} + 1462.45 \text{ MeV}.$$

Nothing is known about the chemistry of a mixed system of matter and anti-matter. But it is sure that microchemistry is an open system both with respect to the methods of investigation and with respect to the objects of interest.

STRATEGY AND TACTICS IN MICRO TRACE ANALYSIS OF ELEMENTS

GÜNTHER   TÖLG
Max-Planck-Institut für Metallforschung,
Institut für Werkstoffwissenschaften,
Laboratorium für Reinststoffe,
Katharinenstraße 17, D-7070 Schwäbisch Gmünd,
Federal Republic of Germany

Abstract. In scientific research determining elements at ng/g levels and
below implies an additional conception in comparison to normal routine analysis.
One of the most important goals must be to look preferably for pertinent methods
and to optimize them in such a critical way, that scientific innovations may be
realized at best. For this the main problem is how systematic errors in micro
trace analysis can be recognized and minimized inherent in the most reliable
procedures.

This contribution is to demonstrate in general and by means of some case studies
the methodical strategy which has been introduced in the author's laboratory
many years ago.

One example is to show the accurate determination of selenium traces in very
different matrices by an optimized universal closely combined decomposition
- preconcentration - AAS procedure, which enables now the determination of
selenium in the ng/g-range in a variety of organic and inorganic samples of
various fields.
A second example demonstrates by means of the analytical characterization of
high-purity beryllium the important role of micro trace analysis in an interdisciplinary research project of materials science, which includes also labour
medical aspects, with respect to the great toxicity of beryllium.
Progress in both fields, which are closely tied to each other is strongly
dependent on the state of the practice orientated micro trace analysis.

1. Introduction

If it is intended to propagate new strategies in trace and micro analysis of the elements one instinctively thinks that the common pattern of thoughts in this field does no longer meet all current requirements. Therefore, to begin with, one expects a brief introduction of the topic which first is to be given more generally and subsequently by way of some case studies from the work of our laboratory.

Although nobody doubts the inseparability of synthesis and analysis in chemistry and materials sciences, however, tremendous problems arise to practice this unity optimally.

There is hardly another discipline of natural science, such as analytical chemistry in general in which the views are differing to such a large extent when its face value compared to adjacent disciplines is to be assessed. On one side, one believes that analytical chemistry is not an independent scientific discipline. It is looked upon as an interdisciplinary auxiliary means - as a tool for sale - which each scientist can make use of. On the other side, the analysts themselves attempt to define and institutionalize their discipline for years. This is done in order to prove the necessity of its independence in teaching and research. Thus, in the opinion of others, the analysts frequently run the risk to surpass their scientific competence if they penetrate other disciplines. These strained relations surrounding the analytical chemistry, which unfortunately only hinders progress in research, must be removed as soon as possible if the analysts' effort is not only to be an end in itself but is also recognized by other scientific fields.

2. The trace analyst and his partners

Therefore, the analyst has to ask the root question of how it is possible to convince colleagues of the neighbouring disciplines - such as chemistry, physics, geo-, bio- and materials sciences - that analysts are indispensible partners.

In order to hasten this process of understanding we have to be aware why these strained relations could arise. Above all it seems to be due to a quick and significant alteration of the analytical problems in the last years, so that an information gap might have occurred with many colleagues of borderline disciplines.

While formerly simply the composition of a material with respect to major
and minor constituents was of interest, which could be determined with
relatively modest means not only by each chemist, but also by scientists
of disciplines contiguous to chemistry, the interest shifted in the mean-
time towards ever decreasing concentration levels, partly down to the pg/g
range. For a long time, not only are statements on integral material
compositions of the samples of interest, but also many questions about the
micro-distribution of definite components in or on a sample or even concerning their chemical type of bonding.

Typical questions are:
- How is the distribution of a trace element in different organs in certain
  compartments of cells or in individual protein fractions of blood serum
  and in which type of bonding does this happen?

- Which importance have trace elements in the eco system soil-plant - animal -
  - man, e.g., as regards carcinogenesis?

- Does selenium have protective power against cancer of special heart diseases as has been found recently in China [1]?

- Is there a serious danger to mankind as a result of a contamination of
  our environment by Hg [2] and other toxical metals or chemicals?

- Is a foreign atom species in a solid matrix segregated or is it substi-
  tutionally or interstitially fixed in the lattice or preconcentrated or
  depleted at internal or external interfaces in certain types of bonding[3]?

The analysis necessary for the elucidation of such and numerous similar
questions for a long time overwhelms many colleagues of adjacent disciplines,
whose scientific statements should base upon analytical data as detailed and
dependable as possible, but also already many specialized analysts. That one
who persists at a time of change-over of analytical chemistry on old concepts,
which he transfers from the time of his education in the present time, certainly contribute only little to progress in research and technology. We
should, therefore, focus our attention on such partners who are already
interested in a cooperation. But we should then consider too, that each
attempt of an analyst to solve scientific problems alone, which exceed his
scope, increases the risk of annoying colleagues. Seeking for a dialogue
and for capabilities and tools for a joint solution of problems in future
should be preferred, instead. In this, we may not forget the solution of
problems we are already encountering today. Also we should preserve enough
play for an analytical basic research which answers not only the purpose
and which is the fundament of all innovations in future.

It is not unjustified that our partners often reproach us, that our research
misses the requirements of praxis. Therefore, we should concentrate ourselves
more on such analytical problems with whose solution we can expect essential
improvements concerning application, accuracy of data, and economy compared
with the present state. While the economical situation can be judged quite
differently depending on the country, detection ability and reliability of
data are defined today on an international basis.

Figures of merit - especially those of accuracy are, however, quite differently
evaluated mainly by the users of methods and data of analysis. This fact may
by no means surprise, since as early as the last century reliable data have
been a matter of course for every careful analyst. Small relative inaccuracies
of 0.1 - 1 % were a standard in classical analysis.

The picture of a great accuracy in macroanalysis has been also first transferred to trace analysis. Well, reproducible results signalized for a long
time also accurate results - that means an agreement with the true concentration in the sample -.

It was not until so-called   interlaboratory comparative control studies
were introduced in the last years that we obtained some information on the
true state of affairs.

In these studies the same samples are analysed by different laboratories.
In spite of excellent reproducible results obtained by respective laboratories large discrepancies from laboratory to laboratory were evident as
witnessed by numerous publications in different fields of trace analysis [4-6].

That is not to say that the accuracy of results has to be brought to perfection in each case. Often we can manage very well with good reproducible
relative values, in order to be able to find the trends in the solution of
problems. But such data ought to be checked critically to avoid the use of
such data as absolute ones by others as is unfortunately often the case.

A comparison of analytical data presupposes standardized procedures in
routine analysis. I am afraid in this it is far too often overlooked -
under stress of conventions and regulations enacted by non-analysts - that
only such methods can be standardized which are already methodically
completely under control. However, this is only very seldom the case in
the field of trace and micro analysis. Here uncritical generalizations are
a very bad strategy. Balanced methods in trace and micro analysis have to
be free from systematic errors, which bias the analytical result - the more

so the smaller are the absolute amounts to be determined. Systematic errors
depend, however, to a large extent on the method to be used, the matrix and
the present concomitant elements if they exceed certain borderline concentrations.

All these parameters depend on the problem of analysis to be solved.
Systematic errors can therefore prevail to such an extent that it is often
absurd to evaluate results statistically because a normal distribution of
errors is lacking [5].

This new situation in trace analysis, which applies to a greater extent to
micro distribution and surface analysis is, however, not only a reason to
direct new strategies outwards, but presupposes also sequential strategies
for the methodical problems which can be summarized in large as follows:
How can systematic errors be recognized?
Which principal sources of error are to be considered?
How can they be reduced to a minimum?

3. Systematic errors

The recognition of systematic errors belongs today to the most important but
also most difficult tasks of the trace and micro analyst. We can choose
various paths (Table 1) [3].

Table 1: Recognition of systematic errors

1. DEVIATIONS FROM STANDARD REFERENCE MATERIALS
2. INTERLABORATORY COMPARATIVE ANALYSES
3. APPLICATION OF INDEPENDENT WORKING SYSTEMS OF ANALYSIS
4. APPLICATION OF ABSOLUTE METHODS OF DETERMINATION
5. APPLICATION OF MULTI-STAGE PROCEDURES EASY TO CALIBRATE
   (CALIBRATION WITH STANDARD SOLUTIONS ACCORDING TO THE
   INCREMENTAL METHOD, CALIBRATION GRAPHS)
6. BY ADDTION OF INCREASING QUANTITIES OF THE SAME SAMPLE
7. YIELD DETERMINATIONS USING RADIO TRACERS WITH THE SAME
   TYPE OF BONDING (E.G. BY ACTIVATION OF THE SAMPLE)
8. STATISTICAL EVALUATION IN NORMAL DISTRIBUTION OF ERRORS
   (OUTLIER TEST)
9. DEPENDENCE OF THE RESULTS ON SAMPLE WEIGHT, TIME, AMOUNTS
   OF REAGENTS, TEMPERATURE ETC.

The safest one is to investigate a sample by at least 2 or 3 different
procedures which show quite different systematic errors characteristic

of the procedure.

In rare cases it will be possible to obtain immediately a statistically significant agreement. Yet the trend towards accuracy is soon recognized.

Is there only one method on hand, we can succeed if the determination proceeds using different sample weights or according to the incremental method using standard solutions.

If there is a radioactive nuclide of the element under study with a radiation energy and half-life suitable for measurement, the systematic error can be established radiochemically and in addition to this controlled and corrected by isotope dilution analysis.

The more extended are the experiences of a laboratory in regard to systematic errors the more it is possible to plan a new procedure with only less sources of error.

In this we may, however, not renounce to statistical yield determinations of the individual steps of the procedure. If possible radiotracers should be used. In most cases ultra trace analytical problems can be solved only by a combined application of the different methods which have been outlined here briefly.

4. Causes of systematic errors

From what has been said so far the question how these systematic errors arise inevitably follows [5], which hardly carry great weight in the analysis of major and minor constituents. The causes of systematic errors in micro trace analysis (Table 2) depend first of all strongly on the methods, the matrix, and the element to be determined. They take increasing effect with decreasing absolute amount of an element to be determined.

Therefore, they can either be dealt with only very generally or by means of case studies.

Generalizations can not be made since each problem practically varies from one another. Therefore each trace analyst has first of all to gather experiences and to learn to excercise self criticism or to be taught to do so very early. He should be pleased if at the beginning he can only prove why a method does not yet reliably work. In contrast to those who believe to produce always accurate results just the critical trace analyst will be rewarded by reliable data in the long run, which - as I believe - are the sole fundament of each science.

Table 2: Causes of systematic errors

1. IMPROPER SAMPLING AND STORAGE OF THE SAMPLE

2. CONTAMINATIONS : SURFACES OF TOOLS AND VESSELS,
OF REAGENTS, AND AUXILIARY SUBSTANCES,
LABORATORY AIR

3. ADSORPTION/DESORPTION:
SURFACES OF TOOLS AND VESSELS,
INTERFACES (FILTRATION, COPRECIPITATION ETC.)

4. VOLATILIZATION :
ELEMENTS (HG, AS, SE U.A.)
COMPOUNDS (OXIDES, HALOGENIDES, HYDRIDES ETC.)

5. CHEMICAL REACTIONS:
CHANGE OF VALENCY OF IONS, PRECIPITATION
EXCHANGE, ION EXCHANGE, COMPLEX FORMATION,
FORMATION OF VOLATILE AND NON-VOLATILE
COMPOUNDS

6. INTERFERENCE OF THE SIGNAL:
MATRIX EFFECTS, TARGET,
OVERLAPPING OF SIGNALS,
SIGNAL BACKGROUND

7. INCORRECT CALIBRATION AND EVALUATION:
INCORRECT STANDARDS, INSTABLE STANDARD
SOLUTIONS, BLANKS, ERRORS DURING MEASUREMENTS,
FALSE CALIBRATION FUNCTIONS,
INADMISSIBLE EXTRAPOLATIONS, ETC.

The different sources of errors can take quite different effects in the individual steps during an analytical procedure (Table 3).

Table 3: Frequency of sources of systematic errors in trace analytical
procedures
± positive/negative error; +++/---: large; ++/--: medium; +/-:small

TYPE OF ERROR	CONTAMINATION REAGENTS, TOOLS LABORATORY AIR	ADSORPTION TOOLS, INTERFACES	VOLATILIZATION	CROSS INTERFERENCES BY ELEMENTS CHEMICAL REACTIONS, SIGNAL-INTERFERENCES, -COINCIDENCE, -BACKGROUND
STEP OF OPERATION				
SAMPLING	+ + +	- -	-	±
SAMPLE PREPARATION	+ + +	-	-	±
DISSOLUTION, DECOMPOSITION	+ + +	-	- - -	±
SEPARATION, PRECONCENTRATION	+ +	- - -	-	± ± ± (+ + +)
DETERMINATION:	+	-	- -	± ± ± (+ + +)
EXCITATION CALIBRATION EVALUATION				

Therefore, assignments can only be made very roughly. However, the experience shows that the most serious errors ensue from contaminations and cross interferences during generation of the signal [5].

5. Avoidance of systematic errors

The elimination of systematic errors thus becomes an important demand in the problem oriented trace and micro analysis of the elements [5,6]. It requires a new strategical mentality for the right choice of an analytical method since as of this no method can claim to be universal. Its advantages and disadvantages can always only be rated in connection with the analytical problem under consideration.

Unfortunately communication problems still exist between specialists of methods, who develop a distinct method and apply it, and the more strategically working analysts who try to solve a given problem optimally with respect to figures of merit and economy. The former ones believe that their method can do everything, the latter ones that this "prestige thinking" is a very bad strategy. For the problem orientated analyst, therefore, the choice of the optimal method unambiguously presupposes a broad knowledge of the capability and limitations of as many as possible analytical tools currently available. In this, to be progressive,does in no case mean to give priority only to new instrumental methods and to reject categorically chemical ones for reasons of being obsolete. Only a balanced symbiosis of both strategies is promising. In consequence of this a generalizing preference of a distinct method is not reasonable [7]. It is simply useless to argue whether, e.g., optical emission spectrometry, AAS with electrothermal atomization (ETA-AAS), X-ray fluorescence analysis (XRFA), differential pulse polarography (DPP), instrumental activation analysis (IAA), or sparc source mass spectrometry (SSMS),can be preferred for bulk analysis or whether, e.g., secondary ion mass spectrometry (SIMS) is more suitable for a depth profile analysis than a combined procedure of ion etching and Auger spectrometry (AES) or a chemical or electrochemical etching procedure in combination with trace analytical determination methods [3].

6. Criteria for the choice of trace analytical methods

All important criteria for the choice of a method (Table 4) are linked very closely together in applied analysis. In the matrix practically each element combination is possible. Only water, air, or gases as well as organic substances are relatively simple matrices, which firstly don't confine the choice of the method essentially.

Table 4: Criteria for the choice of analytical procedures in the micro
and trace analysis of elements

1. SAMPLES - MATRIX

   AMOUNT

   FORM

   NUMBER

2. NUMBER OF THE TRACE ELEMENTS TO BE DETERMINED

3. CONCENTRATION RANGE OF THE TRACE COMPOUNDS
   (DETECTION ABILITY)

4. LOCAL RESOLUTION ON THE SURFACE AND AT DEPTHS
   (MICRO-RANGE ANALYSIS)

5. DEMANDS ON RELIABILITY (ACCURACY, PRECISION)

6. ECONOMIC ASPECTS

Much more complex are geological matrices, ceramics, glass, metals and their
alloys, especially refractory metals, e.g. Mo, W, Nb, Ta, semiconductor
materials, generally high-purity materials. Here the choice of a method
is very different. For instance, the determination of Ta-traces in high-
purity Nb has to be planned quite differently than the determination of
traces of Nb in high-purity Ta. The systematic errors basing on these procedures can be of quite different nature.

In extreme trace analysis mainly the detection ability is of distinctive
interest. It is for the individual methods strongly element dependent least
of all in SSMS. In principle the detection ability for an isolated element
is better than in the presence of concomitant elements, which lead to an
deterioration by cross interferences if they exceed certain borderline
levels. These are again strongly dependent on the method. If we consider,
e.g., in the bulk analysis INAA, which according to text books allows determination of at least 50 elements with high sensitivity, we easily overlook
the limitation that the matrix may have only a small capture cross section
for thermal neutrons, or yield nuclides with a short half-life. This is true
for only a few matrices. On top of that the number of determinable elements
in such matrices varies (Table 5). Thus, e.g., in high-purity Al and Be
about 20 impurities can be detected simultaneously while in high-purity
Nb only the elements Ta and W on account of cross interferences and line
coincidences [3]. This example is to show that general statements of detection
limits - as they are frequently used to compare methods - are of limited

value and can, at best, give some help for orientation. Detection limits of
a method can only be discussed if not only the whole analytical procedure
fixed in all its details is known, but also the other components in the
sample are known at least on an order of magnitude.

Table 5: Limits of detection (DL) of high purity aluminium and niobium
established by INAA

SAMPLE		ALUMINIUM	SAMPLE		NIOBIUM
ELEMENT	[ug/g]	DL [ug/g]	ELEMENT	[ug/g]	DL [ug/g]
Na*	0.006	0.001	W	0.008	0.003
Sc	0.002	≪0.001	Ta	0.9	0.03
Cr	0.010	0.001			
Fe	0.110	0.05	flux of neutrons:		
Co	0.001	< 0.001	9×10^{13} n/s cm^2		
Zn	0.08	0.01			
Ga	0.03	0.002	irradiation time: 24 h		
As	≤ 0.02	0.01			
Zr	≤ 0.1	0.05			
Mo	≤ 0.02	0.02			
Ag	≤ 0.002	0.001			
Cd	0.05	0.02			
Sb	≤ 0.002	0.001			
La	≤ 0.003	0.001			
Hf	≤ 0.001	0.001			
Ta	0.001	0.001			
W	0.005	0.002			
Hg	0.010	0.001			

*irradiation in a thermal column

We may also not overlook that - at least in AA [8] - the detection ability of
an element to be obtained in reality depends less on the sensitivity of a
method than on the relationship of the concentration of the element to be
determined on the sample to its concentration of ubiquity in the sample
environment.

Thus it certainly makes fewer difficulties to determine the smallest levels
of Au, Pt, Co or Re in the pg/g range where this ratio is reasonable, than
e.g. Hg, Cu, Pb, Zn, Cr, elements which are preconcentrated in the environment and can already locally occur in relatively high concentrations of
ubiquity. For the same reason it is practically useless to strive for a
reliable determination of pg-amounts or pg/g levels of Si, Al and elements
abundantly occurringin the earth crust. The actual detection limits may
therefore differ by orders of magnitude from the detection limits theoretically derived by extrapolation from results of higher concentration

levels and so enter literature [9].

Similar methodical limitations exist also in regard to demands on accuracy, which in its turn correlate with the detection ability. Basing on theoretical statistical considerations 10^{-18}g of an element can yet be determined with a precision of 10% [10]. However, even if all systematic errors are under control the attainable precision is practically, governed by the sample inhomogeneity [3], that can be considerable even with single crystals.

In addition to this detection ability and precision of a method determine the minimum amount of a sample to be used. It is naturally obvious that traces of an element can always be preconcentrated on a target area as small as possible from a large sample amount in order to determine still very small levels of an element by methods with relatively low sensitivity, e.g. XRFA. The trend in bulk analysis goes, however, rather toward micro-trace analysis - also as regards micro distribution analysis - in order to be able to determine yet absolute amounts as small as possible. Limiting factors are then again preferably systematic errors and the sample inhomogeneity [3].

For certain instrumental methods, e.g., IAA with fast neutrons, charged particles or photons, SSMS or PIXE (Proton-induced X-ray excitation) a definite size and form of sample is necessary which often confines greatly the use of such methods.

7. Detection ability aspired to

In principle, detection abilities for an element are aimed at such concentrations which are expected to cause still chemical, physical or biological changes in property in a matter [11]. When we consider first naturally existing systems so we have in a first approximation the concentration of ubiquity of the elements as a result of the cosmochemical and geochemical evolution of our planet.

In a similar relationship many elements distribute also in biological substances. For the most part those elements are of interest which are effective in man (Figure 1) and their smallest concentrations which still cause physiological changes [12, 13].

The encroachment by man on naturally existing equilibria focus interest especially on those elements which are essential in very low concentrations, but are toxic at higher ones, e.g., Se, Co, Cr and Mn or on such elements with an ever increasing environmental concentration so that meanwhile - at least locally - serious impacts on man has to be expected as is the case with e.g. Hg, Cd, Pb, As, Ni, Tl and Be.

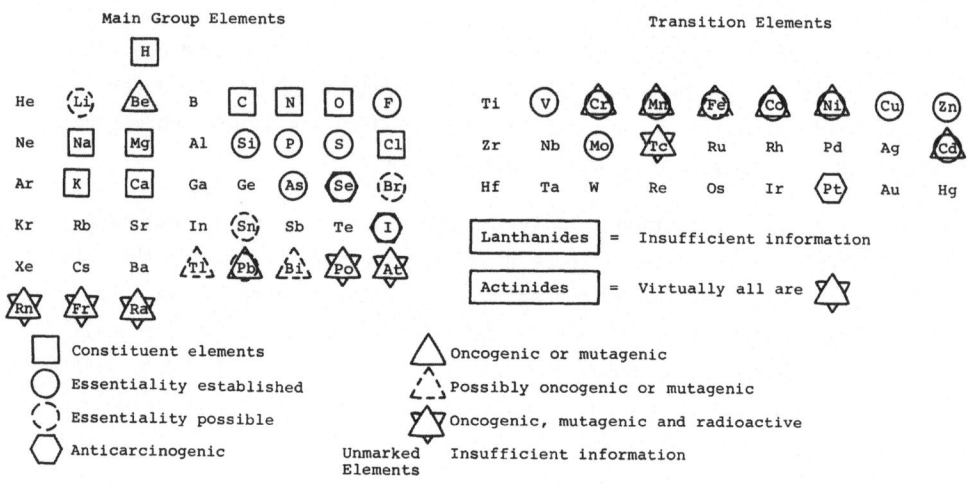

Figure 1: A periodic system of the elements indicating essentiality,
          oncogenicity, or mutagenicity after Schrauzer [1]

Frequently the essential and toxic concentrations differ only by 1-2 orders
of magnitude as e.g. with Se [5]. Furthermore we have to consider that frequently only definite types of bonding of an element are effective as is
conspicuous with Hg where mainly metal organic compounds, which may form,
are biologically active [2]. Only hints exist on the phenomenon that elements
act synergistically or compensate mutually their biological activity as is
the case with Se on one side and numerous heavy metals (e.g. Hg, Cd, Pb)
on the other side. Notwithstanding the complexity of these
problems, nature helps us a great deal. Many of the abundantly occurring
elements in the earths crust are biologically active only at higher concentration levels. A good impression of the detection abilities aimed at
can be obtained from relevant levels of elements in human and animal blood [12]
(Table 6) which provides the transportation of the elements in the organisms
and contains the lowest elemental concentration levels in organs. This applies
equally to all matrices (rocks, soils, foodstuffs with exception of drinking
water) of the superior material ecosystem. The limits of determination lie
practically all on the order $\geq$ 1 ng/g.

If we proceed a step and show interest in the distribution of the elements
in the respective components of the blood and the individual protein fractions

of the serum [14] then detection abilities better by a factor of 10 have to be
aimed at.

Table 6: Detection limits (DL) to be strived for minimal levels of
elements in blood and serum after G.V. Iyengar et al.

DL FOR BLOOD [ng/g]	ELEMENTS	DL FOR PROTEIN FRACTIONS [ng/g]
500	Cu	10
100	AL	5
50	PB, SE, SI	1
10	B, I,	0,5
5	AS, CR, NI, TE, V, ZR	0,1
1	AG, BE, BI, CD, LI, CS, HG, MN, MO	0,05
0,5	CO, TL	0,01
0,05	AU	0,001

A special position in the consideration of natural systems is attributed to
the analysis of the atmosphere as well as to aquatic systems where frequently
already detection abilities on the order of pg/g are relevant [15,16].

Quite different are the conditions with numerous materials which have been
created by man within the scope of modern technology. Here already grades of
purity are strived for and attained, in which the concentration of ubiquity
of the elements - established by nature - are shifted towards extremely low
concentration levels (high-purity metals, semi-conductors, glass for light
conductors, ceramics for implantations and others). The detection abilities
to be aimed here lie on the order of $10^{-7} - 10^{-10}\%$ [3,6]. Mainly electrical
and optical properties and in basic research interaction of solid states
with elementary particles depend strongly on foreign atom levels.

In the analytical characterization of such substances one is practically
always compelled to improve substantially the limits of the currently
attainable detection abilities. Therefore, the experiences gathered just
in this extreme range (high-purity materials analysis) might be of the
utmost importance for all less demanding analytical tasks. Herein general
features occur, especially with regard to systematic errors which last not
least determine the actual attainable limits of detection.

8. Direct instrumental analysis - chemical multi-stage procedures

A further important question in trace analysis is this:
when are direct instrumental multi-element methods or when are multi-stage
procedures that lend themselves at best to oligo-element determinations and have
the reputation to be very laborious to be used. The ideal method would
doubtless be a matrix independent instrumental direct method which is simple
can be simply automated and allows all elements to be determined down to smalles
concentration levels. Solid state mass spectrometry with spark, laser, and
secondary ion excitation, partly also INAA come very close to this ideal
concept. With complex matrices the availability of standard samples which
substantially match the sample under study, is a prerequisite. If there are
already such standards on hand for a compensation of systematic errors, so
a lengthy calibration is only then worth while, when many series analyses
are to be covered. In this, we may, however, never expect extremely low
detection limits.

Normally these require preconcentration steps. Instrumental automatic analysers
are therefore less focal in trace analytical research, they belong already in
the range of commercialized routine analysis.

So the question remains to be answered: what has to happen in all those cases
where no reliable calibration is possible [17]? Then we have to fall back on
multi-step combined or multi-stage procedures which are generally composed
of the steps decomposition or dissolution of the sample, separation and/or
preconcentration of the elemental trace from the matrix and the actual
determination of the isolated elements. After the sample finds itself homogeneously in gaseous phase or in solution a calibration by the standard
addition method is less of a problem. The principal disadvantage of such
multi-stage procedures is doubtless the need of time and staff, which,
however, quickly appears to be less detrimental if we consider how un-
economical are doubtful or even wrong data. On the other side the problem
of systematic errors mainly of contamination problems could in principle
be recognized and partly be solved in recent years [3,5,6,11]. The most important methodical marginal conditions are to be summarized briefly.

a) Choice of suitable working materials for tools such as quartz, glassy
   carbon [18], PTFE, and PP which can be obtained very pure and which are
   thermally and chemically substantially resistant and show after definite
   surface treatment only negligibe adsorption and desorption effects
   against ions [19].

b) Use of only such reagents and auxiliary substances with a concentration level of residual impurities on the order of pg/g which can be achieved by special purification methods [19].

c) Clean working benches and closed-off systems which reduce the contamination by many elements which otherwise might be introduced into the system of analysis by airborn impurities. Such systems improve the actual detection ability by 2-3 orders of magnitude [19].

d) Accommodation of proved working techniques of the classical micro and ultra micro analysis to the new requirements [10].

e) Development of special decomposition, separation and/or preconcentration methods [5,6,20-22].

f) Application of extremely powerful determination methods matching the separation problem in question. Preferably methods for simultaneous multi-element determination, e.g., OES with ICP, CMP, MIP [23], and micro hollow cathode excitation [24], but also for oligo-element determination, e.g., pulse polarography or mono-element determination, e.g., ETA-AAS, spectrophotometric methods, and not to forget micro titrimetric methods, which in special cases, are among the most precise available also in the ng/g range [10].

g) The accuracy of multi-stage procedures, developed with respect to avoiding errors has to be proved significantly by at least a second independent one which can easily be calibrated as well.

After this short survey on the methodical strategical problem range, the last part of my contribution deals with the determination of Se and Be, by means of which I would like to demonstrate our joint strategy which is a fundamental of our interdisciplinary analytical research.

9. Examples of our adopted strategy

Selenium is both in materials science and also in ecological, nutritional, and medical respect an eminently topical trace element [1,5,13]. Therefore, the investigation of a trace analytical procedure to its reliable determination in a broad spectrum of matrices (metals, glass, rocks, soil, plants, animal tissues and others), which are closely tied to each other within an interdisciplinary frame, was an interesting analytical research work. A critical analysis of all current possibilities for determination yielded a real market-gap, for all known methods for Se constitute only partial solutions. Moreover, from results of interlaboratory comparative analyses it was clearly to be seen that Se belongs with the problematic elements which encounters great difficulties in routine analysis as a result of numerous sources of systematic errors. Especially the already very broad-casted hydrid-AAS method is frequently very uncritically used in problem oriented analyses [25,26].

The path for universal solution which we chose [27] consists of a combined decomposition and separation method (Figure 2) where the sample is ashed or heated in a quartz vessel flushed by a stream of oxygen.

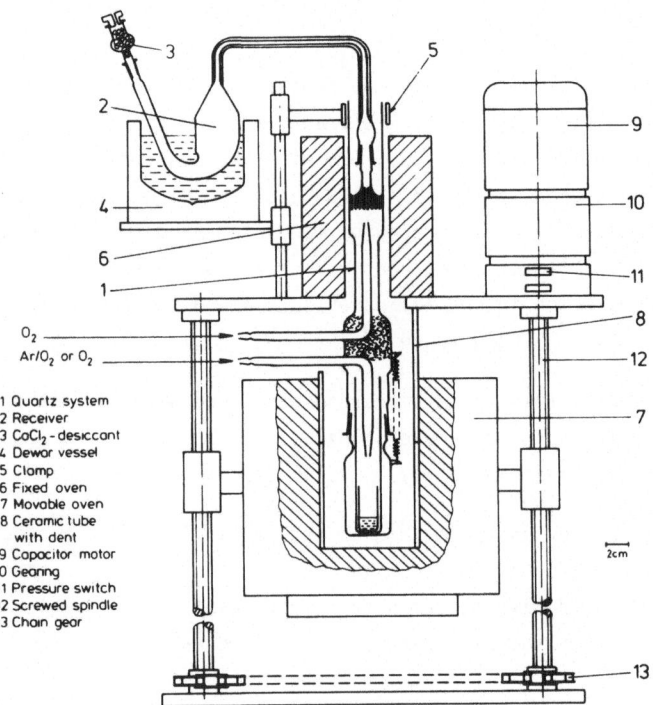

1 Quartz system
2 Receiver
3 CaCl₂ - desiccant
4 Dewar vessel
5 Clamp
6 Fixed oven
7 Movable oven
8 Ceramic tube
  with dent
9 Capacitor motor
10 Gearing
11 Pressure switch
12 Screwed spindle
13 Chain gear

Figure 2: Apparatus of quartz for a combined sample decomposition and separation of selenium via SeO_2 by volatilization for an universal and reliable Se determination in the ng/g range

The volatile SeO_2 which forms is evolved and subsequently trapped together with the oxygen in a receiver cooled with liquid nitrogen. After evaporation of the oxygen the isolated Se can be reliably determined without any problems by ETA-AAS [28], hydride-AAS [25,26], gas chromatography [29], pulspolarography [30] in various matrices (Table 7).

Table 7: Comparison of found Se contents and certified results for
different inorganic and organic matrices

SAMPLE	SAMPLE WEIGHT [mg]	CERTIFICATION [µg/g]	SELENIUM CONTENT FOUND FOR $n \geq 5$ [µg/g]
Granite (U.S.G.S., GSP 1)	200-600	0.059 (INAA)	0.063 ± 0.004
SoGMs 57 (Cu-Zn-Alloy) (BAM,AKP 221)	50	2.3	2.3 ± 0.2
Rg 7 (Cu-Sn-Alloy) (BAM,AKP 227)	10	27	30 ± 2.0
E-Copper (BAM,AKP 360)	150	(0.7)	0.71 ± 0.06
Euro-Cu I	300	(0.1-0.3)	0.205 ± 0.013
River sediment I	300	0.72 ± 0.18	0.66 ± 0.02
River sediment II	60	2.0 ± 0.4	2.32 ± 0.17
Bovine liver (NBS 1577)	15-250	1.1 ± 0.1	1.12 ± 0.07
Wheat flour (NBS 1567)	20-150	1.1 ± 0.2	1.01 ± 0.06
Rice flour (NBS 1568)	30-300	0.4 ± 0.1	0.34 ± 0.034
Orchard leaves (NBS 1571)	100-200	0.08 ± 0.01	0.078 ± 0.008

Similar conditions exist with Be. It is, as is known, the wear metal with
the lowest density and exhibits, therefore, as working material decisive
advantages over other metals, especially in aviation and space operations.
Its use in technology encounters, however, two principal difficulties.
Firstly it is hardly ductile and therefore hardly deformable and secondly
it is very toxic. About its physiological behaviour only very little dependable knowledge exists in regard to labour medicine. The materials part
has therefore to include labour medical aspects in all investigations in
this field. Both problem ranges have therefore been tackled simultaneously [31,32]. The mechanical properties of Be result from a complex interaction of structural and material impurities in its hexagonal lattice. Its
actual mechanical properties we can only know if we examine single crystals
of ultra-high purity. Such a high-purity Be is easily deformable even at
room temperature.

The non-ductility of the technical Be can in the first instance be referred
to an oxygen impurity which preferably leads to a segregation of BeO at

grain boundaries. If it is possible to diminish this effect by reducing the oxygen level by refining of Be, so the ductility increases. As, however, also foreign atoms species (3rd partners) influence the segregation processes in a very complex way we try presently whithin an interdisciplinary high-purity material project to obtain the desired properties by a definite doping of Be with different impurities.

It is unfortunately impossible to go here more deeply into high-purity materials analysis [31] which is closely connected to this project. I want only to say that also here the already demonstrated methodical strategy was consequently preserved. The analytical task could not be solved without a multi-method concept. Directly usable is only INAA for which Be is a convenient matrix. Unfortunately, only about 20 elements can be determined sufficiently powerful by this method in which the most important impurities are only partially included (Table 8).

Table 8: Determination of impurities in high-purity Be by INAA - neutron flux: 9×10^{13} n/sec x cm²; irradiation time: 24 h; evaluation of spectra by Ge/Li detector after 20 - 30 days

ELEMENT	NUCLIDE	T 1/2		STARTING MATERIAL [µg/g]	DL [µg/g]	ELECTROLYTICAL REFININ [µg/g]	DL [µg/g]
Na	Na-24	15.02	h	~ 4	0.1	~ 4.9	0.010
Sc	Sc-46	83.8	d	0.04	0.001	0.0001	< 0.0001
Cr	Cr-51	27.71	d	3.40	0.01	0.100	0.002
Fe	Fe-59	44.6	d	40	1	2.00	0.15
Co	Co-60	5.272	a	0.080	0.005	0.0045	< 0.001
Zn	Zn-65	243.7	d	4.1	0.1	0.940	0.02
Se	Se-75	120	d	~ 0.4	0.1	0.002	0.001
Zr	Zr-95	65.5	d	< DL	1	< DL	0.05
Ag	Ag-110 M	250.4	d	0.250	0.02	0.012	0.001
Sb	Sb-124	60.3	d	0.065	0.005	0.0015	0.0001
Hf	Hf-181	42.4	d	~ 0.2	0.05	0.001	< 0.001
Ta	Ta-182	115	d	0.210	0.005	0.007	< 0.001
W	W -187	23.9	h	0.5	0.005	0.015	0.001
Au	Au-198	2.695	d	0.3	0.005	0.950	0.005
Hg	Hg-203	46.6	d	0.4	0.05	0.015	0.005

The spectrum of methods ranges therefore from INAA via OES with ICP, CMP, and MIP excitation, ETA-AAS and XRFA in combination with the already partly mentioned preconcentration and separation method to the classical spectro-photometric and titrimetric micro methods.

For the especially important determination of small O and N levels special gas analytical methods are necessary [33]. However, it appears to be more interesting to pursue still more the physiological aspects (Figure 3).

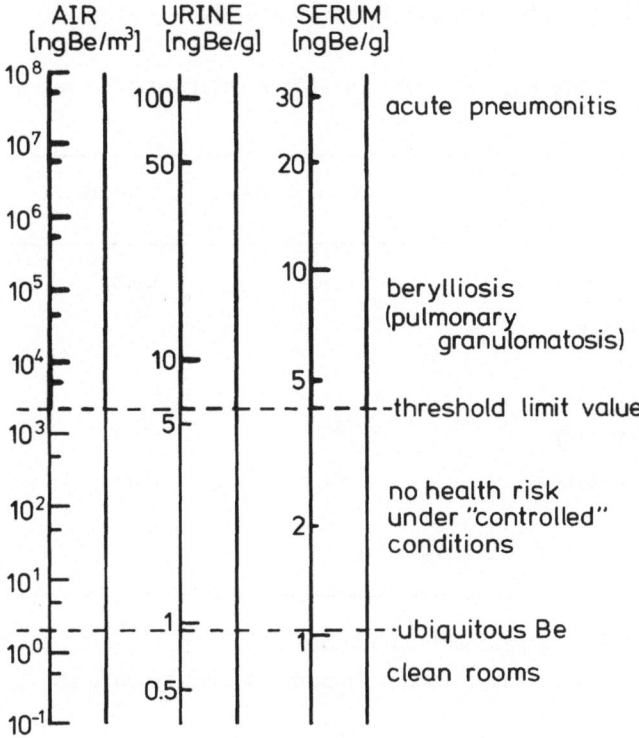

Figure 3: Be-concentrations in air, urine, and serum and their risk for human health

In this are encountered soon the limitations of the present knowledge of the patho-physiological aspect [32]. While we know already quite well the symptomatic part of the beryllosis - which is an offensive pulmonary disease - we are in the dark about the causes.

The first step towards the elucidation begins with the transport mechanism of the Be in human and animal organisms. Since beryllosis occur in the lungs -

- that is at the interface blood/respiratory air. At the onset we focussed
our attention of course together with a labor physician on the question how
distributes Be, which is administered via the respiratory airways, in blood,
Preliminary attempts showed that the Be level in blood of normal test persons
lies at about 1 ng/g. The analytical objective was, however, to detect still
substantially lower Be levels in the cellular blood and in the repective
protein fractions in serum. It has therefore to be strived for a reliable
detection of even pg/g-quantities of Be in a biological matrix. For the
determination of such low Be-levels only chelate GC [34] and ETA-AAS [35] provides satisfactory detection abilities (Table 9).

Table 9: Specification of the most sensitive determination methods for
beryllium [34,35]

Specifi- cation \ Method	AAS with ETA	GC of Be-TFA [*]
Detection limit [ng/g Be]	0.05	0.0001
Relative standard deviation	$\pm$ 5.5 % (5 ng/g)	$\pm$ 4 % (5 pg/g)
Mass balance	85-99 %	90-100 %

[*] TFA : Trifluoracetylacetonat

Moreover, with Be-7 a radioactive isotope with a sufficient specific activity
was available, which aids a lot in elaboration and trouble shooting of the
first step.

This work alone took us about 3 years of man power. Essential prerequisites
for this problem range were now decomposition methods for the biological
matrix (pressure decomposition with HNO_3 in PTFE or glassy carbon vessels)
the quantitative isolation of the traces of Be from the decomposition
solution in the pg-range by liquid-liquid distribution via Be-acetylacetonate
(Be-acac) and still further systematic investigations in order to reduce as
far as possible multiple sources of error in the actual GC and ETA-AAS
respectively (Figure 4).

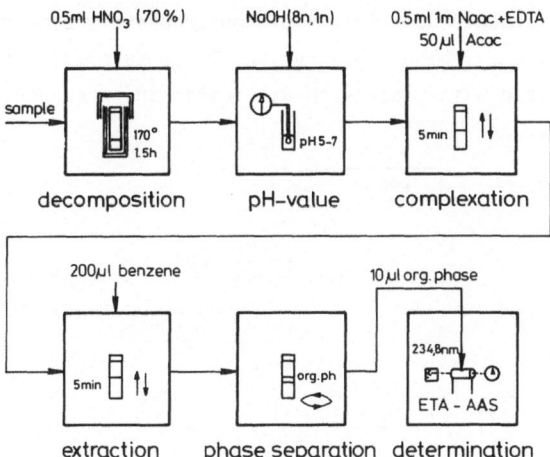

Figure 4: Multi procedure scheme for the accurate determination of ng/g
and pg/g amounts of Be in biological matrices by ETA-AAS

The next step, independent thereof, served to separate the respective
components of the serum down to its protein fraction [36]. It had to happen
under physiological conditions in order not to affect the natural conditions
of bonding of Be in the protein.

Ultra filtration and electrophoresis came into consideration (Figure 5).
The first condition, not to cleave original Be-bonds in the electrophoretic
separation of serum proteins, which contain about 70% of the Be in the
blood serum could only be met by the isotachophoresis. However, it was
not suitable to date for sufficiently large sample amounts being in
compliance with the detection ability of the Be-determination method.
It was not until the apparative problems of the isotachophoresis have
been solved, which now enabled us to use about 3 cm³ serum compared to
300 µl in case of commercially available devices, that we could tackle
the problem. Unfortunately the most important results can also here only
be given very briefly [32].

a) Normally Be is bound largely to γ-globulin which sequesters Be strongly.
   However it shows only a relatively small bonding capacity (Figure 6).

b) On Be-exposed animals and humans a significant increase in the γ-globulin
   portion in serum was established. But the main part of the Be finds itself in the prealbumin (Figure 7) where the bonding is however essen-

tially weaker, so that Be can easily be exchanged in the kidneys. This portion of Be is then practically excreted completely down to the natural concentration of the γ-globulin in urine. These surprising findings are presently evaluated with regards to diagnostic and therapeutic effects and consequences.

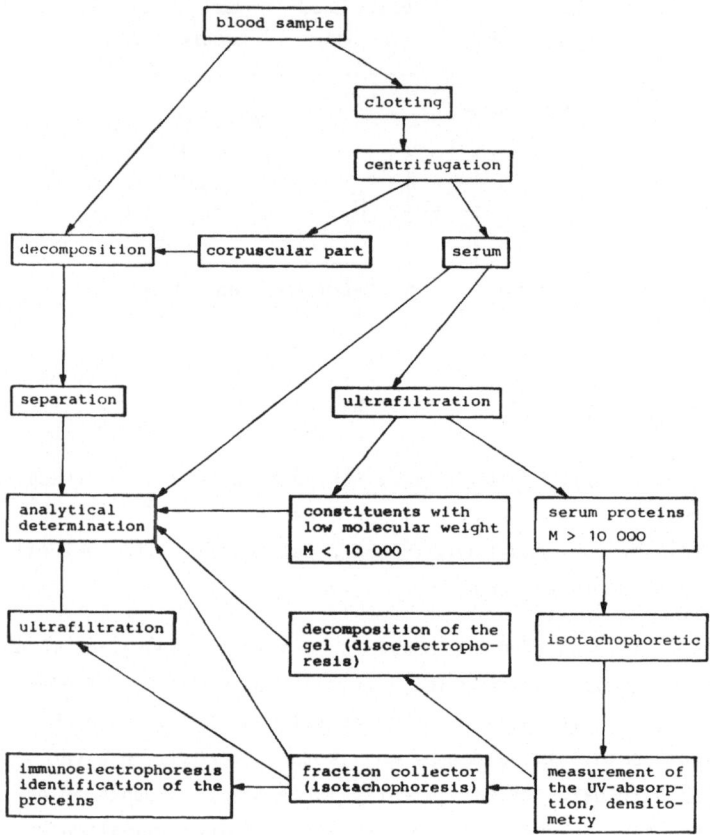

Figure 5: Block diagram of separation and analysis for the determination of trace elements distributed in blood

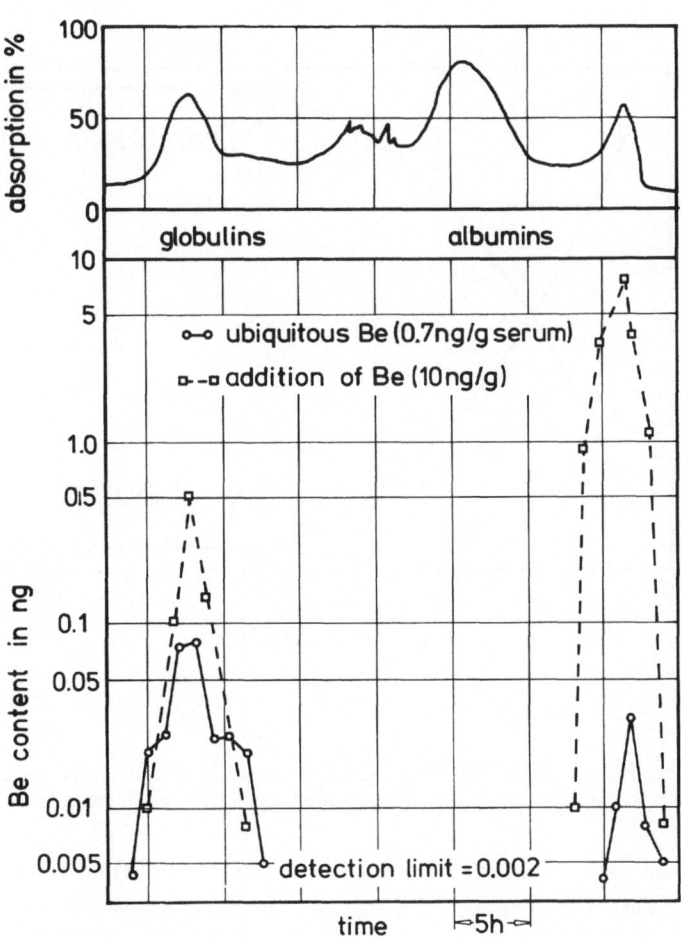

Figure 6: Protein spectrum (UV-diagram 280 nm) and Be-distribution of the isotachophoretic fractions of (3 ml) standard human serum with an ubiquitous Be-content of 0,7 ng/g and after addition of 10 ng/g Be to a normal serum

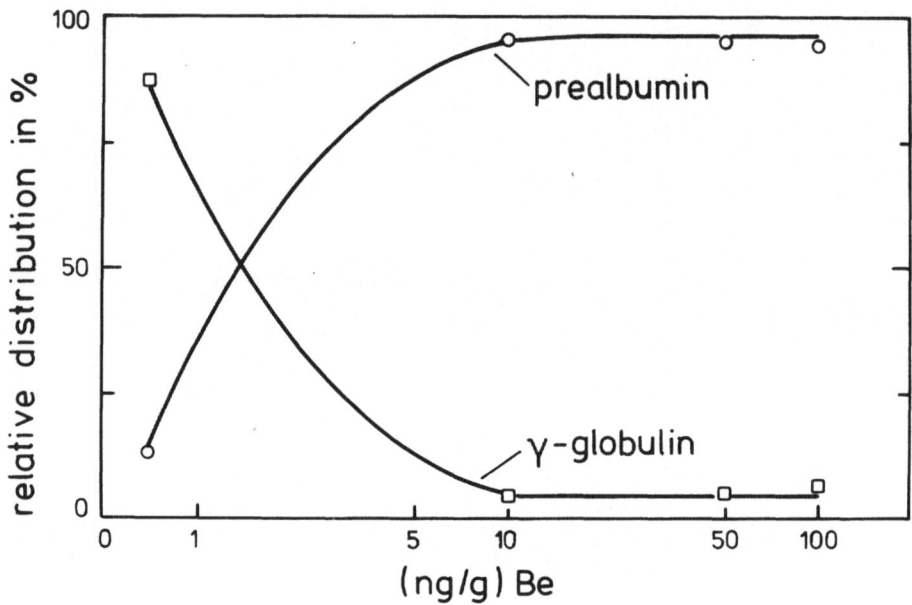

Figure 7: Distribution of the protein-bound Be between pre-albumins and
γ-globulins of standard human serum at different Be-concentrations

Associated with the investigation of the normal distribution of Be in urine
of test persons a further interesting sideeffect occurred. Frequently,
strongly increased Be levels were found and this with statistical significance only with smokers after having considered the phenomenon more closely.

Indeed we found that Be is preconcentrated in tobacco to a relatively large
extent (Table 10), but it occurs only partly in the ashs. From this observation we may infer that it is for the most part incorporated as a volatile
compound via inhaled smoke and again excreted after a certain retardation
time via the urine. In a first preliminary test a group of co-workers and
passive smokers convened to a very extensive smoking party. After having
abstained from smoking subsequently, for days, urine samples were collected.
In both, smokers and passive smokers a considerable increase in the Be-
concentration as compared with "normal urine samples" were found. The incorporated Be is, however, relatively quickly excreted.

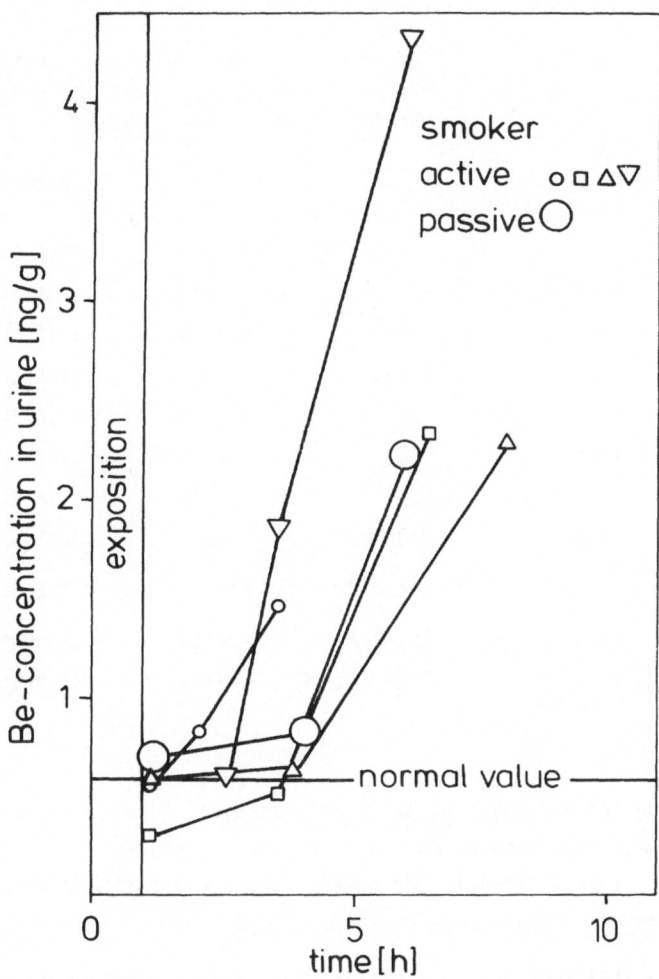

Figure 8: Normal concentration of Be in urine (non smokers) and increasing contents after inhalation of cigaret smoke (active and passive smokers) during an exposition time of 1 h

Table 10: Beryllium contents in different natural materials PAA: Photon
activation analysis, ETA-AAS: Atomic absorption spectrometry
with electrothermal atomisation

SUBSTANCE	BERYLLIUM CONTENT [µg/g]	METHOD	REFERENCE
ROCKS (HOHE TAUERN)	2.7	PAA	JASMUND, NEY
SOIL	1-10	ETA-AAS	STIEFEL
ROAD DUST	CA. 1	PAA	JASMUND, NEY
EGG WHITE	0.008	ETA-AAS	STIEFEL
HUMAN BLOOD	0.001	ETA-AAS	STIEFEL
BLOOD SERUM	0.0007	ETA-AAS	STIEFEL
BREAD	0.01	ETA-AAS	STIEFEL
VEGETABLE	0.02	ETA-AAS	STIEFEL
CORN ASH	0.06-0.67	PAA	JASMUND, NEY
CIGARET TOBACCO	0.1-0.2	ETA-AAS	STIEFEL
CIGARET ASH	0.2-0.7	PAA	JASMUND, NEY
AIR (STUTTGART)	$0.001 \ µg/m^3$	ETA-AAS	STIEFEL

Although these results of our finite marginal investigation can be looked
upon only as an immaterial factor of risk to smokers and passive smokers
so they clearly characterize however the present situation in many sections
of research.

Only an efficient analysis within the scope of a balanced and well operating
interdisciplinary collaboration is the fundament in many branches of modern
sciences. I hope that the conception of an analyst introduced here was a
modest contribution to think more intensively about possibilities how this
goal can be achieved more effectively. I want to finish by expressing
sincere gratitude to all my colleagues and co-workers who enabled me to
deliver these lines of thoughts, and to you for your kind attention.

Acknowledgement. I am grateful to G. Kaiser for his help in translating
the manuscript.

References

1. G.N. Schrauzer; Trace Elements in Carcinogenesis; Advances in Nutritional Research 2, Ed.: H.H. Draper, Plenum Publ. Corp., 1979

2. G. Kaiser and G. Tölg; Mercury in Environmental Chemistry, Ed.: O. Hutzinger, Springer-Verlag, Berlin-Heidelberg-New York, 1980.

3. G. Tölg; Proceedings II, 5th Internat.Sympos.High Purity Materials in Science and Technology, Dresden, 1980, Akademie der Wissenschaften der DDR, Zentralinstitut für Festkörperphysik und Werkstofforschung, Dresden, Helmholtzstr. 20, pp. 57-77

4. F. Ackermann, H. Bergmann and U. Schleichert; Fresenius' Z. Anal.Chem. 296, 270-276 (1979)

5. G. Tölg; Fresenius' Z. Anal. Chem. 283, 257-267 (1977)

6. G. Tölg; Fresenius' Z. Anal. Chem. 294, 1-15 (1979)

7. G. Tölg; Nachr. Chem. Tech. Lab. 27, 250-257 (1979)

8. V. Krivan; Angew. Chem. 91, 132-155 (1979)

9. O.G. Koch, P.D. LaFleur and G.H. Morrison; Pure Appl. Chem., in press

10. G. Tölg; in Wilson and Wilson's Comprehensive Analytical Chemistry, Ed.: G. Svehla, Vol. III, Elsevier Publ. Comp., Amsterdam, Oxford, New York, 1975. pp.6

11. G. Tölg; Pure Appl. Chem. 50, 1075-1090 (1978)

12. G.V. Iyengar, W.E. Kollmer, and H.J.M. Bowen; The Elemental Composition of Human Tissues and Body Fluids, Verlag Chemie, Weinheim, New York, 1978

13. J. Versieck and R. Cornelis; Anal. Chim. Acta 116, 217-254 (1980)

14. Th. Stiefel, K. Schulze and G. Tölg; Internat. Workshop - Trace Element Analytical Chemistry in Medicine and Biology, Neuherberg, FRG, 26th - 29th April, 1980, Walter de Gruyter, Publ. Comp., in press

15. H.W. Nürnberg; The Science of the Total Environment 12, 35-60 (1979)

16. C. Boutron; Anal. Chim. Acta 106, 127-130 (1979)

17. G.H. Morrison; Pure Appl. Chem. 41, 397-402 (1975)

18. L. Kotz, G. Henze, G. Kaiser, S. Pahlke, M. Veber and G. Tölg; Talanta 26, 681-691 (1979)

19. P. Tschöpel, L. Kotz, W. Schulz, M. Veber and G. Tölg; Fresenius'
    Z. Anal. Chem. 302, 1-14 (1980)

20. P. Tschöpel; Aufschlußmethoden, Ullmanns Encyclopädie der Technischen
    Chemie, Band 5, Verlag Chemie, Weinheim, in press

21. A. Disam, P. Tschöpel and G. Tölg; Fresenius' Z. Anal. Chem. 195, 97-109
    (1979)

22. E. Grallath, P. Tschöpel, G. Kölblin, U. Stix and G. Tölg; Fresenius'
    Z. Anal. Chem. 302, 40-51 (1980)

23. P. Tschöpel; in Wilson and Wilson - Comprehensive Analytical Chemistry
    (Rd. G. Svehla) Vol. IX, Elsevier Publ. Comp., Amsterdam,
    Oxford, New York, 1979, pp. 173

24. P. Faßmann; Dissertation an der Univ. Stuttgart, 1980

25. A. Meyer, Ch. Hofer, G. Tölg, S. Raptis and G. Knapp; Fresenius' Z. Anal.
    Chem. 296, 337-344 (1979)

26. S. Raptis, G. Knapp, A. Meyer and G. Tölg; Fresenius' Z. Anal. Chem. 300,
    18-21 (1980)

27. A. Meyer, Ch. Hofer, G. Knapp and G. Tölg; Fresenius' Z. Anal. Chem.,
    in press

28. A. Meyer, Ch. Hofer and G. Tölg; Fresenius' Z. Anal. Chem. 290, 292-298
    (1978)

29. A. Meyer, E. Grallath, G. Kaiser and G. Tölg; Fresenius' Z. Anal. Chem.
    281, 201-209 (1976)

30. G. Henze, P. Monks, G. Tölg, F. Umland and E. Weßling; Fresenius' Z.Anal.
    Chem. 295, 1 (1979)

31. H. Wohlfarth, K. Schulze and G. Tölg; unveröffentlicht

32. Th. Stiefel, K. Schulze, H. Zorn and G. Tölg; Arch. Toxicol. 45, 81-92 (1980

33. E. Grallath; Fresenius' Z. Anal. Chem. 300, 97-106 (1980)

34. G. Kaiser, E. Grallath, P. Tschöpel and G. Tölg; Fresenius' Z. Anal.Chem.
    259, 257-264 (1972)

35. Th. Stiefel, K. Schulze, H. Zorn and G. Tölg; Anal. Chim. Acta 87, 67 (1976)

36. Th. Stiefel, K. Schulze, G. Tölg and G. Zorn; Fresenius' Z. Anal. Chem. 300,
    189-196 (1980)

MICROTRACE ANALYSIS OF NATURAL MINERAL MATERIALS

YURI A. ZOLOTOV

V.I.Vernadsky Institute of Geochemistry and Analytical Chemistry,
USSR Academy of Sciences, Vorobjevscoe shosse 47a
Moscow V-334, USSR

Abstract. Microtrace analysis of geological samples, meteorites and lunar
soil is considered. Particular features of these materials as objects of
microtrace analysis are noted. The main methods for analysis of very small
specimens are described including neutron and gamma activation, mass spectrometry, particularly spark source MS, as well as atomic absorption and fluorescence techniques. Examples of an application of the mentioned methods for analysis
of meteorites and samples from the Moon are given. Attention has been paid to
the representativeness of small samples and standardization. Special techniques for local analysis of natural mineral materials are considered, for
instance X-ray fluorescence microprobe and laser microspectrochemical
analysis.

1. Introduction

By microtrace analysis (MTA) we mean the determination of elements in samples
of several milligrams or less when the concentration of elements is below
10 ppm and the local analysis for trace elements, i.e. the analysis at
different points of the sample.   The usual and unique terrestrial samples
of rocks, ores and minerals, and also meteorites and the lunar soil samples
will be considered as natural mineral materials. Such an analysis of natural
objects has already been considered to some extent in a number of books and
reviews [1-5].

The need for MTA of these materials is generated by several important causes.

Many instrumental methods of analysis demand small amounts of natural materials for single determination. It is known that during neutron activation analysis (NAA) of samples containing highly activating elements, say sodium or chlorine, it is difficult to work with weighed quantities exceeding a few milligrams. Spark source mass spectroscopy (SSMS) deals with samples generally weighing not more than 10-20 mg. Samples weighing 10-30 mg are used in the spectrochemical methods. The analysis in these cases becomes essentially microanalysis although sufficiently large amounts of the material to be analysed can be made available to analysts.

Sometimes it is necessary to analyse very small particles due to the peculiar properties of the object to be analysed, for example the grains of the lunar soil or the chondrules of stone meteorites. The analysis of various mineral inclusions in rocks and ores is also of importance. Finally, the study of spatial distribution of trace elements even in one and the same phase, for example in a mineral grain, is necessary.

These problems can be solved only with the use of diverse instrumental methods having low absolute detection limit and sometimes permitting space resolution.

For MTA the absolute limit of detection should be less than 10^{-7} - 10^{-8} g, or better less than 10^{-9} - 10^{-10} g. These methods belong NAA, SSMS, to some extent spectrochemical analysis, and atomic absorption spectroscopy, particularly the flameless technique.

2. Materials to be analysed

MTA is employed for terrestrial geological materials, particularly for large scale determination of elements. In Table 1 are listed some routine analyses carried out in the laboratories of the USSR Geological Survey [6]. Wide use is also made of local analysis of geological samples by the electron microprobe and laser spectrochemical method. It may however be mentioned that in very rare cases one succeeds in determining trace elements by these methods (see somewhere in this report).

The analysis of meteorites, in particular of traces of rare and scattered elements is important for cosmochemistry [7]. Besides bulk analysis of the meteoritic matter, of importance is the analysis of various inclusions, isolated monomineral fractions, and also of chondrules - congealed glasslike droplets of silicate melt, and the iron-nickel phase, their size being from a few microns to several millimeters.

Table 1: Routine micro trace analysis of geological samples in the USSR
Geological Survey

Analytical method	Elements determined	Materials analysed	Sample consumption, mg	Detection limit, ppm
Emission	B	Rocks, minerals	7 - 10	4
spectrography	Hf*	Silicate rocks	10	2
	Mn	Rocks,minerals	1-10	200
	Bi,Co,Cr, Ga,Ge,La, Nb,Ni, Pb, Sb,Sn,V,Y	Silicate rocks, minerals	10	20**
	REE***	Silicate rocks minerals	10-50	10
X-ray fluorescence spectrometry	Hf	Rocks, minerals	15	500
Absorption spectrophotometry	F****	Minerals raw	10-200	100
	Fe(II)*****	Rocks	5-20	500
Neutron activation	Ce	Apatite-nepheline ores, loparite	5-500	500
	Eu	Apatite, nepheline ores, loparite	3-10	5

   * With chemical pretreatment
  ** Y 6, La and Sb 200 ppm
 *** Rare earth elements
**** Alizarine Complexone
***** 2,2'-Dipyridile.

There are other objects of microanalysis, to which is ascribed meteoritic
or cometary origin. An example is very small silicate and magnetic globules
found in the soils and peat of the region where the Tunguskii explosion had
occurred in 1908.

MTA of these globules each of diameter from 30 to 80 μm was carried out using
NAA [8,9]. For this purpose, 6-17 μg of globules weighed on an ultramicrobalance
and classified according to their external characteristics were taken. The
diabase W-1 was used as standard reference material (SRM). The analysis was
difficult due to small absolute amounts of determinable microelements. However,
the contents of 7 trace elements as well as of 4 main elements was established
(rock-forming elements were also determined by other methods, particularly by
the electron microprobe). The absolute determinable amounts were : for
Eu - 4×10^{-13} g, for Co, Sc, Cs, and Se - $(2-5)\times10^{-12}$ g, and for Zn - 1.5×10^{-10} g

From the contents of elements and the ratios between them one can see that the
globules bear no resemblance to earth rocks. The obtained values confirm the
comet hypothesis of the Tunguskii explosion (the collision of the Earth with
a small comet). A comparison can be made between these globules and the samples
of "rusty soil" and "orange-colour soil" which were found on the Moon by
American astronauts. The variations in the lunar soil can be attributed to
the collisions of comets with the lunar surface [8, 10].

Beginning in 1969, the lunar soil has become the object of microtrace analysis.
The lunar samples brought to the earth by American space ships "Apollo 11, 12,
14, 15, 16, 17" and the Soviet automatic stations "Luna 16, 20, 24" (1969-1976)
have put the task of gathering maximum information about the chemical compositio
of unknown material with the use of small samples. Now sufficient experience ha
been gained in analysing lunar soil and many concrete results have also been ob
tained. These data are available in the proceedings of yearly conferences on
Moon problems held in Houston [11], special issues of some journals [12,13], in
three collections of articles published in Moscow [14-16], and in a number of
papers.

Let us now consider briefly the investigations concerning MTA of the lunar soil
carried out in our institute [14-19]. During preliminary treatment of the soil
it was ensured that the soil would not come into contact with air and the
materials capable of contaminating it. On the basis of the theoretical studies
on sampling, a tentative estimation was made of the minimum amount of finely
ground sample of the lunar soil that can be representative for analysis by
instrumental methods. First qualitative analysis was performed by the spectro-

chemical method with photographic registration; the spectra of samples matched those of earth basalts. The rock-forming elements were determined by the X-ray emission method with relative error of 2% for the confidence level 0.95. For calibration, use was made of standard reference samples of basalts, in particular of BCR-1. NAA was employed as an additional method of determining macrocomponents. Glass globules, microbreccias, various inclusions, and the grains of minerals were studied by the electron microprobe. The trace elements were determined by SSMS and NAA. Additionally spectrochemical analysis and other methods were used. In Table 2 the main methods employed in Soviet and American laboratories for MTA of the lunar soil are given [17].

Table 2: Determination of trace elements in lunar soil

Methods of analysis	Elements determined	Materials analysed
Spark source mass spectrometry	50-60 elements	Regolith, rocks, breccias, glasses, minerals
Neutron activation analysis	30-40 elements	Regolith, rocks, breccias, minerals
Emission spectrochemical analysis	Ag, Au, B, Be, Bi, Cd, Co,Cr,Cu,F,Ga,Ge,In,Li, Mo,Ni,Pb,Rb,Se,Sr,Sn,T1, V,Y*,Yb,Zn,REE*,	Regolith, rocks
Flame emission spectroscopy	Ni,Sr,Zn	Regolith, rocks
X-Ray fluorescence spectrometry	Ag,Au,Ca,Mo,U,Zn,REE*	Regolith, rocks breccias
Atomic absorption spectroscopy flame	Ga,K,Li,Na,Pb	Regolith, rocks
flameless	Ag, Cd	Regolith
Flameless atomic fluorescence spectroscopy	Hg, T1	Regolith
Mass spectrometry with isotopic dilution	Ba,Li,K,Rb,Sr,REE*	Regolith, rocks
Gamma spectrometry	K, Th, U	Regolith
* With chemical pretreatment (REE - rare earth elements).		

A mention may be made of some particular properties of the lunar soil as an object of MTA. No scattered light is seen during atomic fluorescence (AF)

determination of thallium with direct flameless atomization of powders [17,20-23].
For this reason the detection limit of Tl (1×10^{-7} %) was found to be less by two
orders than that for corresponding earth rocks, in the analysis of which significant scattered light is seen. A similar phenomenon is observed when detecting
mercury by this method. On analysing the lunar soil obtained from the marine
region it is observed that the dc arc burns more calmly than in the case of
earth basalts. The curves showing evaporation of elements of the lunar soil
and of the standard reference materials of earth rocks on heating in a dc arc
differ strongly from each other. That is why it is necessary to use buffer
reagents.

The results of analysis of lunar samples carried out in different laboratories
and by different methods are essentially the same, but differ sometimes. This
difference may be due to [17]:

a) insufficient prior information about the composition and properties of the
   soil; that is why not all matrix effects are accounted for when use is
   made of standard reference materials or the matrix effects are not completely
   suppressed by buffers (during spectrochemical analysis),

b) inhomogeneity of the samples which shows up most vividly during MTA,

c) small number of repeated determinations due to the limited amount of the
   samples to be analysed and the effects of random errors,

d) petrological differences of samples.

As a SRM use is generally made of basalt BCR-1 for MTA of the lunar soil. This
enables one to estimate the degree of reliability of obtained data and forms
bases for comparing the results. Perhaps, the time has come to have a SRM of
the lunar soil itself. The errors of determining traces in microsamples of
lunar soils and minerals usually amount to 10-20 %.

3. Analysis of small samples

3.1. Representativity of samples and standardization

Analysis of small samples for trace elements is a most important area of
microtrace analysis of natural mineral materials.

To what extent does a weighed amount of a few milligrams represent the bulk
sample to be analysed? It is clear that the degree of representativity can be
raised by thoroughly grinding and mixing the sample and by increasing the
number of subsequent analyses of the sample, i.e., by increasing the total

mass of the sample used for analysis. A detailed study of this problem as applied to rocks [24] has been made by Belyaev, who determined cadmium in granophyre and gabbro by the atomic-absorption method with flameless atomization of powders [20] and chromium in granite by the spectrochemical method [25]. The consumption of sample for single determination in the first case was 10 and in the second case, 10 or 20 mg. The mass of the laboratory sample was 10 g, the particle size after grinding of specimens was 0.05 mm for granophyre and gabbro and 0.08 mm for granite.

The sample mass (M) which is to be used for the analysis of several weighed amounts is determined from the ratio [24,26]

$$M = \frac{K}{s_r^2}$$

where K is the sampling constant which depends on the particle size [26,27] and s_r is the relative standard deviation (RSD) caused only by sample inhomogeneity. With the increase in the total mass of the sample used for analysis (Figure 1) the average values of analysis results come closer to 1.9×10^{-5} %, the weighed amount being 10 mg. For example, for s_r = 20 % the consumed mass of the sample 50 mg (5 weighed amounts) represents the laboratory sample of 1g.

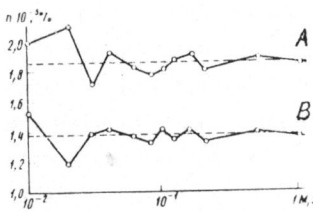

Figure 1: Dependence of average results of cadmium determination in granophyre (A) and gabbro (B) on the total mass of the sample consumed in the analysis. Dotted lines correspond to the limiting value of average results when the total mass of the sample is increased up to 1 g

Determination of chromium in granite represents many difficulties. In granite, chromium is present in the scattered state, but grains of chromite are also encountered; 5-6 grains of chromite are found in each gramm of granite with a particle size of about 0.08 mm. The probability of one grain falling into a specimen of 20 mg is about 10 %. The result of the analysis of single specimen without any grain equals 0.0015 %. Much higher results are obtained in case grains fall into the specimen (for instance 0.01 %). In usual spectro-

chemical analysis, each sample is analysed three times, i.e., 60 mg of the
sample are consumed. The probability of the specimen containing a grain in
this case equals $\sim$ 30 %. Therefore, in some cases there may not be any grains.
Too high results of determination are obtained when chromite grains are
contained in the specimen and such values are generally rejected. These
"rejections" however lead to false results.

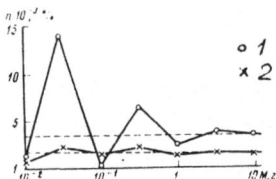

Figure 2: Dependence of average results of determination of chromium in
granite on the total mass of the sample consumed in the analysis.
Dotted lines correspond to limiting average results of determination of total content of Cr (1) and scattered Cr (2) in a
sample of mass 10 g.
1 - by taking into account anomalously high results ("rejections");
2 - without accounting for the "rejections"

Figure 2 shows the dependence of average values of chromium determination
results on the total mass of the consumed sample. Two cases are shown:
(1) when the anomalously high values are rejected and (2) when they are
taken into consideration for calculating the average value. In (2) the
average value approaches 0.03 % for a total sample consumption of 10 g
(500 specimens each weighing 20 mg). In (1) the average value will always
be less. Only the content of scattered chromium is determined when the total
consumption of the sample is 20-60 mg. Actual concentration can be found
out by taking into account the "rejections" and when total consumption of
the substance to be analysed is very large - up to several grams. This is
not feasible practically.

A similar picture may be observed when nickel, tin, and molybdenum in geological materials are determined by the spectrochemical method.

The errors associated with the mineralogical form of trace elements in the
sample can also appear during calibration if use is made of standard reference
materials tested by chemical and other methods of analysis requiring large
weighed amounts. In actual practice, total concentration of an element is
determined using these methods. In case the instrumental method is used,
which requires small amounts of sample, mainly the scattered component is
determined.

With the development of analytical methods requiring small weighed amounts, strict requirements are being placed on homogeneity of the SRM and the selection of methods for their testing. At present, in order to make better use of the analysis results of geological and space objects and to compare the data obtained in various laboratories, it is necessary to mention not only the SRM used for calibration and the assumed concentrations of trace elements present in them, but also the weighed amounts consumed for a single determination and the number of analysed specimens.

It may be mentioned that the number of SRM of geological materials, the concentration of traces for which has been reliably established, is not so large. Recently, IUPAC has published a report [28] on 28 available SRM, in which the concentrations of trace elements for 14 standard reference materials including such wellknown SRM as diabase W-1, granite G-2, basalt BCR-1, andesite AGV-1 prepared in the USA are given. Unfortunately, this report does not contain any data on the USSR made SRM. Information on these SRM is available [29].

For local analysis the term "representativity of a sample" has a slightly different meaning. If analysis is carried out not of a microscopic "point" as such, but of a sample which should basically be homogeneous, then averaging is achieved by subsequent analysis of a large number of points and by scanning the sample surface. For example, for the microprobe method of SSMS (see somewhere else) a 5x5 mm^2 surface is scanned.

3.2. Microtrace analysis techniques

For MTA use is made of only those methods which ensure low absolute limit of detection for specimen of not more than a few milligrams.

3.2.1. Activation analysis

Among different methods of radioactivation analysis NAA is of great importance for MTA. It is widely employed for the determination of trace elements in microsamples of geological and cosmic materials. For determining 30 or more microelements at a concentration of 10^{-7} - 10^{-3} %, it is sufficient to have 1-5 mg of the material. Cases are known where even lesser amounts were used. Morrison [30] published a review of the earlier works on the use of activation analysis in different branches of geology. Other reviews are also available [31, 32], but a large majority of them does not deal especially with microanalysis.

Instrumental NAA is an essential method for MTA of meteorites. It is not an exaggeration if we say that the development of this method determined in many ways the successes of cosmochemistry. The advances made by the reactor technique owing to the use of pneumatic tube conveyors, cadmium protected channels, and others, wide application of Ge(Li)-detectors, the use of computers for processing complex gamma-spectra, and other achievements made in NAA enable one to determine many rare and scattered elements in very small samples of meteorites, in different minerals isolated from them, to increase the sensitivity and accuracy of analysis, to reduce the analysis time and carry out analysis without destroying the sample.

In our institute the neutron activation analysis of meteorites is being developed for about 20 years by Lavrykhina. A review of earlier studies is available [7]. Much attention has been paid to the analysis of impurities in troilite (a ferrous sulphide) - a mineral first detected in meteorites. Thus, the conditions have been found out under which V, Co, Mn, Cs, Cr, and W, and also Fe and Al in troilite of iron meteorite Sikhote-Alin are determined [33].

INAA of stone meteorites and the silicate phases of iron meteorites is hindered by the high induced activity of ^{24}Na with half-life of 15 hours. That is why sufficient time is needed to "cool" the irradiated sample. After cooling, Cr, Ni, Se, Co, Ir, and Fe can be determined using long-life isotopes. Sometimes, when the sample contains sodium in small amounts analysis can be carried out without cooling the sample (for example, olivines of pallasites are analysed in this manner). A large number of chondrules of the stone meteorite Saratov has been analysed and the concentrations of Mn, Fe, Sc, Cr, Co, Ga, Eu, Ce have been determined. The chondrules weighed between 1.6 and 11.3 mg. Use can also be made of the methods involving radiochemical separation of elements to be determined.

Some methods of NAA determination of rare-earth and other rare elements present in small amounts in meteorites have been developed by Kolesov [34,35] who for rare-earth elements achieved the absolute limit of determination equal to 10^{-9} - 10^{-10} g using 3-10 mg of specimen. Verification was done by analysing known geological SRM, for example BCR-1. Use was also made of the radiochemical variant of NAA.

The activation method of determining uranium with registration of fission produ tracks in a specially selected solid dielectric detector is a very sensitive or It can be employed for analysing very small samples of natural materials, parti cularly of meteorites [36-43].

The track method has the following advantages [42]:

a) it is possible to measure exactly the distribution of uranium in very small samples,

b) the determination sensitivity of uranium mounts up to 10^{-3} ppb,

c) analysis can be repeated as the sample is not destroyed,

d) the sample does not receive any uranium impurities in the course of analysis. The maximal relative error of determining uranium amounts to 25%.

The NAA is of great importance for MTA of lunar soil. It was first employed by Morrison et al. for determining a large number of elements present in the lunar samples brought to the earth by "Apollo-11" [44]. Later on, NAA of different lunar samples was repeated both in the instrumental and radiochemical variants. For example, a procedure for the determination of 42 elements present in the samples collected by Apollo-12 was suggested [45]. It involves four irradiations; one of them is followed by instrumental NAA. Radiochemical isolation of radio-active isotopes is carried out after three irradiations.

Dr. Kolesov [46-50] analysed the lunar samples brought by "Luna 16, 20, 24" and "Apollo 11, 12, 14-17". Analysis was carried out mainly of the fine-grain fraction (regolith) containing particles of about 70 μm. The analysed samples weighed 3-6 mg. BCR-1 and the alkali basalt, and the solution of element salts of high purity were used as SRM. The analysis was performed using instrumental and radiochemical variants. The radiochemical variant was used mainly for rock-forming and rare-earth elements. In the case of instrumental NAA the samples and SRM were left standing to decrease the activity of disturbing isotopes and transferred into a test tube for cancelling the background activity of weighing bottles. In the radiochemical variant the samples were fused with Na_2O_2, dissolved in hydrochloric acid, and the elements were divided into several groups by precipitation, extraction, and chromatography. In both cases the samples were measured using gamma spectrometer with Ge(Li)- -detector and 4096-channel pulse analyzer connected to a computer. The composition of elements was calculated by comparing with standard reference samples. The determination errors were

    5-10 % for Co, Cr, Eu, La, Mn, Na, Sc, Sm, Yb
    10-15 % for Ba, Ca, Ce, Fe, Gd, Hf, K, Lu, Nd, Zr
    15 % for Ho, Ta

In Table 3 the results concerning the determination of traces in the soil collected from the continental region of the Moon are presented. They are given in comparison with the data obtained in other laboratories (no data are listed for rock-forming and rare-earth elements).

Table 3: Concentration of some elements in the soil from highland region of
the Moon (ppm)

Element	Sample of "Luna-20"		Sample of "Apollo-16"	
	Soviet laboratories 47-51	American laboratories 13, 52, 53	Soviet laboratories 46	American laboratories 53, 54
Ba	183	103	165	134
Co	25	29	25	27
Cs	0.094	0.07	0.2	0.14
Hf	2.5	2.8	4.5	4.2
Nb	-	8	11.6	11.2
Rb	1.4	1.7	2.9	2.3
Sc	20	16	10.4	9.3
Ta	0.35	0.3	0.65	0.48
Th	0.84	0.84	2.1	1.9
U	0.30	0.29	0.60	0.62
W	(0.1)	(0.1)	2.1	1.9
Zr	-	120	160	151

Among other variants of activation analysis, the gamma-activation method can
be pointed out. It is based on the use of high energy bremsstrahlung obtained
by means of electron accelerators - betatron, microtrons and linear waveguide
accelerators. In the USSR the gamma-activation method is widely used for
analysing geological objects. Selective activation of some elements or a
group of elements can be carried out by making use of the threshold nature
of photonuclear reactions [32]. The value of threshold energy depends on the
type of reaction and the nature of nucleus.

The use of (γ , γ')-reactions holds a prominent place. The Institute of
Nuclear Geophysics and Geochemistry, Moscow, has developed a number of methods
for gamma-activation analysis of rocks, ores, and their treatment products.
The detection limit goes up to 5×10^{-4} % when use is made of betatrons and up
to 1×10^{-4} % with the use of microtrons. Besides, it has developed more sensitive
methods (detection limit up to 1×10^{-6} %) based on the application of linear
electron accelerators.

Activation analysis with the use of protons and alpha particles is also
prominent.

243

3.2.2. Mass spectrometry

Among mass-spectrometric methods, SSMS is most important for MTA. Geological
and space objects are nonconducting materials of complex composition. That is
why the methods of their analysis by SSMS were developed with difficulty. At
present two main variants of such an analysis are known.

One of them is based on powdering the sample to be analysed and mixing it
with a conducting additive mainly with extra pure graphite and obtaining rods
to be used as electrodes. The isotope dilution method can be used for quantitative analysis, particularly for determining numerous elements [55]. References
on earlier publications are given in [56]. Using this method Morrison et al.
analysed meteorites, earth basalts [57,58]. Many elements present in the lunar
soil have also been determined [59,60]. Jackson and Strasheim [61] have analysed
the international geological SRM, lunar soil, and other natural materials.
Recently, Knab and Hintenberger [62] have improved the analysis procedure by
making the measurements independent of calibrated standards. They determined
some elements with 10^{-4} - 10^{-7} % detection limit in meteorites and with 10^{-4} %
in the standard sample of diabase W-1. This method has found application in
many other investigations.

The second method was developed in our institute by Chupakhin, Ramendik et
al. [63-66]. It does not call for the use of conducting additives. A small amount
of powdered material (5-15 mg) is pressed into a tablet covered with an aluminium
film. As a result, a thin layer of dielectric ($\sim$ 0.3 mm in thickness) is formed.
The tablet containing the sample is fixed in one of the supports of the ion
source and serves as an electrode. The probe made of tantalum wire about 70 µm
in diameter (Figure 3) acts as second electrode.

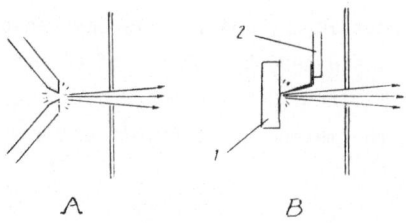

Figure 3: Location of electrodes in the spark ion source of mass spectrometer for usual (A) and probe (B) analysis techniques:
1 - sample;
2 - antielectrode (probe)

The probe is rigidly fixed relative to the outer slit of the ion source. The tablet can move in three directions. This enables scanning of the surface and obtaining an average composition of the sample by repeating the analysis [67]. The distance between the probe tip and the sample surface is a few tenths of a millimeter. Radio-frequency voltage is applied on the probe in relation to the tablet which results in sample-dielectric and vacuum breakdown, sputtering and ionization of the sample (Figure 4).

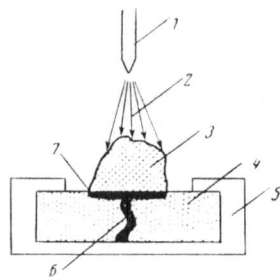

Figure 4: Plasma formation upon vacuum breakdown of a thin layer of dielectric:
            1 - probe;
            2 - electron beam;
            3 - plasma;
            4 - a pressed layer of the sample;
            5 - aluminium tablet;
            6 - conducting channel (it is formed after the breakdown of the
                dielectric);
            7 - area of plasma contact with the sample surface

The sample surface lies parallel to the surface of the first slit of the ion--optical system; this permits the ion current to be increased by 4 times (up to 4×10^{-9} A) compared with the first method of analysing geological samples and the analytical accuracy to be raised. Tantalum and aluminium do not produce a significant effect on the mass spectrum. The spark discharge in vacuum destroys mainly the anode; that is why analysis is carried out with negative voltage of the probe. This is achieved by the use of a special device [68]. The analysis with detection limit $\sim 10^{-7}$ % needs about 5 mg of the substance, the absolute sensitivity of the method being 4×10^{11} atoms which for iron impurities amounts to 3×10^{-11} g.

In this method the final phase of spark discharge is also cut off. This is done to ensure stable ion-formation conditions. As a result, the mass spectrum is simplified and the superposition of complex ion lines on the analytical lines is decreased. This permits enlarging the range of elements to be determined.

A special device is used to cut off the spark discharge [65].

Such a microprobe method was employed for analysing lunar soil [69]. In the analysis use was made of internal standards of Mn, Ba, and Zr for three groups of determinable elements.

Standard samples of rocks were not used; lunar soil samples thoroughly analysed by other methods were used for calibration. A portion of the surface (4x5 mm^2) was processed for spark discharge analysis of lunar soil; 2-3 mg of the sample was consumed in the experiment which continued for 2.5 hours.

The microprobe method of mass-spectrometric analysis was employed for solving many other analytical problems [70-75].

Other variants of mass spectrometry, i.e. secondary ion mass spectrometry, laser mass spectrometry, and mass spectrometry with ionization by electron impact were also used for MTA of natural mineral materials. The latter was employed for determining gas components of the earth and space samples. It was also used for determining inert gases in regolith and fragments of lunar materials [76].

3.2.3. Other methods

Among other methods of microtrace analysis of natural mineral materials, mention may be made of spectrochemical analysis, particularly with the use of new spectrum excitation sources (ICP and others), flameless atomic absorption and atomic fluorescence methods, different chemical and physico-chemical methods employed for micro- and ultramicroanalysis. Without discussing these and other methods, we shall consider only the works on AA and AF analysis of powder samples and the methods of "chemical" ultramicroanalysis.

Belyaev and Koveshnikova [20, 21] developed sensitive AA and AF methods for the determinion of volatile elements (Hg, Tl, Cd, Ag, Bi) with pulse electrothermal atomization during evaporation of elements into the argon atmosphere from small powdered samples. The AF method of determining Hg and Tl and the AA method of determining Ag and Cd were applied in the analysis of lunar soil [17, 22, 23]. For determining mercury, 2-5 mg of the soil was used. In samples of 5 mg the detection limit of this element equals 1×10^{-7} % (s_r = 33%). Heating of sample up to the required temperature results in the appearance of a single light pulse of atomic fluorescence of mercury. The amplitude of this pulse serves as an analytical signal.

Using another device for AF determination of mercury [77], the solution to the so called "mercury paradox" has been obtained in our institute. The point is that the mercury content in minerals of coaly chondrites is always found to be many times more than expected according to the laws governing distribution of elements in the solar system. It has been found that the samples of meteorites displayed in museums and stored in laboratories capture mercury vapours from the air; the more the storage period the greater is the content of mercury in them [78].

In many cases MTA can be carried out with the use of fine techniques of chemical decomposition of samples, separating and concentrating necessary components, and determining their concentration by methods having low absolute limit of detection. In doing so it is not necessary to use those methods of analysis which ensure high concentration sensitivity. Tölg is quite successfully developing the ideology of similar methods by paying special attention to the study of sources of errors and the ways of eliminating them [78-84]. The initial samples of the substance to be analysed must not necessarily be small in this case, but the obtained concentrate has usually a small mass. Prior-to-analysis decomposition and concentration has a number of advantages [85, 86]: this almost solves the representativity problem of the sample to be analysed, makes the calibration easier to an appreciable extent and enables the internal standards to be introduced readily. Besides, it has other advantages also. Here it is important to have a rational combination of the decomposition, separation, preconcentration, and determination methods [79, 86, 87]. MTA involving chemical preparation is of great importance in the analysis of ultra pure substances; it is used also for the analysis of natural mineral materials. For example, a method has been developed for determining nanogram amounts of fluorine in rocks using weighed amounts of a few milligrams [88]. All operations are carried out in one closed system: fluoride is separated from the matrix by diffusion and determined by the kinetic method.

Ultramicroanalysis where use is made of a microscope holds a specific position. The ultramicrochemical technique first developed with the aim of utilizing atomic energy employs essentially the usual chemical and physico-chemical methods, but uses samples a few micrograms in mass and a few milliliters in volume. It is however directed mainly to the determination of main components and very rarely employed for trace analysis. Besides this owing to several reasons this method has not found wide application. One of the few teams working on ultramicrochemical analysis is functioning in the Vernadsky institute. The results of methodological investigations made by this team are available in the monograph by Alimarin and Petrikova [89]. In the course of investigations, this team has developed an ultramicrobalance capable of weighing samples up to 10^{-7} g with an accuracy of about 2×10^{-9} g.

3.2.4. Combination of methods

MTA is generally used in combination with other methods of determining rock-forming elements and the elements of intermediate concentrations. Not one but several methods are usually used for trace analysis of most important materials. That is why it has become necessary to find a rational combination of different analytical methods, in any case to decide the sequence of their application. For instance, a complete schematic of analysis of microscopic natural materials, say of lunar soil, different minerals, and others, can be represented [65]:

a) Instrumental NAA without destroying the sample, determination of rock-forming elements and microelements.

b) Powdering of sample (if it is necessary), pressing into an aluminium tablet.

c) X-ray (emission or fluorescence) analysis, determination of rock-forming elements.

d) Use of spark source mass spectrometry for trace analysis.

e) Removal of the leftover sample from the tablet; neutron activation analysis with radiochemical separation.

When compiling such schemes care should be taken that the involved methods supplement and controll each other. The idea of developing such combinations of methods is being prosecuted at the Vernadsky institute by Belyaev.

4. Local analysis

Local analysis of natural mineral materials is of great importance also for geochemical and cosmochemical investigations. It is used in those cases when grains, phases, and inclusions are very small to be analysed by "usual" micro-analysis methods, and also for knowing how microcomponents are distributed. Thus, it is necessary to determine the composition of microsections observed under the microscope for diagnosing microinclusions of minerals; this enables to know how different elements are distributed both in mineral grains and the spaces between them, and also in solid mineral solutions. For local MTA, methods having still lower absolute limits of detection, for example 10^{-11} - 10^{-15} g or even less should be used. In this case, calibration and quantitative analysis in general present difficulties.

4.1. Electron microprobe

This technique is the main method used for local analysis of geological objects.

Despite its high absolute sensitivity (up to 10^{-15} g), it does not ensure concentration sensitivity sufficient for trace analysis. Its relative detection limit is 10^{-1} - 10^{-2} %; this pertains also to the present day modifications of this method [90-93]. Only in some cases, by taking special measures one succeeds in determining concentrations of the order of 10^{-2} - 10^{-3} ; Il'in and Kolomeitseva [94, 95] determined the traces of Ni, Cr, and Co in olivines of meteorites, lunar soil, and terrestrial ultrabasic rocks. The analysis could not be carried out by usual methods, particularly in the case of meteorites, due to the difficulty of collecting the monomineral fraction.

The detection limit of germanium in iron meteorites was calculated to be 2×10^{-3} ; but for this pulse counting should continue for 30 minutes [96].

4.2. X-ray microprobe

Only the sections of samples which withstand electron bombardment can be investi gated by electron microprobe. A large variety of objects such as powders, micro-samples of arbitrary configuration, fusible and volatile minerals, samples of sedimentary rocks cannot be practically analysed by electron microprobe. That is why for local analysis the application of X-ray microprobe seems to be alluring, which gives rise to an X-ray fluorescence spectrum of the sample. X-ray fluorescence analysis is nondestructive, contactless, and can be applied to samples in any aggregate state. To the development of X-ray fluorescence analysis methods for small amounts of substances and of local analysis are dedicated a number of investigations and design elaborations.

Il'in of the V.I.Vernadsky institute has developed an X-ray fluorescence micro-analyzer [97]. Its working principle is based on the design of electron micro-probe analyzer, in which the electron microprobe of increased intensity initi-ates in the target primary X-ray radiation that reaches the studied sample through a limiting diaphragm.

The analytical characteristics are being improved. At the first stage it ensures the following: concentration sensitivity - 3×10^{-3} %, absolute sensitivity - 3×10^{-9} g, and localization - 100 μm.

4.3. Ion microprobe

The ion microprobe (microprobe variant of SIMS) offers significant possibilities as far as sensitivity is concerned [98-100]. This method can be used for deter-mining all elements, the consumption of sample being very small : 10^{-15} g for static ion microprobe and 10^{-6} g for the dynamic one. Its localization is higher than that of the electron microprobe - up to 1 nm compared to 1 μm. The detection limit of secondary ion mass spectrometry is 10^{-3} - 10^{-4} % for the static variant and 10^{-8} - 10^{-9} % for the dynamic one; absolute sensitivity is as great as 10^{-18} g.

The ion microprobe is not yet applied on a large scale for the analysis of natural materials. This involves the use of not only sophisticated and very costly equipment, but also is associated with difficulties in obtaining quantitative results. This method was employed in [102]. In plagioclase, the elements having atomic number from 3 to 82 were determined with relative sensitivity of about ∿ 1×10^{-4} % and absolute sensitivity ∿ 10^{-18} g. The ions $^{16}O^{-}$ having an energy of 20 keV served as primary ions.

4.4. Laser microspectrochemical analysis

The laser microspectrochemical analysis is used on a relatively large scale in the USSR [103-105]. In general, the sensitivity and reliability of this method are not very high, but in some cases its sensitivity can be increased to a level sufficient for trace analysis. For example, relatively low detection limits [105] were obtained by applying a spark discharge of 8-10 thousand volts and using spectrograph of increased resolving power on evaporating mineral grains of diameter up to 0.7 mm. Use was made of a powerful laser (5-6 J). The absolute sensitivity reached 10^{-8} - 10^{-10} g.

The sensitivity of the method depends on the amount of substance that evaporates, and hence, on the crater size. All other things being equal, the crater size is determined by the properties of the mineral to be analysed. For example, with the rise in mineral hardness the crater depth decreases (Figure 5) [3].

In the USSR, for laser microspectrochemical analysis use is made of LMA-1 and LMA-10 instruments (GDR) [103, 106] and other devices also [104].

4.5. Other methods of local and surface analyses

The surface analysis for natural mineral materials is not so important as for material science, electronics, and the study of catalysis or corrosion processes. Nevertheless, numerous rapidly developing methods of local and surface analyses [101, 107, 108] have interesting features which may finally find application in studies of geological and space objects. The main methods are : X-ray photoelectron spectroscopy (XPS), Auger electron spectroscopy (AES), ion scattering spectroscopy (ISS), and SIMS.

Recently, a proton microprobe has been announced. Comments on this methods have been given by Laitinen [109].

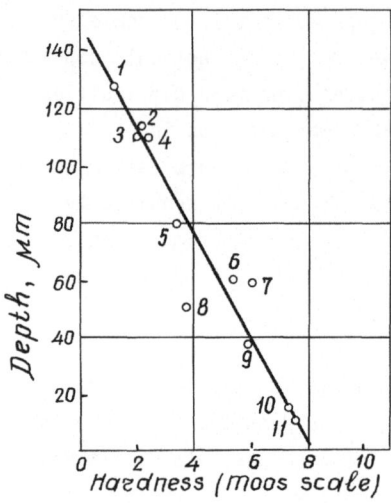

Figure 5: Depths of craters formed during laser spectrochemical analysis vs. hardness of minerals to be analysed :
1 - molybdenite;
2 - galenite;
3 - cinnabar;
4 - bismuthinite;
5 - sphalerite;
6 - ilmenite;
7 - pyrite;
8 - chalcopyrite;
9 - haematite;
10 - zircon;
11 - beryl

References

1. R.D. Reeves and R.R. Brooks, Trace Element Analysis of Geological Materials, Wiley, New York, 1978.

2. Analytical Chemistry in the Exploration, Mining and Processing of Materials. Eds.A.Strasheim and T.W.Steele, Butterworths, London, 1978.

3. A.K. Rusanov, Principles of Quantitative Spectrochemical Analysis of Ores and Minerals, 2nd ed., "Nedra", Moscow, 1978 (Russian);

4. M. Pinta. Modern Methods for Trace Element Analysis. Ann Arbor Science Publishers, Ann Arbor, 1978.

5. W. Schrän, Wiss.Z.Karl-Marx-Universität, Leipzig, Math.-naturwiss. R., 28, No. 4, 433 (1979).

6. Reference Data on Methods for Determination of Chemical Composition of Rocks, Ores and Products of their Treatment. Comp. by G.A.Volkova and E.P.Osiko, "Nedra", Moscow, 1979 (Russian).

7. A.K.Lavrukhina, in: Progress in Analytical Chemistry, "Nauka", Moscow, 1974,p.131 (Russian).

8. E.M.Kolesnikov, A.Yu.Lyul and G.M.Ivanova, Astronomicheskii Vestnik, 11, 209 (1977).

9. E.M. Kolesnicov, A.Yu.Lyul and G.M.Ivanova, in: Cosmic Substance on the Earth (Problem of Tunguskii Meteorite), "Nauka", Novosibirsk 1976, p. 87.

10. G.M.Brown, J.G. Holland and A.Peckett, Nature, 242, 515 (1973).

11. Proceedings of Annual Lunar Sci. Conferences, Houston, Texas.

12. Earth and Planet Sci. Lett., 17, No. 1 (1972).

13. Geochim. Cosmochim. Acta, 37, No. 4 (1973).

14. Lunar Soil from Sea of Fertility, "Nauka", Moscow, 1974 (Russian).

15. Regolith from the Highland Region of the Moon, "Nauka", Moscow, 1979 (Russian).

16. Cosmochemistry of the Moon and Planets. Proc. of the Soviet-Amer. Conf. on Cosmochem. of the Moon and Planets, 4-8 June 1974, Moscow, "Nauka", Moscow, 1975 (Russian).

17. Yu. I. Belyaev, Zh. Anal. Khim., 32, I88 (1977).

18. Yu. A. Zolotov, "Pravda" Newspaper, No. 249 (1976).

19. Yu. A. Zolotov, Essays on Analytical Chemistry, "Khimia", Moscow, 1977 (Russian).

20. Yu. I. Belyaev, A.M. Pchelintsev, N.F. Zvereva and B.I. Kostin, Zh. Anal. Chim., 26, 492 (1971).

21. A.M. Pchelintsev, Yu. I. Belyaev, A.V. Karyakin and T.A. Koveshnikova, Zh. Anal. Chim., 26, 1355 (1971).

22. Yu. I. Belyaev and T.A. Koveshnikova, in: ref. 14, p. 335.

23. Yu. I. Belyaev and T.A. Koveshnikova, in: ref. 15, p. 468.

24. Yu. I. Belyaev, V.I. Shcherbakov and L.A. Makarova, Zh. Anal· Khim., 31, 230 (1976).

25. Yu. I. Belyaev and L.M. Khitrov, Zh. Anal · Khim., 18, 310 (1963).

26. C.O. Ingamells and P. Switzer, Talanta, 20, 547 (1973).

27. A.D. Wilson, Analyst, 89, 18 (1964).

28. O.G. Koch, Pure and Appl. Chem., 50, 1531 (1979).

29. S.V. Lontsikh and A.K. Parshin, Zh. Anal · Khim., 34, 2446 (1979).

30. G.H. Morrison, J. Radioanal. Chem., 18, 9 (1973).

31. V.V. Sulin, in: Nuclear-physical Methods for Analysis of Materials, Atomizdat, Moscow, 1971, p.206 (Russian).

32. D.I. Leipunskaya, S.I. Savostin and V.V. Sulin, Vestnik Akademii Nauk SSSR, No. 1, 44 (1972).

33. A.Yu. Lyul, Izv. Akad. Nauk Latv. SSR, No. 2, 14 (1972).

34. G.M. Kolesov, J. Radioanal. Chem., 30, 553 (1976).

35. G.M. Kolesov, Meteoritika, 35, 59 (1976).

36. P.B. Price and R.M. Walker, Nature, 196, No. 4856, 732 (1962).

37. R.L. Fleisher, P.B. Price, R.M. Walker, Annual Rev.Nucl.Sci., 15, 1 (1965).

38. R.L. Fleisher, C.W. Nalser, P.B. Price, R.M. Walker and U.B. Marvin, Science, 148, 629 (1965).

39. R.L. Fleisher, Rev. Scient. Instrum., 37, No. 12 (1966).

40. R.L. Fleisher and P.B. Price, J. Geophys. Res., 69, 331 (1964).

41. L.I. Genaeva, L.L. Kashkarov and A.K. Lavrukhina, Meteoritika, 28, 102 (1968).

42. L.I. Genaeva, L.L. Kashkarov, A.K. Lavrukhina and L.V. Yukina, Meteoritika, 31, 137 (1972).

43. L.L. Kashkarov and V.L. Koshkin, in: Low-Radioactivity Measurement and Applications, Proc. of Int. Conf. 6-10 Oct. 1975, The High Tatras, Czechoslovakia, Slov. Pedag. Nakladat., Bratislava, 1977, p. 341.

44. G.M. Morrison and R.A. Nadkarni, J. Radioanal. Chem., 18, 153 (1973).

45. A.O. Brunfelt and E. Steinnes, Talanta, 18, 1197 (1971).

46. Yu.A. Surkov and G.M. Kolesov, in: ref. 15, p. 345.

47. G.M. Kolesov and Yu.A. Surkov, Radiochimiya, 21, 138 (1979).

48. Yu.A. Surkov, G.M. Kolesov, I.N. Ivanov and A.P. Shpanov, Kosmichesk. Issledovaniya, 16, 107 (1978).

49. I.N. Ivanov, F.F. Kirnozov, G.M. Kolesov, B.N. Rivkin, Yu. A. Surkov and A.P. Shpanov, Radiatsionnaya Technika, No. 11, 215 (1978).

50. Yu.A. Surkov and G.M. Kolesov, Kosmichesk. Issledovaniya, 15, 261 (1977).

51. R.A. Kuznetsov, V.B. Pankratov and B.G. Lurê , Zh. Anal. Khim., 34, 1564 (1979).

52. Geochim. Cosmochim. Acta, 37, 869, 909, 927, 953, 963 (1973).

53. Proc. IV Lunar Sci. Conf., Houston, Texas, March 1973, 2, Chemical and Isotopes Analysis, 1973, p. 1127, 1209, 1275, 1297, 1349, 1379, 1399, 1415, 1427, 1445, 1461.

54. LSPET (The Lunar Samples Preliminary Examination Team), Science, 179, No. 4068, 23 (1973).

55. J.F. Jaworski and G.H. Morrison, Anal. Chem., 47, 1173 (1975).

56. Trace Analysis by Mass Spectrometry. Ed. by A.J. Ahearn, Academic Press, New York and London, 1972.

57. G.H. Morrison and A.T. Kashuba, Anal. Chem., 41, 1842 (1969).

58. H. Hinterberger, Naturwissensch., 56, 262 (1969).

59. G.H. Morrison, J.T. Gerard, A.T. Kashuba, E.V. Gangadharam, A.M.Rothenberg, N.M. Potter and G.B. Miller, Science, 167, 505 (1970). Geochim. Cosmochim. Acta, 2, 1383 (1971).

60. G.H. Morrison, Anal. Chem., 43, 22A (1971).

61. P.F. Jackson, A. Strasheim, Analyst, 99, No. 1174, 26 (1974).

62. H.J. Knab and H. Hinterberger, Anal. Chim., 52, 390 (1980).

63. M.S. Chupakhin and A.L. Polyakov, Zh. Anal. Khim., 27, 523 (1972).

64. G.I. Ramendik, M.S. Chupakhin, Yu.G. Tatsii and V.I. Derzhiev, Zh. Anal. Khim., 29, 238 (1974).

65. G.I. Ramendik, Zh. Anal. Khim., 32, 1990 (1977).

66. M.S. Chupakhin, O.I. Kryuchkova and G.I. Ramendik, Analytical Prossibilities of Spark Source Mass Spectrometry, Atomizdat, Moscow, 1972 (Russian).

67. G.I. Ramendik, Yu. G. Tatsii and M.S. Chupakhin, Zh. Anal· Khim., 28 736 (1973).

68. T.V. Babushkina, G.G. Sikharulidze and P.K. Nikolaev, Zh. Anal. Khim; 30, 172 (1972).

69. N.J. Hubbard, G.I. Ramendik, S.I. Gronskaya, I.Ya. Gubina and V.N. Gushchi Geochimiya, No. 6, 803 (1975).

70. M.S. Chupakhin, G.I. Ramendik and O.I. Kryuchkova, Zh. Anal· Khim. 24, 965 (1969).

71. M.S. Chupakhin, Kazakov I.A. and O.I.Kryuchkova, Zh.Anal.Khim., 24, 165 (1

72. F. Konishi, Int.J.Mass-Spectrom. Ion Phys., 9, 33 (1972).

73. Yu.A. Zolotov, N.V. Shakhova, O.I.Kryuchkova, S.I. Gronskaya, B.Ya. Spivak G.I. Ramendik and V.N. Gushchin, Zh. Anal· Khim., 33, 1253 (1978).

74. O.M. Petrukhin, Yu.A. Zolotov,V.N.Shevchenko, O.I.Kryuchkova,S.I.Gronskaya Gushchin,G.I.Ramendik,V.V.Dunina and E.G.Rukhadze,Zh.Anal.Khim.,34,33

75. G.I. Ramendik and A.E. Lisitsin, Geokhimiya, No. 4, 566 (1979).

76. A.P. Vinogradov and I.K. Zadorozhnii, in: ref. 15, p. 547.

77. Yu.I. Stacheev and S.A. Stacheeva, Zavodsk. Labor., 42, 144 (1976).

78. Yu.I. Stacheev, A.K. Lavruchina and S.A. Stacheeva, Geokhimiya, No. 9, 1390 (1975).

79. G. Tölg, Talanta,19, 1489 (1972); 21, 327 (1974).

80. G. Tölg, Naturwissensch., 63, 99 (1976).

81. G. Tölg, Pure and Appl. Chem., 44, 645 (1975).

82. G. Tölg, in: Comprehensive Analytical Chemistry, Vol. III, Eds.C.L.Wilson and D.W. Wilson, Elsevier, Amsterdam-London-New York, 1975.

83. G. Tölg, Z. Anal. Chem., 283, 257 (1977).

84. G. Tölg, in: Spurenelemente - Analytik, Umsatz, Bedarf, Mangel und Toxikologie. Symposium in Bad Kissingen 1977. Georg Thieme Verlag, Stuttgart, 1979, S.1.

85. Yu. A. Zolotov, Pure and Appl. Chem., 50, 129 (1978).

86. Yu. A. Zolotov, Analyst, 103, 56 (1978).

87. Yu. A. Zolotov, Zh. Anal. Khim., 32, 2085 (1977).

88. Yu. Auffarth and D. Klockow, Anal. Chim. Acta, 111, 89 (1979).

89. I.P. Alimarin and M.N. Petrikova, Qualitative and Quantitative Ultra-micro Chemical Analysis, "Khimia", Moscow, 1974 (Russian).

90. Physical Aspects of Electron Microscopy and Microbeam Analysis. Eds. B.M. Siegel and D.R. Beaman, Wiley, New York, 1975.

91. Practical Scanning Electron Microscopy. Electron and Ion Microprobe Analysis. Eds.J.I.Goldstein and H. Yakowitz, Plenum Press, New York, 1975.

92. S.J.B. Reed, Electron Microprobe Analysis, Cambridge Univ. Press, Cambridge a.o., 1975.

93. I.B. Borovskii, F.F. Vodovatov, A.A. Zhukov and V.T. Cherepin, Lokal Methods for Analysis of Materials, "Metallurgiya", Moscow, 1973 (Russian).

94. L.N. Kolomeitseva and N.P. Il'in, Geokhimiya, No. 4, 521 (1978).

95. A.P. Vinogradov, N.P. Il'in and L.N. Kolomeitzeva, in ref. 16, p. 97

96. J.I. Goldstein, J. Geophys. Res., 72, 4689 (1967).

97. N.P. Il'in and F.I. Bochkaev, in: Progress in Analytical Chemistry, "Nauka", Moscow, 1974, p. 74.

98. G.H. Morrison and G. Slodzian. Anal. Chem., 47, 932A (1975).

99. V.T. Cherepin and M.A. Vasil'ev, Secondary Ion Emission of Metals and Alloys, "Naukova Dumka", Kiev, 1975.(Russian).

100. F.P. Viehböck, Mikrochim. Acta, Suppl. No. 5, 385 (1974).

101. C.A. Evans, in: Proc. III, Int. School on Cristal Growth, New Hampshire, USA, 1977, p.1-51. Anal. Chem., 47, 818A, 855A (1975).

102. M.K. Pavicevic and J.R. Hinthorne, in: 6 Jugosl. Kongr. za cistu i primi jen. hem., Sarajevo, 1979. Sinop., Sarajevo, 1979, p. 185.

103. H. Moenke and L. Moenke, Laser Microspectrochemical Analysis, Adam Hilger Lt., London, 1973.

104. N.V. Korolev, V.V. Ryukhin and S.A. Gorbunov, Emission Spectral Micro-analysis, "Mashinostroenie", Moscow, 1971 (Russian).

105. Laser Local Spectrochemical Analysis of Minerals. Compilers: M.V. Bobrova, V.M. Paikova and N.G. Suslova, Alma-Ata, 1975 (Russian).

106. L. Moenke-Blankenburg, H. Moenke et al, Ienscoe obozrenie, No. 3, 107, (1975).

107. S. Hofmann, Talanta, 26, 665 (1979).

108. Quantitative Surface Analysis of Materials. Spec. Techn. Publ. 643, Ed. N.S. McIntyre, 1978.

109. H.A. Laitinen, Anal. Chem., 50, 545 (1978).

MICROANALYSIS OF POLYMERS
WITH SPECIAL EMPHASIS ON SURFACE CHARACTERIZATION

REIMER HOLM
Bayer AG, Applied Physics Department
D-5090 Leverkusen, Federal Republic of Germany

Abstract. When small amounts of a substance are enriched in small volumes
or in monolayers at the surface of a solid body, they can be directly detected by instrumental analysis methods without any chemical separation.
In the polymer field surface analysis with methods like ESCA, AES, SIMS,
and ISS is still in its infancy. Examples of application are the investigation of surface changes, such as the effects of weathering, etching and
corona treatments, and the detection of surface deposits consisting of
materials like antistatic agents, lubricants, and exuded constituents.
Because of their high surface sensitivity and ability to differentiate
between surface and bulk phenomena, surface analysis methods are particularly suitable for the investigation of interfaces between polymers and
metals, e.g. in radial tires or cable insulations. It is pointed out that
in these cases the inadequacies of the preparation methods limit the value
of the analytical findings more than the inadequacies of the analysis
methods.

1. Introduction

High polymers are encountered in many forms in our daily lives, e.g. as
fibres, foams, engineering plastics, and coatings. As the emphasis in
polymer research has shifted from the polymerization process towards the
processing of polymers, microanalysis methods have gained increasing importance.

Analysis of polymer matrices is routinely done by IR spectroscopy.

For the identification of additives, pigments, and fillers there are well-

tried analytical procedures, most of which combine suitable detection
techniques with separation or accumulation. But if small amounts of a
substance are enriched in small volumes or in monolayers at the surface
of a solid they can be directly detected by instrumental analysis without any chemical separation. In cases where these methods can be used
they provide, more rapidly, information that is more accurate and detailed (e.g. in respect of the spatial distribution) than that provided
by the classical techniques. Several examples of the application of
instrumental microanalysis and surface analysis are given below. They
relate to the analysis of specks, to the investigation of surface changes,
such as the effects of weathering, etching, and corona treatments, and to
the detection of surface deposits consisting of materials like antistatic
agents, lubricants, and exuded constituents. These methods are also particularly suitable for investigation of interfaces between polymers and metals,
e.g. in radial tires.

2. Analysis of specks

The inorganic elements are generally identified by electron X-ray microanalysis (Fig. 1).

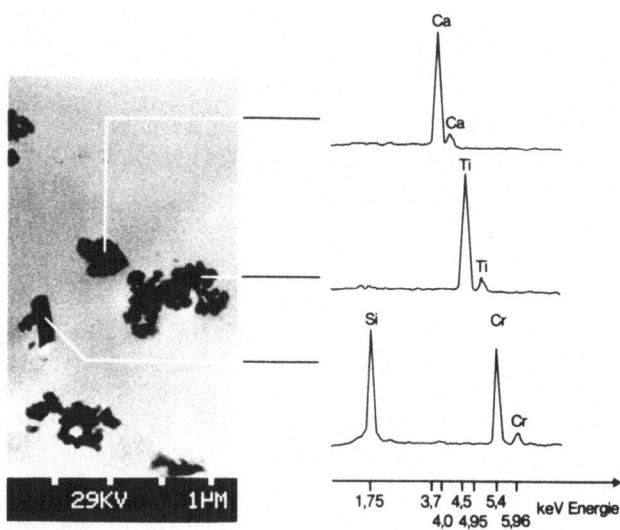

Figure 1: Identification of inorganic pigments in a polymer by energy-
dispersive X-ray microanalysis (STEM image of a thin section)

If the specks are by nature organic, analysis is frequently possible by
infrared-spectroscopic micromethods. Information about any compounds
present can also be obtained by LAMMA [1,2]. In this case the zone of
interest is evaporated from a thin section with the aid of a high performance laser, the position of the zone being selected with a light
microscope, and the charged fragments leaving the plasma are detected
with a time-of-flight mass spectrometer. An example is given in Fig.2,
where it was desired to find out whether light-coloured specks in a
rubber sample were attributable to the formation of crystallized
agglomerates by an accelerator (tetramethyl thiuram disulphide).

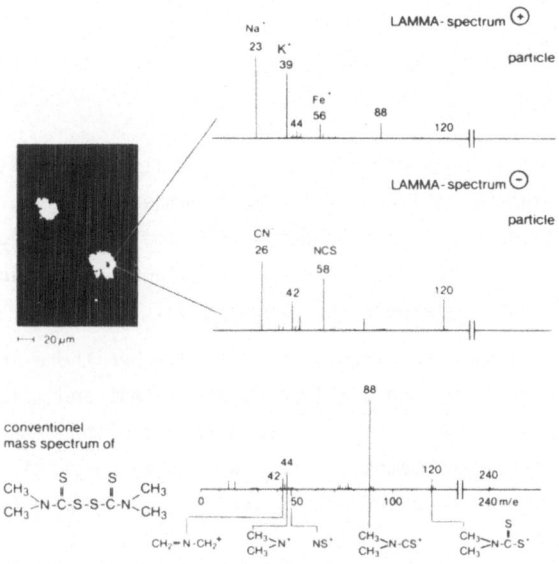

Figure 2: LAMMA spectra of specks of tetramethyl-thiuram-disulfide

X-ray microanalysis permits no definite conclusion in this respect because the sulphur detectable by this method is also present in the
vicinity of the specks and, apart from this, may belong to other additives. Accordingly the lightcoloured areas and the accelerator likely
to be responsible for them

were analyzed by LAMMA.

As the mass spectra of the specks and of the substance with which they were compared are almost identical there is no doubt that the specks consist of crystallized particles of the accelerator. The characteristic ions in the negative LAMMA spectra are

$$\text{mass 26:} \quad (CN^-), \quad 42: (CH_2 = N - CH_2^-), \quad 58: (N - C - S^-) \text{ and}$$

$$120: \quad \left(\begin{array}{c} CH_3 \\ CH_3 \end{array}\!\!>\!N - \overset{S}{\overset{\|}{C}} - S^-\right).$$

In the positive LAMMA spectrum there appear, in addition to ions from impurities in the rubber (23: Na^+, 39: K^+, 56: Fe^+), the characteristic ions of the mass

$$44: \left(\begin{array}{c} CH_3 \\ CH_3 \end{array}\!\!>\!N^+\right), \quad 46: (NS^+), \quad 58: (N - C - S^+) \text{ and } 88: \left(\begin{array}{c} CH_3 \\ CH_3 \end{array}\!\!>\!N - CS^+\right).$$

The conventional mass spectrum of the accelerator, as excited by electron impact, shows intensive lines at the mass numbers 88, 120, 44 and 42. It follows that laser excitation produces the same fragment ions as conventional mass spectrometry does (though the intensity distributions are somewhat different as between the two methods).

Under the most favourable circumstances (i.e. when there is a high probability of ionization and no overlapping of element and molecular fragment ions) the mass spectrometric detection obtained with LAMMA makes this an exceptionally sensitive method, a fact which may be particularly useful in the investigation of extraneous processing techniques (e.g. detection of catalyst traces).

3. Surface analysis

Many methods can be used to obtain information on the surfaces of solids. But only a few of them, namely ESCA, AES, SIMS and ISS, as characterized in Fig. 3, are suitable for extensive analytical use. A review is given in ref. 2.

All these methods have their advantages and disadvantages. But it is a fact that at the present state of the art ESCA is the leading surface analysis technique for the study of chemical problems, especially in the organic field, because

- ESCA is in general non-destructive,
- ESCA is normally free from charging problems in measurements on insulators,
- ESCA permits the most reliable detection of compounds,
- ESCA permits the most reliable quantitative measurements.

The surface is irradiated with X-rays (e.g. Al Kα) and as a result of the photoeffect and the Auger effect electrons are emitted and their kinetic energy is measured. On the basis of existing tables the emitting elements can be easily identified. Fig. 4 gives some wide scan spectra of polymers.

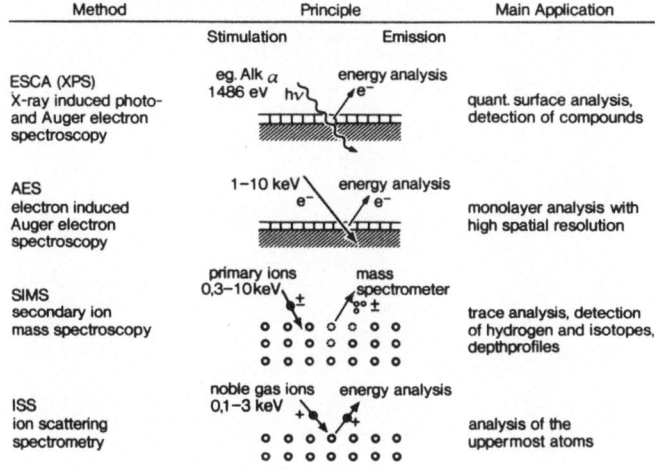

Figure 3: Methods for surface analysis

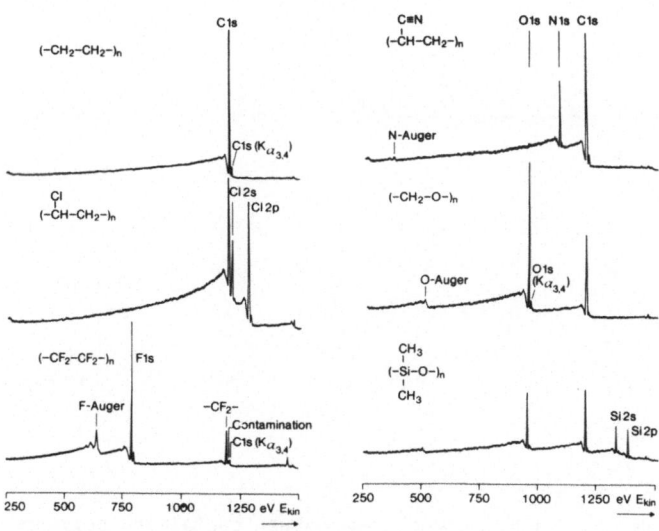

Figure 4: Wide scan ESCA spectra of some polymers

If the electron lines are investigated in more detail with expanded abscissa, chemical shifts are observed which give us information about the binding states of the emitting elements.

In Fig. 5 typical C1s and O1s spectra of some oxygen containing polymers are shown.

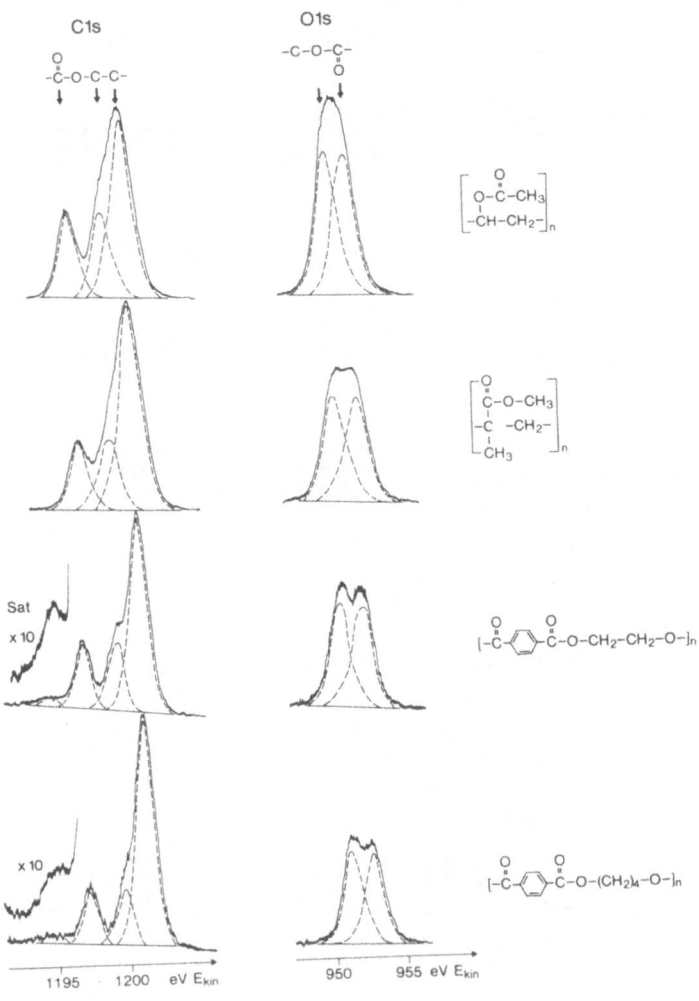

Figure 5: C1s and O1s spectra of some oxygen containing polymers

Aromatic rings can be identified by shake up satellites, as can be seen for PET and PBT in Fig. 4. A special advantage of ESCA is the quantitative evaluation of the spectra. If there are lines separated by chemical shifts each line directly reflects the number of atoms in the corresponding binding state (similar to chemical shifts in NMR). If different elements are concerned, sensitivity factors can be taken from existing tables [4] with a precision of 5-10 % in the case of polymers. The surface sensitivity of ESCA is given by the escape depth of the photo and Auger electrons [5]. The inelastic mean free paths of electrons with a kinetic energy of 1000 eV are approximately 1-2 nm. Thus about 99 % of the intensity of the lines of an ESCA spectrum is provided by electrons escaping from a surface layer which is less than 10 nm thick.

If information is to be obtained from lower depths, surface layers must be removed. The most widely used technique is that of sputtering. With ISS and SIMS this removal is effected through the nature of the excitation itself; in ESCA and AES an additional ion gun is required.

The energies used in ion bombardment (500 eV - 5 keV) are several orders of magnitude greater than the binding energies of the molecules and therefore sufficient to break chemical bonds and form new ones. Where organic compounds are concerned such ion-induced reactions are particularly easy to observe because in this case there is no restoration of the initial state through annealing processes (Fig. 6) [6]. It may also be assumed, however, that it is no longer possible to identify inorganic compounds reliably in areas opened up by ion bombardment.

Some examples will now be given to demonstrate the state of the art in surface analysis of polymers.

Several publications (e.g. ref. 7) describe investigations concerned with the etching of polytetrafluoroethylene (PTFE) as a means of facilitating the bonding of this material. Satisfactory sticking of adhesives can be obtained by, for example, treating the surface with Na in liquid ammonia. In the etched layer neither N nor Na, and no F, but only C and O in one binding state each, were detected by ESCA. It follows that a layer of the PTFE thicker than the escape depth of the photoelectrons of F or of C in $C-F_2$ groups was decomposed. The less severe treatment with Na/naphthalene/ tetrahydrofurane gives a considerably thinner etched layer; signals from the PTFE below it are still obtained. The adhesiveness is lost if the plastic is irradiated with ultraviolet light in the presence of air. The growth in the intensity of C1s signals from CF and CF_2 groups can be explained by the formation of gaseous products during the deactivation and the associated chemical removal of the etched layer. Heating to 400°C in air or irradiation with UV light in air removes the etched layer entirely with the result that the ESCA spectrum again shows the signals of the pure PTFE.

Another way in which the surface energy of polymers can be increased is to subject them to corona discharge treatment [8]. This treatment is applied to polyethylene, for example, to make the material easier to print on, easier to bond, etc. ESCA reveals that this treatment results in oxidation of the carbon atoms at the surface, even acid or acid anhydride groups being formed (Fig. 7).

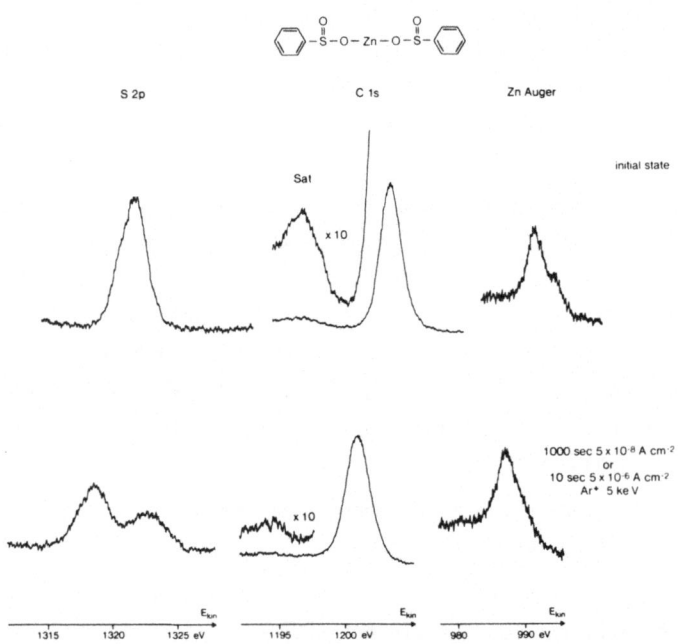

Figure 6: Decomposition of zincphenylsulfinate under Ar[+] ion bombardment

Other oxidation products with smaller chemical shifts (keto, aldehyde, alcohol, and peroxide groups) are represented by almost equal intensities, with the result that no minimum is formed in the region of 1,196 to 1,201 eV. Long-lived radicals may also result in the formation of lines within this shift range. Fig. 8 shows how the intensity for the oxidation products is reduced during the storage of the treated samples. Volatile products are formed by rearrangement processes at the surface, and some of the oxidized material may migrate into the bulk. Ozone treatment of polybutadiene, on the other hand, results mainly in the formation of ketone groups (Fig. 7). The oxidation products can also be identified by infrared and electron spin resonance spectroscopy; to obtain sufficient intensities, however, very high radiation doses are necessary. When ESCA is used as a surface analysis method, effects can not only be detected, but also determined quantitatively, at technically normal irradiation doses. ESCA is also not restricted to a special sample geometry as, e.g., ATR-IR is.

In view of the opportunities offered by ESCA for the early recognition of oxidation processes this method is also used to study the effects of weathering.

Another application of surface analysis methods is the detection of surface deposits (of antistatic agents, fibre finishing agents etc.) and of surface exudations by substances added to polymers (e.g. lubricants).

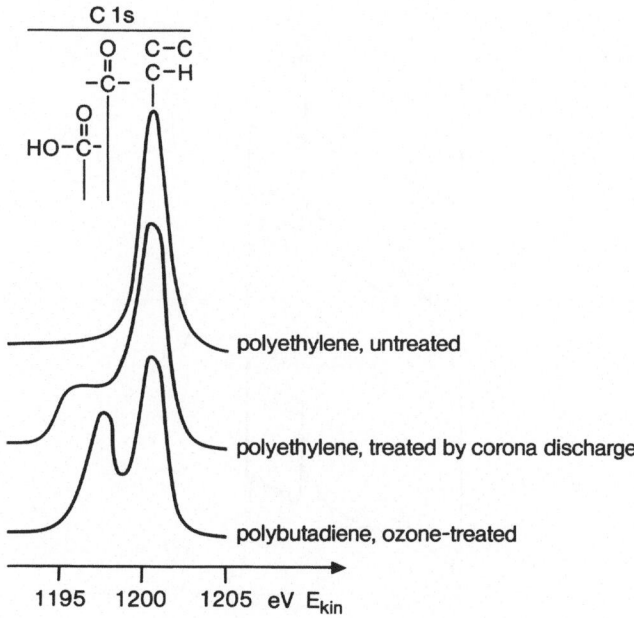

Figure 7: Effect of surface treatments of polyethylene and polybutadiene

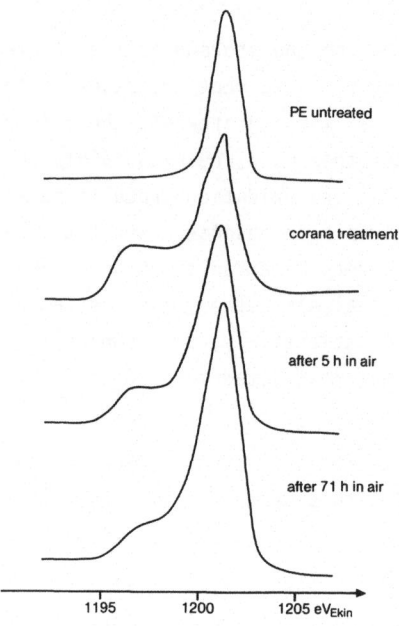

PE untreated

corana treatment

after 5 h in air

after 71 h in air

1195　　　　1200　　　　1205 eV$_{Ekin}$

Figure 8: C1s spectra of corona discharge treated polyethylene

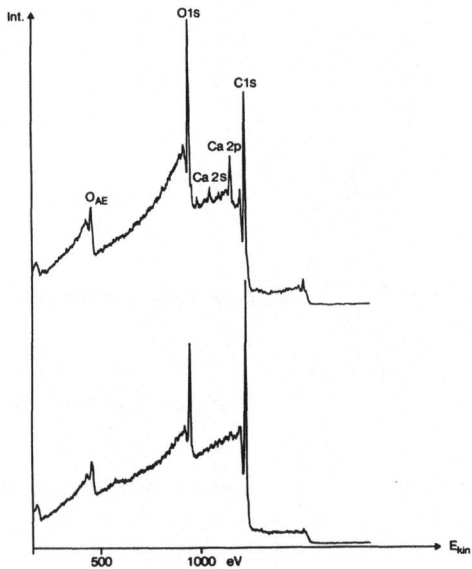

Int.

O1s

C1s

Ca 2p

Ca 2s

O$_{AE}$

500　　　1000　eV

E$_{kin}$

Figure 9a: ESCA spectra of polyethylene with exuded Ca stearate

An example is the detection and even semi-quantitative determination of calcium stearate, which is often added to polyethylene as a lubricant. An excessive calcium stearate concentration at the surface impedes welding. This problem has been successfully studied both by SIMS (Fig.9a) (observation of the masses 40, 44 (Ca^+), 56 (Ca O^+), 57 ($CaOH^+$)) and by ESCA (Ca 2s, 2p lines) (Fig.9b).

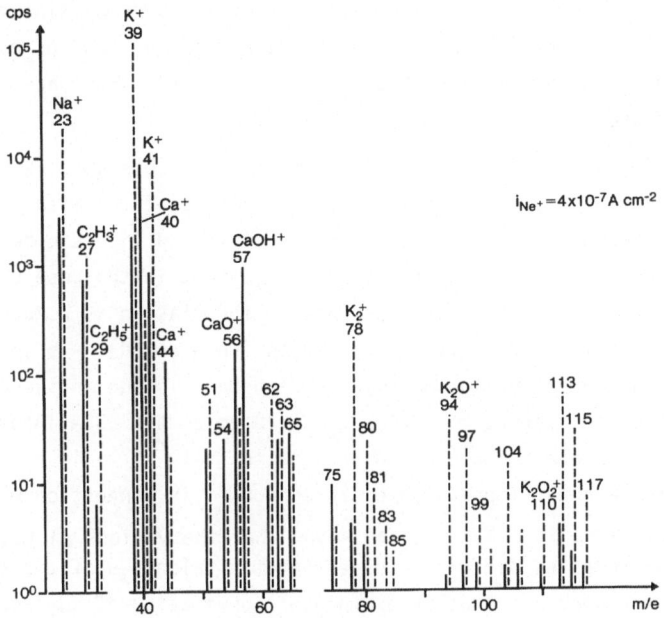

Figure 9b: SIMS spectra of polyethylene with (-) and without (---) excuded Ca stearate

The oxygen and COOH signals are not unambiguous, because the migration of the Ca stearate is observed only at elevated temperatures, where at least a partial oxidation of the PE cannot be excluded. The sensitivity of SIMS is limited in this case by the overlapping of elemental ions (e.g. Ca^+) and organic fragment ions.

4. Interface analysis

A wide field for the application of microanalysis and surface analysis methods is the investigation of the adhesion of polymers to metals.

Polyethylene (PE) that is melted onto aluminium undergoes oxidation - the more so if it is high density, rather than low density, polyethylene [9,10]. Thus the usually good adhesion of extruded PE to aluminium

is probably attributable to oxidation processes of this kind. By virtue
of its greater surface sensitivity ESCA enabled the oxidation of PE to be
detected, whereas the ATR-IR technique failed to reveal it.

It should be pointed out in this connection that, in cases of inadequate
adhesion, ESCA - as well as other techniques - are able to detect possible
foreign matter at the Al-PE interface and to reveal whether separation
has occurred exactly at the interface or whether Al or aluminium oxide
have been transferred to PE or vice versa. But in the case of good
adhesion getting information about the polymer-metal interface is a
problem of preparation.

The situation is similar in the study of the processes involved in the
galvanization of plastics, as well as of the possible blemishes. For
example, pickling and activation products at the surface can be identified [11]. The N1s intensity (from the SAN content in the case of ABS)
reveals the extent to which the surface of the plastic is homogeneously
covered by pickling or activation products (such as Pd). A problem that
arises in the determination of relative coverage is the influence of
metal particle diameters of the order of 10 nm on the ESCA intensities.
Such influences are being studied very intensively at various laboratories
in view of the great importance of this problem in connection with supported catalysts. It has also been found that the various pickling and
activation processes differ in respect of the valencies of the metals
deposited for seeding. Another example of application of surface analysis
methods is the study of galvanizing blemishes such as blisters. They can
be caused by oxidation, by dirt or by the formation of undesirable intermediate layers, even if these are only a few monolayers thick.

Surface analysis methods have been applied with great success in model
experiments to gain a fundamental understanding of the bonding of brass-
coated steel cords to rubber in radial tires [12,13]. As the adhesion is
brought about by a very thin interfacial zone (Fig.10) it is reasonable
to study this zone by surface analysis methods. For this purpose, however, preparation techniques are needed which enable the interface to
be uncovered. In this way a problem of interface analysis becomes one
of surface analysis. Unfortunately no existing technique allows that to
be done in a controlled manner. In general the cord and rubber are separated under liquid nitrogen. The separation line can occur more or less
as a matter of chance either at the bond layer/brass interface or at
the bond layer/rubber interface, or it may be in the adjacent rubber.
As the freshly ruptured surface reacts in ambient air the sample must

be prepared under a protective gas. These preparative difficulties, considerably more than the inadequacies of the methods themselves, may reduce
the value of the analytical findings.

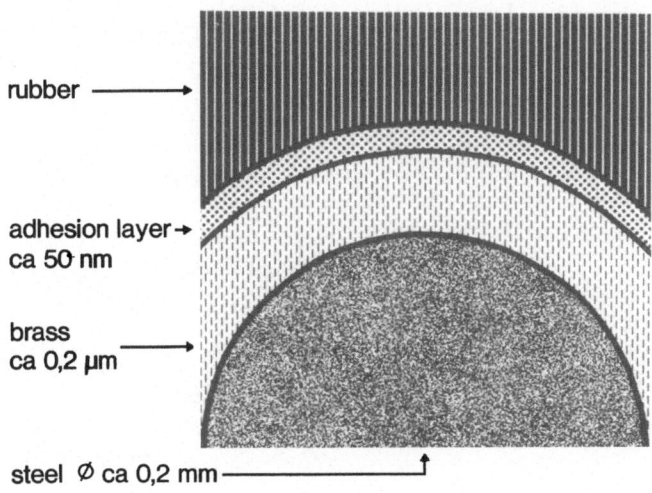

rubber

adhesion layer→
ca 50 nm

brass
ca 0,2 µm

steel ⌀ ca 0,2 mm

Figure 10: Schematic drawing of the interface tire cord/rubber

In view of the difficulties encountered in this specific case, none of the
methods by itself permits comprehensive conclusions. So it is necessary to
combine several surface analysis methods and to supplement them by such
other methods as SEM, TEM and the electron microprobe.

Fig. 11 shows ESCA depth profiles obtained in a model experiment with a brass
plate after separation of the bond under liquid nitrogen. Starting from the
surface of the break, depth profiles were obtained both in the direction of
the brass and in that of the rubber; they were then put together at zero
depth in Fig. 15. This gives the concentration profile of a number of elements in the interface between the brass and rubber. As can be seen, the
adhesion layer consists essentially of an approximately 80-nm thick Cu(I)-
sulphide layer with a small excess of sulphur and, on the brass side, it
has an underlayer consisting of some ZnS + ZnO, which acts as a barrier to
the diffusion of the copper (thus preventing excessive copper sulphide

formation).

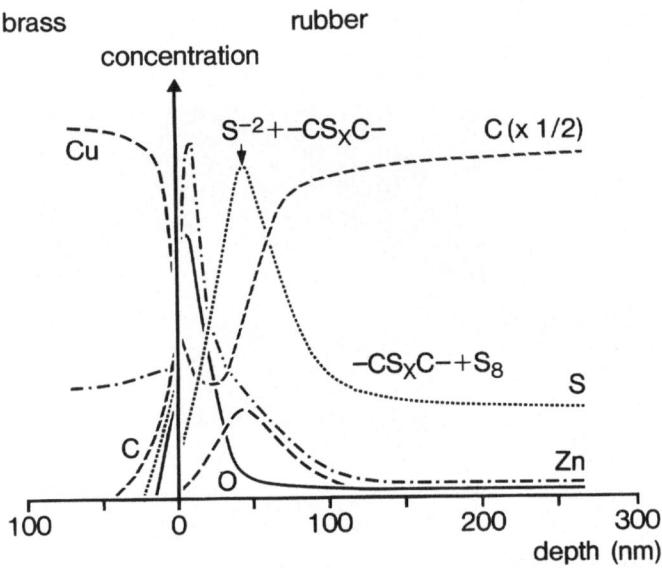

Figure 11: ESCA in-depth profile of a rubber-to-brass sample broken
at liquid nitrogen temperature (after W.J.van Ooij [13])

Because of charging and decomposition, corresponding profiles cannot be
obtained by AES on the rubber side; they can, however, be obtained on the
brass side. In this case micro-AES can be used to advantage to obtain
depth profiles of residual islands of the Cu_2S layer on cord samples
(Fig. 12). As the C signal declines, the S signal is seen to increase
before the Cu and Zn signals. The S intensity passes through a maximum,
which may be situated before those of Cu and Zn. The depth to which an
accumulation of sulphur occurs, and the relative positions of the maxima
of S, Cu and Zn, differ considerably according to the bonding system
employed. Unexpected and striking is the fact that Fe lines appear rather
early in the depth profile. Their intensity increases linearly until
saturation in the steel core is reached. It may be concluded that there
is no ideal brass layer on the cord but a ternary Cu-Zn-Fe alloy. Thus
Fe has also to be considered as a reaction partner during the formation

of the adhesion bond.

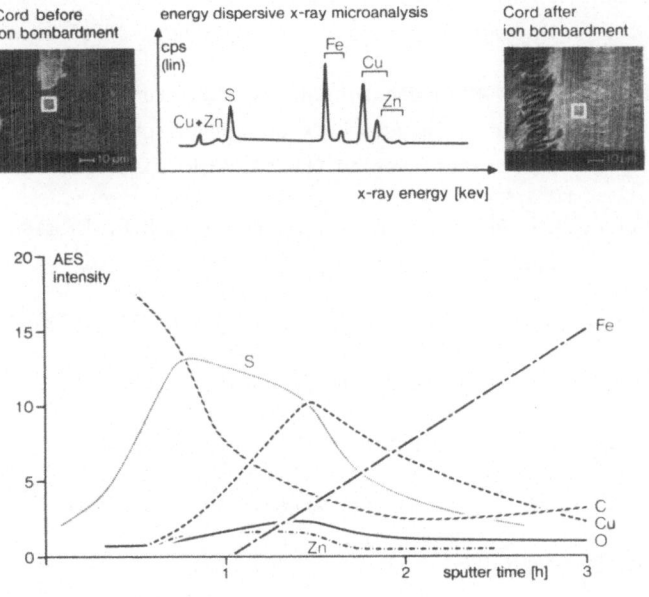

Figure 12: AES in-depth profile of a tire cord surface with a combined
SEM/AES apparatus.
The tire was broken at liquid nitrogen temperature

The routine use of surface analysis methods in the polymer field is still
in its infancy. The examples discussed should have demonstrated, though,
that polymer surfaces and - if suitable preparation techniques are used,
interfaces, too - can be characterized. The last example, in particular,
was intended to show that, where complex problems are involved, surface
analysis methods should be used, not in isolation, but together with other
microscopic and spectroscopic techniques. It is also essential to develop
adequate preparation techniques, especially in the case of good adhesion
in interface analysis.

References

1. R. Wechsung, F. Hillenkamp, R. Kaufmann, R. Nitsche, E. Unsöld,
   H. Vogt, Microscopica Acta, Suppl. 2 (1978) 281.

2. R. Holm, S. Storp: Methoden zur Untersuchung von Oberflächen, in
   Ullmanns Encyclopädie der technischen Chemie, Band 5,
   4. Auflage, Verlag Chemie, Weinheim, 1980.

3. D.T. Clark, D.B. Adams, A. Dilks, J. Peeling, H.R. Thomas, J. Electron
   Spectr. 8 (1976) 51.

4. C.K. Jorgensen, H. Berthou, Disc. Faraday Soc. 54 (1972) 269.

5. M.P. Seah, W.A. Dench, Surf. Interf. Anal. 1 (1979) 2.

6. S. Storp, R. Holm, J. Electron Spectr. 16 (1979) 183.

7. H. Brecht, F. Mayer, H. Binder, Angew. Makromol. Chem. 33 (1973) 89.

8. R. Holm, S. Storp, Phys. Bl. 32 (1976) 342.

9. D.T. Clark, W.J. Feast, K.R. Musgrave, J. Ritchie, J. Polymer Sci.,
   Polymer Chem. Ed. 13 (1975) 857.

10. D. Briggs, D.M. Brewis, M.B. Konieszko, J. Mat. Sci. 12 (1977) 439.

11. K. Richter, G. Kley, J. Robbe, J. Gähde, J. Löschke, Z. Chem. 18
    (1978) 390.

12. W.J. van Ooij, Surface Sci. 68 (1977) 1.

13. W.J. van Ooij, Kautschuk + Gummi, Kunststoffe 30 (1977) 739.

VI. INSTRUMENTATION, METHODS AND AUTOMATION IN MICROCHEMISTRY

CLUSTERING FOR MICROANALYTICAL DATA

DESIRÉ L. MASSART
Vrije Universiteit Brussel, Farmaceutisch Instituut,
Laarbeeklaan 103, B - 1090 Brussels, Belgium

Abstract. Clustering is a chemometrical technique, used for the grouping or classification of objects characterized by multivariate analytical data. The hierarchical and non-hierarchical algorithms are discussed. Among the hierarchical methods the average linkage and Ward's method are preferred, while the preferred non-hierarchical method is MASLOC. The latter is the only method with a build-in rule for the selection of significant clusters.
The importance of combining clustering with other multivariate techniques, such as factor analysis, is stressed.

1. Introduction

Clustering is part of a new field of analytical chemistry, called chemometrics. Chemometrics has been defined as the application of mathematical and statistical tools to chemistry. It is a collection of techniques from fields such as information theory, pattern recognition and clustering, correlation techniques, etc. Its development started among other places, in Austria, where the possibility of applying information theory was perceived perhaps 10 years ago and it is now achieving full status. One of the symptoms is that one of the two-yearly fundamental reviews of Analytical Chemistry is now devoted to Chemometrics [1].

Chemometrics is, however, still a very new science to many analytical chemists. For this reason, this article will consists mostly of an introduction to clustering technology. Some recent results obtained by us will also be discussed.

The usual starting point of a clustering problem is a large and multivariate data set. The interpretation of the data set is often impossible without putting some order into the data. Clustering is the collective name of a set of techniques to uncover the structure residing in the data set. This is achieved by a classification or grouping of objects (for instance samples that have been analysed) characterized by similar patterns of analytical data or other characteristic data.

2. Preliminary steps

One starts with a data set $x_{i,j}$. These raw data are usually transformed. In many instances this transformation consists in scaling the data. One computes $z_{i,j} = \dfrac{x_{i,j} - \bar{x}_i}{s_i}$, where $\bar{x}_i$ and s_i are the mean and the standard value of the ith object.

The clustering procedure then continues by measuring the similarity between all pairs of objects i. The larger the similarity, the "closer" or the more similar these objects are. Very often one uses the Euclidian distance $\Delta_{i,k}$ as similarity value. When J measurements have been carried out on each object, this is given by:

$$\Delta_{i,k} = \sqrt{\sum_{j=1}^{J} (z_{i,j} - z_{k,j})^2}$$

$\Delta_{i,k}$ being the distance between i and k.   The smaller the distance is, the larger the similarity. Other transforms such as the log-transform and the Γ-transform [2], are also possible and other distances such as Mahalanobis distance or the correlation coefficient can also be employed. All these similarities are gathered in a similarity matrix such as the one given in Table Ia. It now remains to group in some way the more similar objects. Two kinds of algorithms allow this, namely hierarchical and non-hierarchical clustering algorithms.

3. Hierarchical clustering algorithms

The most typical result of hierarchical clustering is the dendrogram (see figure 1). A hierarchy of clusters is then observed : E is an object, part of cluster EF; cluster EF is part of the larger cluster EFGH, which is a part of the still larger cluster E to M , which in turn is a part of the largest possible cluster namely the one which encompasses the whole data set. Biological taxonomy has such a structure, for individual living objects constitute species, groups of species constitute genera, etc.   For this reason hierarchical clustering is called numerical taxonomy by biologists. This term was also introduced in analytical chemistry by us [3], but the term hierarchical clustering is to be preferred.

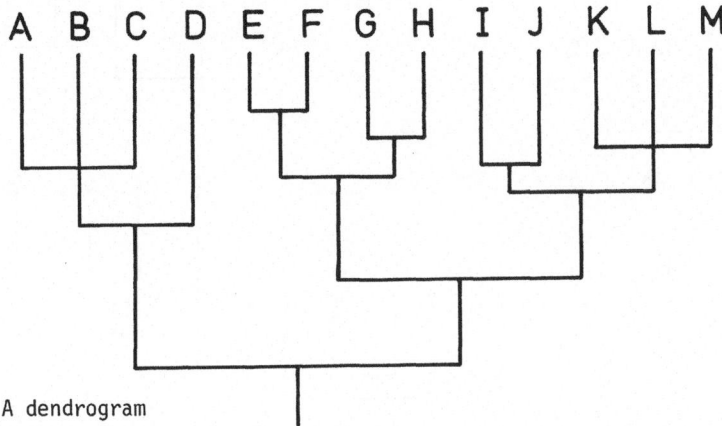

Figure 1: A dendrogram

Most scientists seem to prefer hierarchical to non-hierarchical clustering methods, probably because the mathematics and the concept are extremely simple. Analytical chemists are no exception, so that most applications of hierarchical clustering in analytical chemistry are of the hierarchical type.

We will introduce hierarchical clustering here with a small example taken from ref. [4]. It is supposed here that five objects A, B, C, D and E must be clustered. The distances between them are given in the similarity matrix of Table 1.

Table 1: Example of hierarchical clustering

a)	A	B	C	D	E
A	0				
B	40.0	0			
C	38.7	17.3	0		
D	110.4	70.7	78.1	0	
E	111.4	72.1	80.6	14.1	0
b)	A	B	C	D	E
A	0				
B	40.0	0			
C	38.7	17.3	0		
D*	110.9	71.4	79.3	0	

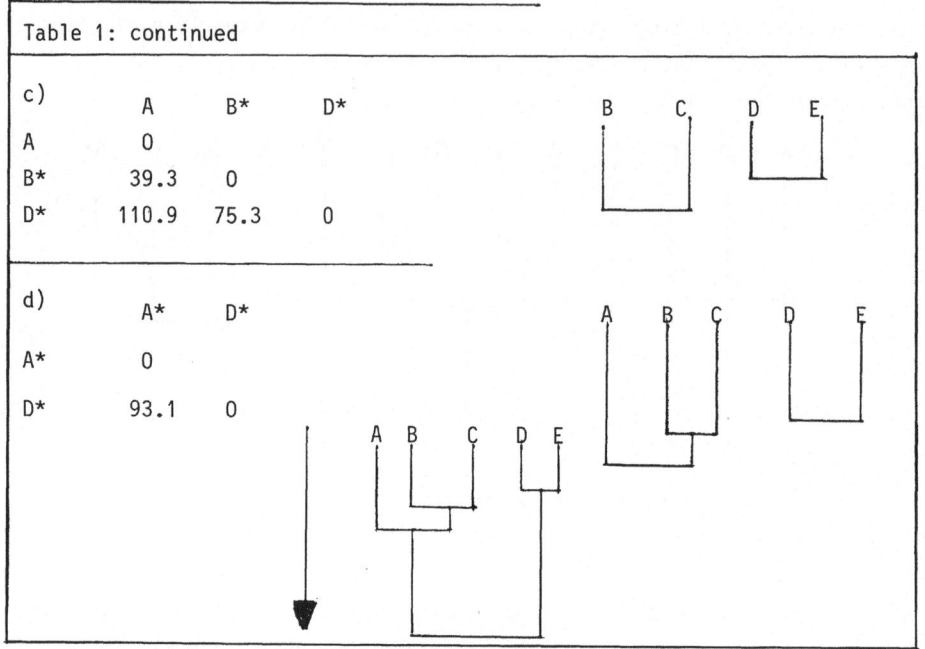

Table 1: continued

c)

	A	B*	D*
A	0		
B*	39.3	0	
D*	110.9	75.3	0

d)

	A*	D*
A*	0	
D*	93.1	0

Since D and E are clearly the most similar objects in the set, it is logical, that if one cluster is created, it should contain D and E. A cluster D* containing D and E is created and replaces D and E in the similarity matrix. To do this one must now be able to determine the distance between D* and the remaining objects. One supposes that this distance is the average between the distances to D and E. For example, $\Delta DA = 110.4$, $\Delta ED = 111.4$, $\Delta D*A = (110.4 + 111.4)/2 = 110.9$. One can now replace D and E by D*, which results in Table 1b.

This procedure is now repeated and B and C are replaced by B*, which results in Table 1c. In this table, the distance AB* is the smallest one, so that one joins A to the already existing cluster B*, which results in a larger cluster A*. The resulting similarity matrix now consists of only one distance, so that the two remaining clusters A* and D* are now joined. All elements have now been linked one to another and the clustering algorithm stops. The sequence of joining objects or clusters is given by the dendrogram in Table 1d.

It remains now to isolate clusters from this dendrogram. One can do this in two ways namely by cutting successively the highest links (resulting in the clusterings ABC/DE, A/BC/DE, etc.) or all links representing a higher distance then some threshold distance. If this distance would be set at 30, this would result in the clustering A/BC/DE.

Many different hierarchical clustering methods exist. They differ mostly in
the determination of the distance between pairs of clusters or between
clusters and objects. In the example the distance between D* and A is the
average between D and A and E and A. This is called an average linkage method.
One can also declare that $\Delta D*A$ = min (ΔDA, ΔEA) = 110.4 (which is called
single linkage) or that $\Delta D*A$ = max (ΔDA, ΔEA) = 111.4 (which is called
complete linkage). One can also determine the centroid of cluster D* and
$\Delta D*A$ is then equal to the distance between this centroid and A (centroid
linkage, also called the median method). A last possibility, which we will
discuss here is Ward's method or the error sum of squares. Ward defines the
heterogeneity E_p of a cluster p as the sum of the squared distances of each
of the objects to the centroid of the clusters. In deciding which clusters
should be joined at a certain stage, one does not consider distances between the candidate clusters as do the other methods but an increase in
heterogeneity after linkage. All possible linkages between pairs of clusters
or clusters and objects are considered and one carries out that one for which

$$E_{(p,q)} - E_p - E_q = min$$

($E_{p,q}$ is the heterogeneity of the cluster obtained by joining p and q).
At this point, the layman who reads this article will want to know which
of these methods should be employed. Unfortunately, it is quite impossible
to state that one method is always and in all aspects better than another
and it requires some knowledge to select the best one for a particular
application. However, some general conclusions may be made. Single-linkage
and complete linkage show so-called space distorting effects. In single
linkage, for instance, one joins two clusters when one element of one
cluster is near to an element of the other cluster.

This may lead to the "chaining" of poorly separated clusters. For this
reason single-linkage is said to be space-contracting. Centroid linkage
has another disadvantage : it is a non-monotonous method. This means
that in the dendrogram the distance does not increase in a monotonous
way for successive fusions : reversals may occur, leading to cross-over
and dendrograms that are difficult to interpret. It results that average
linkage and Ward's method are the methods with the least disadvantages.
On the other hand, single linkage is computationally the simplest method.

4. Non-hierarchical clustering algorithms

To explain hierarchical clustering, biological classification was used as
an analogue (see higher). Here one can use an operations research analogue [5].
This concerns the optimal location of supermarkets. Consider figure 2 which
is a map of 10 towns A to J. One needs to locate p supermarkets in such a
way that the sum of the distances from the towns to the nearest supermarket
is minimal. Suppose first that the supermarkets must be located in some of
the towns.

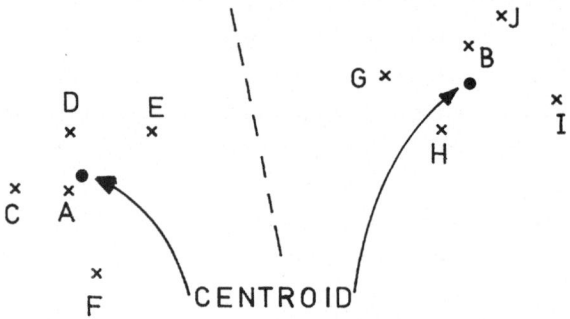

Figure 2: Nearest centrotype and centroid sorting.
The centroids are shown by the arrow and
the centrotypes are A and B

If 2 supermarkets are wanted, they should be located in A and B. A and B are
called centrotypes. By observing which towns shop in A and which in B one
creates two clusters or, in operations research language, a 2-median. To
find p clusters in a data set it is therefore sufficient to find the p-median.

Instead of requiring that the supermarkets be located in one of the towns
one can also permit the location of these supermarkets somewhere in between
the towns. The optimal locations are then the centroids of the clusters of
towns (see also figure 2). Since the objects are classified in clusters
according to their distance from these centroids, one calls these methods
nearest centroid sorting methods.

The minimization must be carried out for a fixed number of clusters p. The
basic procedure is the folowing (see figure 3):

a) give an initial list of p seed points and a corresponding initial
   classification of the N objects in clusters by grouping the objects
   with the seed point to which it is nearest. Determine the centroids
   of these clusters

b) for each object, ask the question: "is the object more similar to the
   centroid of any other provisional cluster than the one with which it

has been located or not?"
c) for the objects for which the answer is yes, relocate the object in the
   correct cluster and recalculate the centroid of the gaining and the
   losing cluster
d) scan the list of objects repeatedly until the location of objects does
   not change anymore or until a fixed number of scans has been carried out.

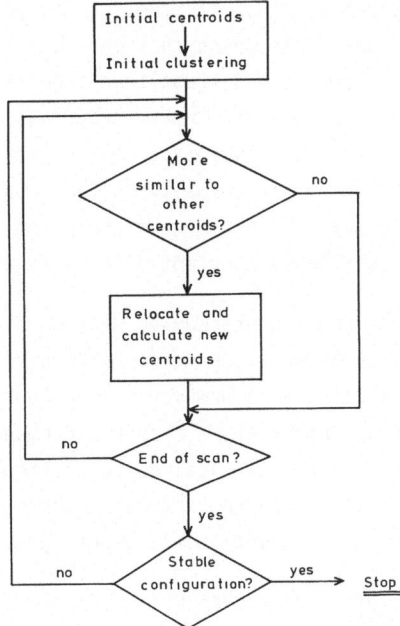

Figure 3: Flow schema of a typical nearest centroid sorting
          procedure for a fixed number of clusters

The clustering reached is then considered as optimal. The simplest of the
non-hierarchical centroid sorting procedures is Forgy's method [6]. In its
simplest form the initial list of seed points consists of the p first objects
in the list and relocation steps are carried out after a complete scan of
all the objects. McQueen's k-means method [7] is the same except that new
cluster centroids are calculated without waiting until a complete scan has
been carried out.

Quite clearly, the principal disadvantage of these methods is that one has
to specify beforehand a certain number of clusters. RELOCATE and ISODATA

start out more or less as McQueen's k-means method but go on to find another number of clusters than that specified initially. RELOCATE, which is to be found in the computer package Clustan [8] computes the similarity between all pairs of clusters and joins the two most similar ones so that p-1 clusters result. This is used as the initial clustering for carrying out McQueen's or Forgy's method, to obtain a p-1 optimal clustering. Again the two most similar clusters are joined which results in p-2 clusters, etc.   ISODATA [9] is a much more sophisticated, but not necessarily a better method. It lumps clusters together or splits them up according to criteria which must be given by the user. This means that the user has to state for instance that if two clusters are less distant than a certain threshold, one should fuse them. In some versions of ISODATA not less than 7 parameters must be furnished by the user. Since clustering is used to make sense of a data set, the properties of the set are usually not well known and it is therefore not easy at all to make an intelligent choice of these 7 parameters. This is felt by us as a very important disadvantage of the method.

MASLOC, which is a centrotype sorting procedure, developed by Massart and Kaufman [10] very much resembles McQueen's k-means methods in its first stage. The choice of the initial clustering is however more elaborate and much nearer to the final clustering. Moreover, a branch and bound optimization procedure permits to obtain with mathematical certainty the optimal configuration. This is not the case with the centroid methods, where different solutions may be obtained with different initial configurations.

To obtain different numbers of clusters MASLOC is simply rerun for all possible prespecified cluster numbers. Up to this stage, it offers only two minor advantages compared with other non-hierarchical methods, such as RELOCATE, nl.

1°) if the branch and bound option is used, it always gives the optimal solution

2°) the centrotype is one of the classified objects and may be considered as a representative object while the centroid is an abstract quantity. This was found to be useful. For example, in a study on the optimal choice of representative substances (probes) for the characterization of the chromatographic behaviour of stationary phases, we [11] clustered about 70 substances according to their chromatographic behaviour on 25 stationary phases. If the object is to obtain 4 probes then one asks for 4 clusters and selects the centrotypes of each cluster as the probes.

The major advantage of MASLOC is the introduction of a hierarchical element
in this non-hierarchical technique. This is obtained by considering the
sequence of clusterings obtained from the 2-clustering (2 clusters) to
the N-clustering (the clustering whereby each of the N objects forms a
cluster on its own) and its permits a solution to two of the more important
problems in clustering, namely the detection of outliers and of significant
clusters.

One of the big problems of pattern recognition in general and of clustering
in particular is the detection of outliers, objects which do not belong to
any of the clusters.

In centrotype sorting, outliers soon form a cluster on their own, with one
element only, this one element being its own centrotype. MASLOC finds outliers by looking for elements which form such clusters at low p-level.

Another extremely important problem of clustering is which particular
clustering is significant. One can always cut a dendrogram at a certain
level and obtain clusters or one can ask for a non-hierarchical clustering
containing p clusters. This does not mean that these clusters are significant, are natural clusters.

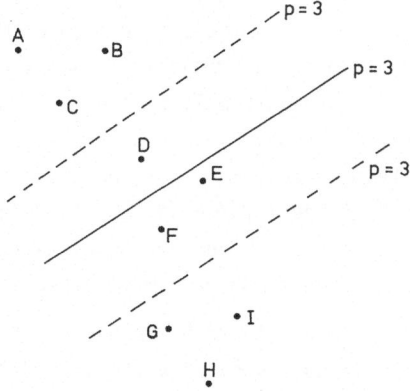

Figure 4: Clustering of 9 objects. The p = 2 clustering is not
"natural" or significant; p = 3 significant

Consider figure 4, which represents a simple two-dimensional clustering
problem. Clearly, the correct solution would consist of 3 clusters A-C,
D-F and G-I. If one doesn't know this and one supposes that one is not
able to observe it visually, then one will ask for the 2-clustering, the

3-clustering, etc. The 2-clustering would yield clusters A-D and E-I. This is correct from the computational point of view but not significant, since the natural cluster D-F is cut in two. In the 3-cluster solution one would obtain the three correct clusters and therefore a significant solution. When going from the 2-clusters to the 3-cluster solution one would, if one is not able to view a map such as figure 4, expect that some of the samples grouped together in one of the two clusters, would be separated in the 3-cluster solution but not the inverse such as is the case in this example. When this does occur, it means that the clusters concerned were not significant. This gives rise to the following rule [10]:

"A certain cluster, existing at level p is considered significant when at higher p-values no clusters are formed containing one or more of its elements mixed up with elements of the other cluster existing at p-level."

Applications of this method have shown its value. The following examples will show this.

- In a study on the metabolism of branched chain fatty acids in the mammary gland, ten goats were milked and 14 branched-chain fatty acids were determined in the milk [12]. The 14 acids were clustered using MASLOC with their procentual concentration in the milk as the variables. Four major clusters were identified and it was found that these four clusters coincide completely with different metabolic pathways. These results are really remarkable. One should compare the enormous experimental efforts which have gone into the differentiation of the metabolic classes using classical biochemical techniques with the extremely simple experimental set-up used here: simply 10 goats milked during five successive weeks! The difference is made by the much better use of the data in the simple set-up. Clustering has given the information for the obtention of which biochemists have worked many years with more sophisticated and more elaborate experiments but without adequate multivariate data reduction.

- A cosmochemical data base was developed by Esbensen and Buchwald [13]. In the form studied by us it consisted of a set of 493 iron meteorites for which 33 variables, mostly concentrations of major and trace elements were determined. The extensive literature on the subject has led to a classification in 15 classes with some 40 or 50 border cases and 75 outliers. The dataset is not a well behaved one for several reasons. The first reason concerns missing data. These are a serious handicap for pattern recognition methods and this is certainly the case for some of the otherwise very good methods using statistical techniques such as

factor analysis. Techniques to fill in missing data exist   and are the
object of some fine recent research. However, the quantity of missing
data may never be too large. For clustering purposes missing data do
not seem to be too large a problem. In the present case, variables with
up to about 50% of missing data were admitted, so that the data set
consisted finally of 493 samples and 13 variables. Another reason why
this data set is not well behaved is that the clusters overlap or are
not at all well separated. Moreover these clusters are elongated which,
reputedly, is not a good shape for most clustering methods, since these
tend to prefer round clusters.

The method used is a two-stage one. The first one consists of clustering
the whole data set using MASLOC and the object of this first stage is
triple, namely:
-- to compare, object by object, the existing classification with the
   clustering
-- to try and break up the data set into subsets for more detailed inspection
-- to eliminate clear outliers.

Detailed results will be reported elsewhere but an example can be considered here. According to Esbensen, group IA consists of 87 meteorites.
MASLOC-first stage clusters 85 of these in one significant cluster, together with some meteorites which according to Esbensen are outliers
and two meteorites from other groups. In the second stage, all the
objects declared IA by either Esbensen or MASLOC-first stage are re-
scaled and reinvestigated with a new MASLOC. The result is that a
significant cluster is now formed containing 86 of the 87 meteorites
considered IA by Esbensen plus two outliers also according to Esbensen.
One may conclude that IA is indeed a significant entity. Three meteorites
are reassigned and detailed analysis of the data shows this relocation
to be logical. The important thing to us is of course not whether the
group IA consists of 87 or 86 iron meteorites, but rather that a
classification technique yields very rapidly a classification, which
took a long time to be derived by other means.

An industrial application for the photographic industry was also developed.
One of the requirements when one produces film rolls for the cinemato-
graphic industry is that some of the color parameters vary as little as
possible in each lot which is sold. If, for example, a production batch
consists of 40 rolls, then MASLOC allows optimal grouping into clusters

with the same colour characteristics.

All these and other applications show that clustering allows in many instances better use of experimental data so that the data yield more information in a faster way than is usually the case at present. Analytical chemists are known to go to much trouble to assemble very accurate data in an optimal way. Once these data are assembled, they very much have the reflex to hand them over to other people. One of the ways they can go further is to help in the interpretation of the data, for instance by the use of chemometric techniques such as clustering. Readers of this article who might be convinced of the truth of this statement and who would like to carry out clustering on their own datasets, will then very probably ask the question which of the described methods is the best one. Unfortunately it is impossible to say which clustering method is the best one. The impression of the author is that the following methods are most likely to produce the best results: the average linkage and Ward's methods among the hierarchical methods and RELOCATE and MASLOC among the non-hierarchical ones. In any case, it is always best to try and approach a classification problem from at least two sides. One of the best combinations is MASLOC or another clustering procedure plus factor analysis on the whole data set or on the separated clusters as proposed in SIMCA [14].

References

1. B.R. Kowalski, Anal. Chem. 52, 112 R, 1980.

2. O.H.J. Christie and S. Wold, Anal. Letters, 12, 979, 1979.

3. D.L. Massart and H. De Clercq, Anal. Chem. 46, 1988, 1974.

4. D.L. Massart, A. Dijkstra and L. Kaufman. Evaluation and Optimization of Laboratory Methods and Analytical Procedures, Elsevier, Amsterdam 1978.

5. D.L. Massart, L. Kaufman, Anal. Chem. 47, 1244A, 1975.

6. E.W. Forgy, cited by M.R. Anderberg, Cluster Analysis for Applications, Academic Press, New York, 1973.

7. J.B. McQueen, cited by M.R. Anderberg, see ref. 6

8. Clustan, developed by D. Wishart and obtainable from The Clustan Project, Computer Centre, University College of London, 19 Gordon Street, London WC1HOAH

9. G.H. Ball and D.J. Hall, Isodata, A novel method of Data Analysis and Pattern Classification, AD 699616, Stanford Res. Inst., Menlo Park, California (1965).

10. D.L. Massart, L. Kaufman and D. Coomans, Anal. Chim. Acta, in press.

11. H. De Clercq, M. Despontin, L. Kaufman and D.L. Massart, J. Chromatogr. 122, 535, 1976.

12. A.M. Massart-Leen and D.L. Massart, to be published.

13. H.K. Esbensen and V.F. Buchwald, to be published.

14. S. Wold, Chapters 19 and 20, of ref. 4.

FLOW INJECTION ANALYSIS AND ITS FUTURE DEVELOPMENT

JAROMIR   RUZICKA

Chemistry Department A

The Technical University of Denmark

Building 207, 2800 Lyngby, Denmark

Abstract. A large number of analytical methods can be accommodated in a
simple FIA system. FIA technique does not require a complete mixing of the
sample and the reagent solutions or reaching of chemical equilibrium, and
the reaction product does not even have to be stable in time. The controlled
dispersion of the sample zone is the most important aspect of the techniques
as its concept has led also to the design of automated FIA, stopped-flow
FIA, merging zones FIA, extraction FIA as well as of an FIA scanning method.

1. Introduction

The advantages of processing microlitre sample volumes have been amply demonstrated in classical works of Pregl [1], Emich [2], and Benedetti-Pichler [3],
who originated and continued activities of the famous Graz school of
microchemistry. Yet today, fifty years later, microtitrations, micro-
colorimetry and other microanalytical techniques are seldom used, and the
majority of routine analytical assays is still performed at millilitre
rather than microlitre scale, though a microchemical approach would undoubtedly save sample and reagent materials, laboratory space and even
energy. Yet the reason why the majority of analysts have so far preferred
to work in a semimicroscale is simple: it requires much less skill to
pipette, measure, mix, heat, filtre or otherwise handle reproducibly the
millilitre than the microlitre volumes of liquids. Where the sample
material is scarce, like in clinical chemistry, the manual operations

have been successfully replaced by automated instruments, but the cost in
terms of instrument complexity has been high, and the simplicity of
classical microchemical methods based on volumetric analysis in the
capillary tube [2,3] has all vanished.

2. Present status

It may very well be so that the flow injection analysis [4,5] will allow
the final step towards miniaturization in analytical chemistry to be made.
There are several reasons for such belief, as the FIA apparatus is simple
to design, and the new method has already shown a wide range of applications.
For the present titrations, optical and electrochemical measurements as well
as chemical separations based on solvent extraction, ion exchange and gas
diffusion can be performed usually within ten to twenty seconds after sample
injection. The range of detectors used so far is wide. Since the first FIA
method was developed with the aid of a colorimetric detection and a gas
sensing electrode in 1974, fluorimetry, nephelometry, atomic adsorption,
voltametry and lately also chemiluminescence have been applied as detecting techniques. The method allows, besides simple automation of existing
assays, a design of entirely new approaches based on the chemical reactions
taking place on the interface between two liquids in motion.

The principle of a simple FIA method is best explained with an example of
a colorimetric method such as determination of chloride:

$$Hg(SCN)_2 + 2Cl^- \rightarrow HgCl_2 + 2SCN^-$$
$$2SCN^- + Fe^{3+} \rightarrow Fe(SCN)_2^+$$

where the increase of intensity of the red colour of the iron(III)
thiocyanate complex is measured at 480 nm. (Figure 1)

The flow chart of the FIA apparatus (Figure 1a) shows that a chloride
sample (volume 25 µl) is injected into a carrier stream of a mixed reagent
which is being pumped at a rate of 0.8 ml/min. During the passage through
the reactor coil, which has an internal diameter of 0.5 mm and a length of
50 cm, the injected sample zone is mixed with the components of the carrier
stream and the colour resulting from the above chemical reaction is
monitored in a flow-through cell (volume 18 µl) and recorded. A series of
chloride samples, covering the range 5 to 75 ppm Cl, has thus been analysed
(Figure 1b) by injecting each sample four times (left). Further, the 30 ppm
as well as the 75 ppm Cl samples were injected and the response curves

were recorded at a higher paper speed (right). This simple experiment, which is included in the compulsory part of the undergraduate training programme in instrumental analysis at our Department [6], clearly confirms the high reproducibility of FIA determination, its unusual speed and simplicity of instrument design.

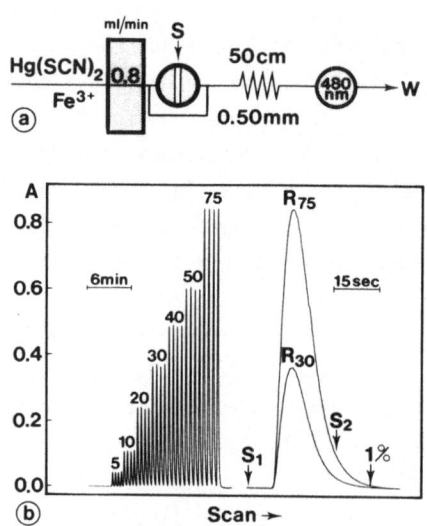

Figure 1 (a): Flow diagram for the spectrophotometric determination of chloride. S -point of sample injection, W - waste.

(b): Record output showing chloride analysis in the system depicted in (a). (Reproduced with the permission of Wiley from ref.5)

Surprizingly, flow injection analysis has not been confined to automation of such simple tasks as the above described colorimetry. The schematic flow charts of FIA methods designed so far (Figure 2) illustrate the versatility of this new approach and it is useful at this stage to try to make some future projections. Before attempting this, however, the concept of sample zone dispersion, which is the key to understanding FIA has to be discussed briefly.

Flow injection analysis is in contrast to all other techniques of continuous flow analysis, based on a combination of sample injection, controlled dispersion and exact timing, which allows to obtain a reproducible readout in spite of the fact that the mixing between sample and carrier solution is incomplete, the chemical reactions do not necessarily reach equilibrium and the recorded signal is transient.

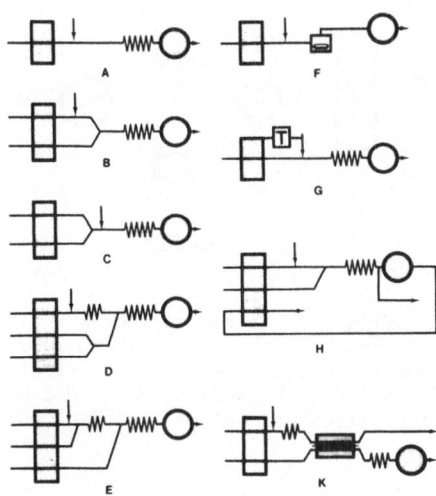

Figure 2: Various types of FIA flow designs employing one (A, B) or several reagents (C to E), stoppedflow (G), solvent extraction (H), gas diffusion or dialysis (K). (Reproduced with the permission of Wiley from ref. 5)

In other words, all events which yield the analytical read-out do not need to reach a steady state, but must be strictly repeatable. In order to achieve this the movement of all liquids have to be reproducible and the geometry of the flow path between the injection port and the flow detector has to be properly designed, so that the dispersion of the sample zone exactly suits the chemical assay to be performed. In the simplest case of a single line FIA system, a choice can be made between four different geometries of a flow path (Figure 3 A to D), each of the reactors yielding different dispersion and residence time. Because the analytical read-out in the majority of FIA methods is derived from peak height (Figure 1b, Figure 4), the dispersion D has been defined as a ratio between the original concentration of the analyte (i.e. prior to injection) $C^\bullet$ and the analyte concentration at the maximum of the peak C^{max}. (This means that for D=2 the sample solution has been diluted 1:1 by the reagent carrier stream). The residence time, available for chemical reaction to take place, is then the time span between sample injection (S) and appearance of peak mamimum (shown as the full point in Figure 4).

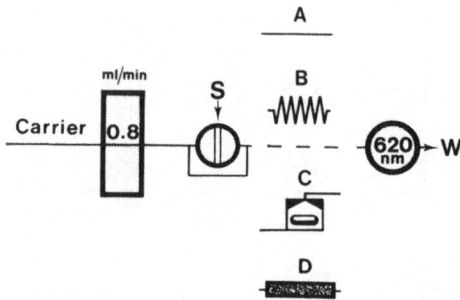

Figure 3: Choice of the flow-through reactor for a FIA system. A - 15 cm
of 0.5 mm I.D. tube, B - 60 cm of 0.5 mm I.D. tube, C - mixing
chamber volume 1.0 ml, D - pearl string reactor 2 m long,
0.86 mm I.D. filled with glass beads of 0.6 mm diameter. Sample
volume 25 microlitres, S - point of injection, W - waste

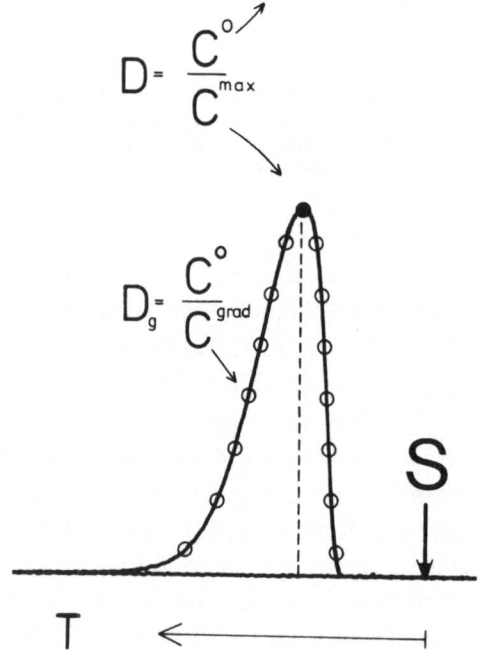

Figure 4: A typical FIA curve. S is the start of recording at the moment
of injection, T is the time

These parameters, together with a full shape of the response curve may be, for any FIA system, investigated by injecting a dye into a colourless carrier stream and by recording via a colorimetric flow through cell the dispersion curve.

In a short narrow tube (reactor A), the dispersion of the sample zone will be limited (D = 2) and the sample maximum will appear within seconds after sample maximum will appear within seconds after sample injection (Figure 5A).

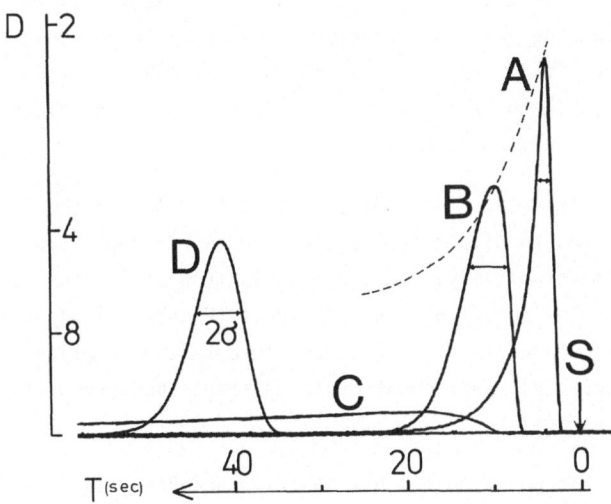

Figure 5: The dispersion curves obtained by injecting 25 microlitres of dye into a colourless carrier stream which was then lead through reactor types A to D (cf. Figure 3 and text). The scale on the left shows dispersion D. Note that the open tube reactor (A, B) would yield for long T values much larger dispersion (broken line) than the pearl string reactor (D)

Consequently, such system cannot accommodate assays based on chemical reactions, but it is well suited as a transport system for assayed material into the detector with the aim of obtaining a reproducible measurement, as the sample material is not mixed with the carrier stream (D can be further decreased by increasing the injected volume). Measurement of pH by a glass electrode, of pCa by a calcium ion-selective electrode or conductivity measurements are typical applications of a FIA design using limited dispersion [4].

In a longer narrow tube (reactor B), the dispersion of the sample zone has a medium value (2 to 10) and the residence time may be adjusted up 10 seconds without undue broadening of the sample zone (Figure 5B). In such conditions a great majority of FIA assays has been performed so far [1,2], either in a single line system (cf. Figure 1) or in more complex flow designs (Figure 2). The element of fluid, corresponding to peak maximum has, for the curve B, a dispersion 3,6 which gives sufficient excess of reagent to react with the sample materia. The residence time is still short (approx. 10 seconds) and the sample material entirely clears the system within 20 seconds. This allows three samples to be processed each minute, while the analytical readout is available within 10 seconds after sample injection. This amazing speed of response and short working cycle is unique for FIA methods and it is most surprizing that over sixty different species were determined by FIA so far [5], though the reaction time has been restricted to a fraction of a minute.

The mixing chamber (reactor C) yields a large dispersion of the sample zone and unless one wishes to dilute the sample material strongly and prolong the sample cycle substantially, the use of this type of reactor is not recommended, because the resulting response curve (Figure 5C) occupies an undue length of time and its maximum is so close to the baseline that one would tend to rescue the measurement by integrating the curve area rather than to measure peak height.

It is natural at this point to discuss shortly the often neglected, yet essential step of any analytical assay, the mixing of sample and reagent solutions. In manual operations the fast and homogeneous mixing of reacting components is an undoubtful necessity, but in automated operations where the flow of liquids can be precisely controlled the homogeneous, and complete mixing of entire volumes of sample and reagent solutions is not a prerequisite for performing a reproducible assay. This is exactly how FIA differs from all previous methods of continuous analysis, which were designed so that sample and reagent solutions were mixed by means of pulsers[7] or mixing chambers [8,9] or where the individual segments of liquid, separated by air bubbles were treated so that a complete mixing was achieved within each segment [10,11]. By abandoning the desire of mixing homogeneously and by introducing the concept of controlled dispersion, the sampling frequency and sensitivity of measurement are dramatically improved, as clearly demonstrated by response curves A, B and D compared with curve C, obtained with a mixing chamber (Figure 5).

It is quite surprizing that many assays could have been accommodated in the FIA system so far in spite of the fact that the time available for chemical reactions involved has been as short as 10 to 15 seconds (cf. Figure 5 curve B). Further prolongation of the residence time beyond this short period, was not possible so far, because longer tube lengths lead to greater dispersion and thus to a decrease of sampling frequency. An alternative, the stopped flow FIA combined with a parallel storage of sample zones [5], is a feasible, yet more demanding approach requiring a more complex instrument design.

3. Future developments

The use of packed reactor (D) is the most recent development in the area of controlled dispersion for a FIA system. Its most successful version, the pearl string reactor, recently designed in the Netherlands [12], allows three to four samples to be stacked following each other in a narrow tube (0.9 mm I.D.) filled with glass beads of slightly smaller diameter (0.6 mm). The linear dispersion in such reactor which has a tortuous path with a very low flow resistance, is surprizingly low (σ = 2.5 sec. for curve D on Figure 5), which allows the individual samples to be injected with high frequency (i.e. 6 x 2.5 sec. = 18 seconds, cf. ref. 5) so that three to four samples may be processed simultaneously at a time, without any mutual intermixing. This ingenious design will undoubtedly find application whenever slower chemical reactions are to be accommodated in a FIA system, or when a higher sensitivity of determination is desired.

So far, only the FIA methods where peak height is the basis of the analytical readout have been considered. It would be, however, a great pity to disregard all possibilities which offer themselves when the vast number of all points constituting the FIA curve is considered (Figure 4).

The gradient FIA techniques utilize the concentration gradient formed at the raising (front) and falling (tailing) edge of the sample zone, as identified by the peak shape. The average composition of each of these (imaginary) elementary elements of fluid can be formally described by the gradient dispersion D_g, which is derived in a similar manner as the dispersion D (Figure 4). Thus the ratio of the original analyte concentration to the analyte concentration in an element of fluid on the gradient is:

$$D_g = \frac{C^\bullet}{C_{grad}} \quad \text{i.e.} \quad \frac{D}{D_g} = \frac{C^{grad}}{C^{max}}$$

thus offering a choice of continuous dilutions of sample solution from the most concentrated one at peak maximum (where $D_g = D$) to a small fraction of $C^\bullet$ at the foot of the peak.

The simplest use of gradient profiles is in the field of kinetic measurements, performed by means of the stopped-flow FIA [4,5]. Considering, say a spectrophotometric assay, one may choose to stop a certain section of the dispersed sample zone in the flow cell and then measure the increase of absorbance while the chemical reaction take place. By increasing the time interval between the moment of injection and the stop of the pump (T_1 T_2 T_3, Figure 6), the ratio between the concentrations of the sample and reagent solutions is changed, due to various D_g values, the excess of reagent being lowest at the peak maximum.

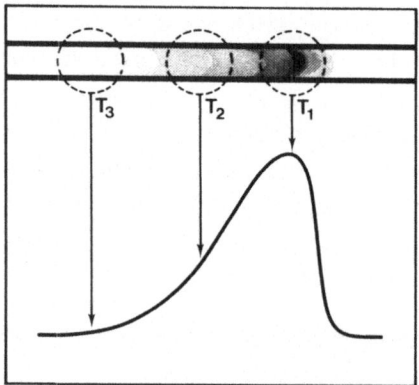

Figure 6: Dispersed sample zone and corresponding recorder output. By stopping sections of the sample zone with increasing delay time (T_1 to T_3) different sample/reagent ratios can be monitored. (Reproduced with the permission of Wiley from ref. 5)

Therefore, the recorded curves (Figure 7 a to f) will have a decreasing slope, as with an increase of T the analyte concentration is decreased, but as at the same time the reagent excess increases, also the linearity increases as seen on curves a to e. This allows the measuring range to be chosen by a simple adjustment of the electronic timer which controls the pump stop delay, and such approach is certainly more practical than

the usual manual serial dilution of sample and reagent solutions. It is
expected that the method will find applications in enzymatic analysis of
substrates and vice versa in the assay of catalytic activity of enzymes.

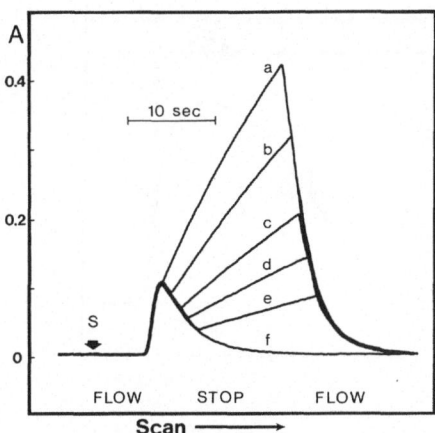

Figure 7: Stopped-flow analysis of glucose by means of glucose
dehydrogenase-NADH reaction showing the influence of
increase of delay time on the slope of reaction rate
curve. Stop period of 15 sec. started after injection
of 25 microlitre sample within (a) 7.8 s, (b) 8.8 s,
(c) 9.8 s, (d) 10.0 s and (e) 11.8 s, while (f) is
continuous pumping. (Reproduced with the permission
of Wiley from ref. 5)

FIA titrations are also based on the use of the concentration gradient
formed at the interface between the carrier and sample solution. Carrier
solution is the titrant, while the injected solution is the analyte to be
titrated. The concentrations of the titrant and analyte are so chosen that
in contradistinction to direct FIA measurements the sample solution is
prior to injection (C°) stoichiometrically more concentrated than the
carrier stream. Thus, following the injection, the carrier stream reacts
with the dispersed, and thus less concentrated, layers of the sample
solution formed at the leading and tailing edges of the zone during injection and on the way towards the detector. Thus, instead of measuring
peak height, peak width is being measured, as it is proportional to the
amount of analyte which reacted with the carrier stream in the dispersed
layers in the interfaces between the sample zone and the carrier stream.

All what is needed is to indicate the time span Δt between the two equivalence points located in the two imaginary elements of fluid of identical D_g
values (Figure 8).

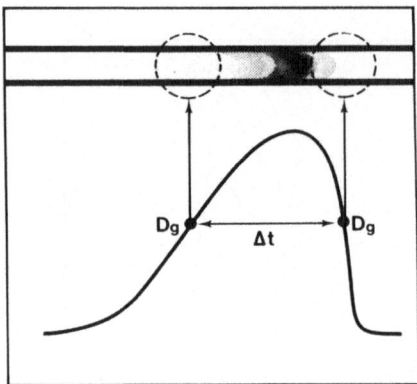

Figure 8: In each dispersed sample zone two elements of fluid can be
located, which formally have identical D_g values. This
phenomenon can be utilized in FIA titrations (Reproduced
with the permission of Wiley from ref. 5)

As indicator one may use an ion-selective electrode [13] or other potentiometric detection, or an optical detector if a colour indicator is being
used. So far acido-basic, complexometric [4,5,13] and redox titrations [14]
have been successfully tried. In all cases the time interval Δt is the
function of the concentration of the injected analyte

$$t = (f) \, C_{analyte}$$

and the type of function depends on the concentration profile of the dispersed sample zone (Figure 5 A to D). In the case of a purely exponential
profile, which can be closely approximated by the one tank model [4,13] it
can be shown that

$$\Delta t = V/v \ln 10 \, \dot{} \log C_{anal} - V/v \ln 10 \log C_{titr}$$

where V is the volume of the mixing stage, v is the pumping rate and C refer
to the concentrations of the analyte and titrant. The most recent development
in this area is the replacement of a mixing chamber, with its relatively
large holdup volume((1 ml) curve C, Figure 5) by a reactor with one, or many

mixing stages. Such an imaginary mixing stage may have formally a much smaller V thus allowing the TIA titrations to be performed at so far unpreceded speed. As the sample zone is no longer stretched in time, a rapid readout device, such as a microcomputer is used to obtain sufficient reproducibility of recording Δt which may be as short as five to ten seconds, thus allowing the high speed FIA titrations to be performed at a rate of two to four determinations per minute [14].

It follows from the foregoing that the concept of dispersion and its understanding is a key to the design of all types of FIA techniques. It is easy to construct a simple FIA system which would yield reproducible results. It is another matter to optimize the performance of the system and this is especially true for a gradient FIA technique. The theoretical models describing dispersion in a FIA system [5] serve as a useful guide. In the past much attention was paid to the diffusion model, based on the Taylor-Aris theory, while the tank-in-series model was somewhat left aside. This attitude is however to change in future, as both pearl string reactors and high speed FIA titrations are best explained on the basis of the mixing stage approach.

The main future challenge lies undoubtedly in the adaptation of chemical assays based on colorimetry, potentiometry, chemiluminescence, fluorescence and calorimetry to FIA, either as direct measurements or as titrations. In contrast to manual methods, the FIA procedure does not require reaching of chemical equilibrium or a long term stability of the product to be measured. Reaction conditions and temperature and time can be strictly controlled in a FIA system where microlitre volumes of sample and reagent solutions are protected from ambient atmosphere while they move, or stand still in a tube of a narrow bore. With a little imagination one may see the traditional tool of microchemists - the capillary tube - gaining finally the central position in the analytical laboratory which is still dominated by batch operations carried in test tubes. The simple reason for such development is replacement of tedius and often clumsy manual handling by strictly controlled mechanized movement of liquids.

Acknowledgement. The author wishes to express his appreciation to Elo H. Hansen and to Anders Ramsing for their cooperation, criticism and valuable discussions.

References

1. F. Pregl, "Quantitative Organische Mikroanalyse", 5. Aufl., Springer-Verlag, Wien 1947.

2. F. Emich, "Mikrochemisches Praktikum", Von J.F. Bergmann Verlag, München 1931.

3. A.A. Benedetti-Pichler, "Introduction to the Microtechnique of Inorganic Analysis", J. Wiley, New York 1942.

4. J. Ruzicka and E.H. Hansen, Anal. Chim. Acta, 114 (1980) 19.

5. J. Ruzicka and E.H. Hansen, "Flow Injection Analysis", J. Wiley, New York 1981.

6. E.H. Hansen and J. Ruzicka, J. Chem. Educ. 56 (1979) 677.

7. W.J. Blaedel and G.P. Hicks, Anal. Chem. 34 (1962) 338.

8. G. Nagy, Zs. Feher and E. Pungor, Anal. Chim. Acta 52 (1970) 47.

9. V.V.S. Eswara-Dutt and H.M. Mottola, Anal. Chem. 47 (1975) 357.

10. L. Skeggs, Am. J. Clin. Pathol., 28 (1957) 311.

11. L.R. Snyder and H.J. Adler, Anal. Chem., 48 (1976) 1022.

12. J.M. Reijn, W.E. van der Linden and H. Poppe, SAC 80 Meeting, Paper B 25, Lancaster 1980.

13. J. Ruzicka, E.H. Hansen and H. Mosbaek, Anal. Chim. Acta, 92 (1977) 235.

14. A.U. Ramsing, J. Ruzicka and E.H. Hansen, Anal. Chim. Acta (in press).

DETERMINATION OF MICRO-COMPONENTS IN MICRO-VOLUMES USING ELECTROANALYTICAL FLOW SYSTEMS

K. TÓTH, ZS. FEHÉR, G. NAGY and E. PUNGOR
Institute for General and Analytical Chemistry,
Technical University,
Gellért tér 4, H - 1502 Budapest, Hungary

Abstract. A brief survey is given on the trends of development and possible applications of electroanalytical sensors and techniques under flow-through conditions. Attempts are made to demonstrate the advantages of miniaturization of sensors such as ion-selective electrodes and voltammetric detectors, as well as to discuss special techniques ensuring an increase in sensitivity and decrease in sample volume. Authors discuss in detail the realization, the advantages and the limits of flow-through titration techniques for micro-analytical purposes.

1. Introduction

The steadily increasing demands for the determination of micro-components in micro-volumes have directed the attention of researchers towards the need for the miniaturization in electroanalytical chemistry also which resulted in new electroanalytical micro-techniques on one hand, and micro-sensors and measuring cells on the other.

This involves for example "in vivo" measurements, intracellular process monitoring, clinical analysis and metal surface analysis.

In respect of measuring small concentrations some stationary-solution electroanalytical techniques, i.e. differential pulse polarography, stripping methods as well as potentiometry with ion-selective electrodes are to be considered. However, in stationary electroanalysis surface reactions, such as adsorption, dissolution processes and in addition to this, at voltammetric electrodes the charging of the electrical double layer may set a limit for scaling down of macro-techniques.

Such problems have been discussed in detail among others by Hulanicki and coworkers [1] as well as Pungor and coworkers [2] in connection with the application of ion-selective electrode for the determination of ion-concentrations in small volumes, as well as by Adams [3] and Bishop [4] for voltammetric electrodes.

The electroanalytical flow-through methods are considered of special importance in the analysis of micro-constituents. This can be attributed to the following main factors:

a) more sensitive determination can be attained
   - with ion-selective electrodes, as the sample being in contact with the electrode surface is moving continuously, the electrode-sample interaction does not allow appreciable change neither in the sample solution nor at the electrode surface;
   - with voltammetric sensors, under hydrodynamic conditions, the signal is increased as an effect of convective mass transport. Furtheron, in constant voltage voltammetry at solid electrodes, which is most frequently used in flowing systems, the residual current is reduced, moreover, it can readily be compensated electrically or by differential electrochemical technique, which further improves the signal to noise ratio;

b) the dynamic characteristics of electroanalytical sensors are improved under flowing conditions, as a relatively thin and well-defined diffusion layer is established at the electrode-solution interface.

Moreover, it was found that the application of continuous flow systems does not necessarily increase the sample size. Examples taken from the field of segmented [5] and unsegmented [6-9] continuous flow analysis show that even a few µl sample volumes can be sufficient for analysis with flow-through techniques in general, and with electroanalytical flow-through techniques also.

In the following,as an illustration of the trends of development in micro-electroanalysis some attractive examples taken from recent literature will be shown, without aiming at completeness.

Remarkable results were obtained in designing neutral carrier type liquid ion-selective membranes optimized for monitoring ionic constituents of biological fluids, such as sodium, potassium, calcium and chloride in blood serum and whole blood by Simon and his group [10]. The multi-ionic flow-through micro-cell incorporating the appropriate micro-electrodes, and measuring device, for example, enables highly selective and accurate measurement of blood constituents to be carried out. Besides, the micro-electrodes (tip size 1 μm) and the combined ion-selective reference micro-electrodes have been suggested to be used for other extracellular and intracellular measurements also e.g. in brain cell environment, in neurones, in rat cerebellum etc. [11].

Similarly, Brown, Pemberton and Owen [12] have developed PVC matrix organo-phosphoric acid based micro-electrodes for use in intracellular measurements with a tip-size of 1 - 2,5 μm.

As another example,the flow-through tubular type micro-ion-selective electrodes are to be mentioned. Such cells were worked out and employed by Blaedel and Dinwiddie [13] for monitoring copper ion concentrations, furtheron by Van der Linden and Oostervink [14] for flow-injection analysis of copper (II) ions in the millimolar to micromolar concentration range.

Adams [15] has directed the attention towards a new field of application of voltammetric micro-electrodes, that is "in vivo" monitoring of processes in rat brain cells. For studying rat neurons of a total volume of 30 μl, miniaturized graphite paste and calomel electrodes (diameter between about 0,5 mm and 0.08 mm) have been employed for taking voltammetric re-cordings in the range of -0,2 to + 1,2 V vs. SCE. By chronoamperometric measurements the metobolism of organic compounds such as catecholamines can be followed.

Pungor and co-workers [16] have designed a voltammetric micro-cell for studies on the drug concentration level and flow-rate of blood in living organisms. Several types of voltammetric cells have been designed for flow injection analysis. Thus, Pungor and co-workers [6,7] have used mainly silicone rubber based graphite electrodes,while Strohl and Curran [17] have employed RVC detector in amperometric and coulometric flow through cells for flow injection analysis.

Pihlar, Kosta and Hristovski [18] have developed a cylindrical microampero-
metric flow through electrode for the determination of absolute amounts of
less than one nanogram of cyanide in volumes as small as 10 µl with flow
injection analysis. The amperometric flow through electrode consisted of
a 0.9 mm diameter silver wire, the effective length was 6.0 cm and its area
1.7 cm^2;   the dead volume 130 µl.

Another important application of the flow-through electroanalytical cells
is high pressure liquid chromatography.

The combination of flow through electroanalytical methods with HPL
chromatography enhances the performance of electroanalytical techniques
with respect to chemical resolution. In this area first of all flow-through
amperometric and coulometric cells, and somewhat less frequently conducti-
metric and different potentiometric cells have been designed. With voltammetric detectors, e.g. 10 pmol of total analyte is sufficient to provide
analytical information.

The low-volume, high stability and sensitivity flow-through electroanalytical
cells developed for chromatographic detection, however, can advantageously
be used in other hydrodynamic techniques,too.

2. Flow-trough titration techniques

The potentialities of flow through direct electroanalytical techniques in
micro-analysis in respect of both sample size and concentration have been
discussed in earlier review papers [5-9, 19].

In practice, however, the titration technique is considered to have certain
advantages over direct methods, namely that:

- it is more reliable, which is due to the fact that changes in the
  detector characteristics have a smaller effect on the concentration
  determination,as the evaluation is based on chemical equivalence;

- the evaluation does not require calibration against high purity standards;

- the matrix or medium effect is less than in the case of direct methods.

As known,conventional batch-type titration techniques using electroanalytical
sensors as end-point detectors have had a widespread application in microanalysis, too. In this respect, pioneering work was done by Alimarin and
co-workers [20] in developing a batch-type coulometric titration technique

for metal ion determinations on the ultramicro scale, in 3-5 µl volumes. As
an end-point detection biamperometry, and as reagents different complexing
agents generated by constant current electrolysis were successfully employed.
This excellent ultramicro titration technique, however, requires skill and
practice.

A considerable amount of effort has been made to develop flow-through
titration techniques and as a result titrations to a fix point - which
may be the equivalence point - as well as complete titrations have
been realized in streaming solutions.

In the former case the mass flow of the reagent is controlled so as to
bring the titration to a pre-determined degree of titration - mostly up
to the chemical equivalence point. The output parameter of the system is
the reagent mass flow [21, 22].

Griepink and co-workers [23], employing chemical engineering principles, have
optimized the control of the reagent addition. Accordingly, the rate of
analysis has remarkably increased.

In the latter case, however, the mass flow of the reagent is programmed, thus
allowing a strict time-dependent reagent addition. As known, different
types of programs have successfully been applied for reagent addition [24-31],
i.e. linear, triangle and exponential concentration profiles, enabling
complete titration curves to be recorded.

In recent work the possible application of the triangle programmed titration
technique developed in our laboratory [26-30] have been investigated to solve
micro-analytical problems.

The fundamentals of the technique are:

- the sample solution is continuously streaming in a so-called analysis
  channel at a constant rate;

- the reagent is added according to an appropriate isosceles triangle
  program resulting in a certain degree of over-titration at a given point
  of the analysis channel. This means that the mass flow of the reagent
  increases for a time period (τ) from zero up to maximum value, after
  which it decreases to zero. Thus, during one program the mass flows of
  the reagent and the sample reach stoichiometric equivalence twice;

- with an appropriate detector system placed at a given point of the analysis
  channel/right at the confluence point or at a given distance/ the con-

centration change of the sample, reagent or both is followed resulting
in two complete titration curves being the mirror image of each other.

The mass flow of reagent equivalent to that of the sample, V_{Re}, and from
this the sample concentration is determined on the basis of the two
equivalence points, showing up clearly on the recorded titration curve.
The sample concentration can be determined either from the time elapsed between the appearance of the two equivalence points, Q, using the following
equation:

$$c = (\tau - \frac{Q}{2}) \frac{b}{a} \cdot \frac{n}{v}$$

where τ is the half time of a whole reagent addition program;
a and b are the stoichiometric constants of the titration reaction;
n is the slope of the reagent addition program; V_R = nt;
v is the flow-rate of the sample;
or from a Q vs. c calibration curve.

This possibility of evaluation based on two equivalence points can be considered as a special advantage of the triangle programmed titration technique [27].

Programmed reagent addition can most conveniently be carried out by current-
-programmed electrolysis, by coulometric reagent generation (Figure 1). This
principle has been employed throughout our work.

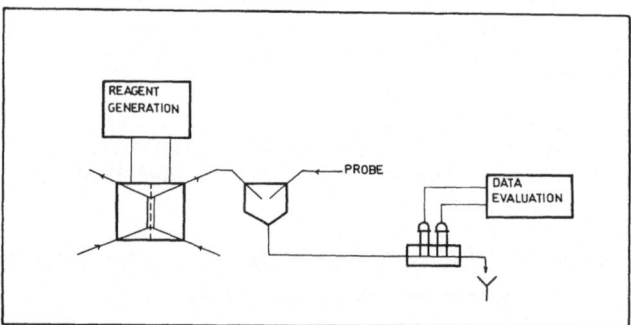

Figure 1: The schematic diagram of the experimental set-up used for
triangle programmed titration

In practical applications three problems had been faced:

a) Programmed reagent generation under flow-through conditions with 100% current efficiency;

b) Continuous detection under hydrodynamic conditions;

c) Data acquisition and processing;

In our work the generation of silver, mercury, hydrogen, hydroxide ions as well as bromine and iodine has been solved with 100% current efficiency for determining various anions, cations, acids, bases and organic compounds [32].

Flow-through cells have been developed for potentiometric, amperometric and photometric detection [7].

Data processing involves the determination of the location of the titration end points, which can be done manually or using an on-line desk-top computer [30,33] This is fairly simple in the case of amperometric and in some instants with photometric detection, since the end point is the intersection point of two straight lines. However, the location of the equivalence points of potentio-metric and some photometric titration curves is more difficult and needs some previous transformation.

In optimizing the triangle programmed titration technique for micro-analytical studies the following experimental variables must be considered.

The three experimental variables are:

$$i_{max} \quad , \quad 2\tau \quad \text{and} \quad v.$$

In selecting i_{max} it should be considered that a certain but not too high degree of overtitration is desirable. According to our experience, in the case of a given program i_{max} is selected to attain a 30% overtitration of the most concentrated sample solution, which obviously means a higher degree of overtitration for more dilute samples. Moreover, one titration program enables titrations to be carried out with appropriate accuracy within a concentration range of one order of magnitude.

The effect of 2τ is twofold:
The larger the value of 2τ, the steeper is the Q vs.c calibration graph, the longer is the duration of a titration and the larger is the sample volume required for one determination.

On the other hand, too short 2τ does not result in well defined titration curves, i.e. the shape of the curve is distorted (Figure 2).

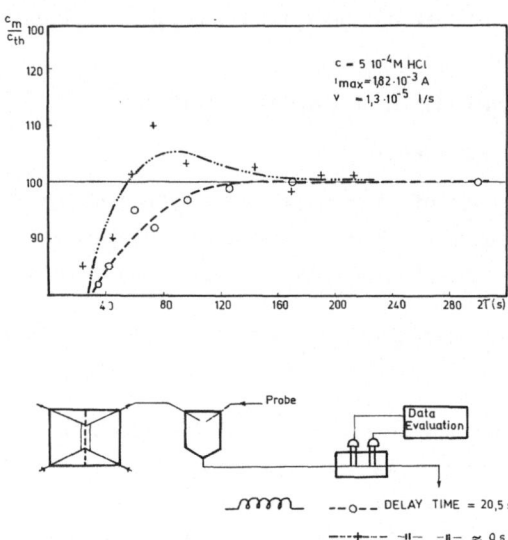

Figure 2: The effect of 2τ on the analytical results obtained at the
titration of HCl with electrically generated OH⁻ ions employing
a flow-through glass electrode as a detector

The distortion of the shape of the titration curves is determined by the
value of 2τ as well as by the hydrodynamic parameters, mainly the flow
rate of the sample solution.

With instantaneous reactions, v only affects the mass flow of the analyte,
i.e. the higher the flow rate, the steeper is the Q vs.c calibration graph.
Accordingly, higher flow rates are favoured. On the other hand, assuming
a constant titration time, 2τ , an increase in the flow rate means larger
sample volumes. As already mentioned, the flow rate affects also the shape
of the titration curve. At extremely low flow rates, the titration curves
are distorted, which is especially marked at short titration times (2τ)
(Figure 3).

With slow titration reactions, however, changes in the flow rate have a
pronounced effect on the reaction time. In this case, the otherwise
favourable increase in flow rate, under similar conditions allows shorter
reaction times which is reflected by a reduction in the slope of the
ϱ vs.c calibration graph.

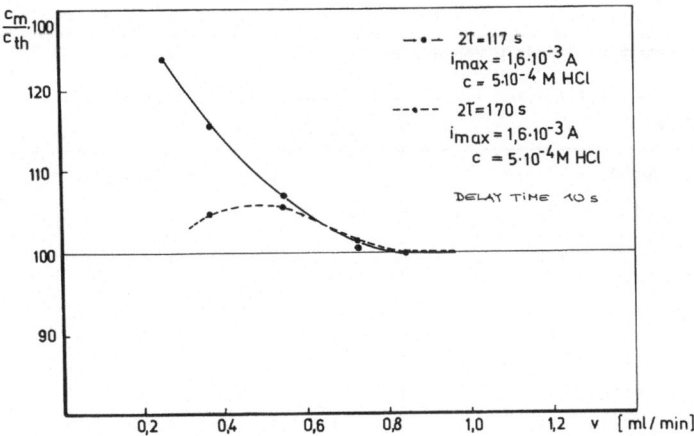

Figure 3: The effect of flow-rate on the analytical results obtained at
the titration of HCl with electrically generated OH⁻ ions
employing a flow-through glass electrode as a detector

The triangle programmed titration technique can be used advantageously in
different fields of analytical chemistry such as pharmaceutical analysis,
water analysis, environmental analysis etc. Table 1 gives a survey on the
types of compounds determined with the technique. Among the possible
applications two fields i.e. acid-base and bromimetric titrations will be
mentioned to show the trends of our latest research in this area.

Flow-through potentiometric [35] and photometric detectors [34] have been
designed for the accurate determination of various strong and weak acids
and bases with the programmed titration technique. Thus, for example
hydrochloric acid and other strong and weak acids or bases could easily be
titrated in 10^{-4} M concentration range employing a flow-through glass
electrode based potentiometric or photometric detection (Figure 4).

Efforts have been made to improve and simplify our data processing method
by software and hardware transformation using for example differentiating
measuring systems [33,34] and computer-aided curve smoothing and transformation (Figure 5) [35]. This holds promises to improve the analytical
results in respect of both sensitivity and accuracy.

Table 1: A survey on the types of compounds determined by triangle programmed titration technique

DETERMINATION	DETECTION
Hydrochloric acid	Potentiometric, Photometric
Perchloric acid	" "
Sulphuric acid	" "
Phosphoric acid	" "
Boric acid	" "
Acetyl salicylic acid	" "
Ammonium hydroxide	" "
Sodium hydroxyde	" "
Halides (Cl^-, Br^-, I^-)	"
Cyanide	"
Thiocyanate	"
Prostaglandin compounds	Voltametric
Chlorpromazine	"
Promethazine	"
Guanine	"
Amidazophen	"
Ascorbic acid	"
Phenol	"
p-Cresol	"
Arsenous acid	"
Sulphide	"
Cyanide	"

The flow-through triangle programmed titration technique has been found
especially suitable for bromimetric determination of various compounds. This
can be attributed to the following factors:

- the time of the reaction between the sample and reagent in flow-through
  circumstances is rigorously controlled by the flow rate and the geometry
  of the system;

- the product of the titration leaves the system, thus no secondary reaction
  may take place;

- the effect of the reaction product on the detector system which may results
  in ill-defined end-point is negligible since the time of contact of sample
  with the sensor is small and uniform.

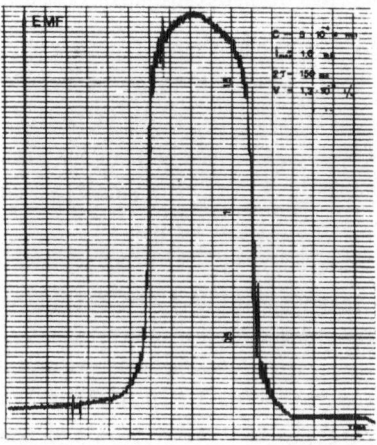

Figure 4: Triangle programmed titration graphs of HCl using electrically
generated OH as titrant with potentiometric detection employing
a flow-through glass electrode [34]

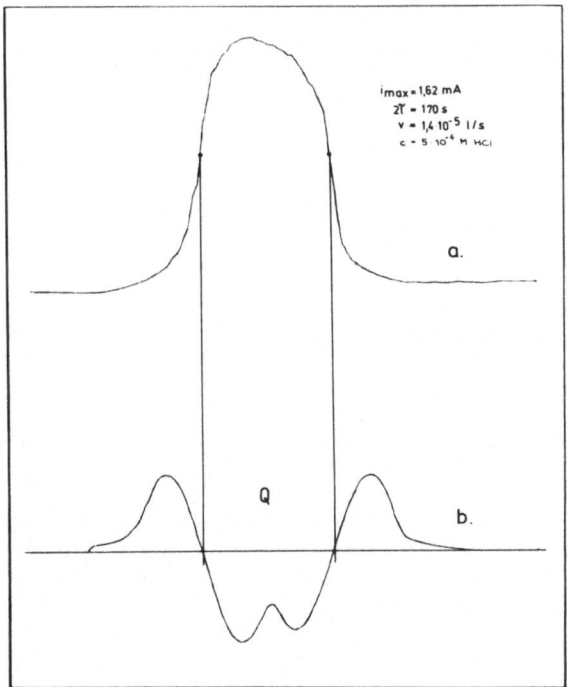

Figure 5: Potentiometric triangle programmed titration curves obtained after
computer aided curve smoothing (a) and second derivative formation (b)

The triangle programmed titration technique has been successfully employed for the bromimetric determination of prostaglandin compounds such as $PGF_{2\alpha}$, (7-{3α,5α -dihidroxy-2β-{(3S)-3-hydroxy-trans-1-octenyl}-1α-cyclopentyl}- -cis-5-heptenoic-acid). The $PGF_{2\alpha}$ molecule takes up bromine in a fast reaction [36]. Biamperometric detection was employed to follow the course of titration, using two platinum electrodes polarized with 400 mV. The titrimetric determination of $PGF_{2\alpha}$ in 10^{-5} M concentration range could easily be carried out with appropriate accuracy.

This flow-through titration technique allows the determination of $PGF_{2\alpha}$ in various pharmaceutical preparations, e.g. in tablets also.

Summing up it may be concluded that flow-through electroanalytical techniques have opened up new possibilities in the field of micro-analysis among which flow-through titration techniques stand out with their reliability and accuracy.

References

1. A. Hulanicki, A. Lewenstam and M. Maj-Zurawskai. Anal. Chim. Acta, 107 (1979) 121.

2. E. Gräf-Harsányi, L. Pólos and E. Pungor. Investigation of the membrane layer of copper (II) selective electrode, Paper presented at 8th International Microchemical Symposium, Graz, 1980.

3. R.N. Adams. Electrochemistry at Solid Electrodes, Marcel Dekker, New York, 1969.

4. E. Bishop. Report on the activation and deactivation of solid electrodes, with particular reference to platinum, Paper presented on IUPAC meeting Commission V-5, Electroanalytical Chemistry, 1975.

5. L.R. Snyder, J. Levine, R. Stoy and A. Conetta. Anal. Chem. 48 (1976) 942A.

6. G. Nagy, Zs. Fehér and E. Pungor. Anal. Chim. Acta, 52 (1970) 47.

7. E. Pungor, Zs. Fehér, G. Nagy, K. Tóth, G. Horvai and M. Gratzl. Anal. Chim. Acta, 109 (1979) 1.

8. J. Ruzicka and E.H. Hansen. Anal. Chim. Acta, 99 (1978) 37.

9. D. Betteridge. Anal. Chem., 50 (1978) 832A.

10. P.C. Meier, D. Ammann, H.F. Osswald and W. Simon. Med. Progr. Technol. 5 (1977) 1.

11. D. Ammann, P.C. Meier and W. Simon. Design and use of calcium-selective microelectrodes "Detection and measurement of free Ca^{2+} in cell" , C.C. Ashley and A.K. Campbell (Eds) Elsevier (North-Holland, 1979.)

12. H.M. Brown, J.P. Pemberton and J.D. Owen. Anal. Chim. Acta, 85 (1976) 261.

13. W.J. Blaedel and D.E. Dinwiddie. Anal. Chem., 47 (1975) 1070.

14. W.E. Van der Linden and R. Oostervink. Anal. Chim. Acta, 101 (1978) 419.

15. R.N. Adams. Anal. Chem. 48 (1976) 1126 A.

16. Zs. Fehér, G. Nagy, E. Pungor. Hungarian Scientific Instruments, 1973, 15.

17. A.N. Strohl and D.J. Curran. Anal. Chem., 51 (1979) 1045.

18. B. Pihlar, L. Kosta and B. Hristovski. Talanta, 26 (1979) 805.

19. W.J. Scott. On line analysis in clinical chemistry using ion-selective electrodes "Electroanalysis in hygiene, environmental

clinical and pharmaceutical chemistry", W.F. Smyth (Ed), Analytical Chemistry Symposia Series Vol. 2 Elsevier Sci. Publ. Comp. 1980.

20. I.P. Alimarin, M.N. Petrikova, T.A. Kokina. Mikrochimica Acta No 1-2 (1973) 1.

21. O. Aström. Anal. Chim. Acta, 105 (1979) 67.

22. G. Horvai, K. Tôth, E. Pungor. Anal. Chim. Acta, 107 (1979) 101.

23. G.W.S. van Osch and B.Z. Griepink. Anal. Chem., 284 (1977) 267.

24. D.L. Eichler. Technicon Symposium, 1969, Vol. 1, Mediad, New York, 1970, p. 51.

25. B. Fleet and A.Y.W. Ho. Anal. Chem. 46 (1974) 9

26. G. Nagy, K. Tôth and E. Pungor. Anal. Chem. 47 (1975) 1460.

27. G. Nagy, Zs. Fehêr, K. Tôth, and E. Pungor. Anal. Chim. Acta, 91 (1977) 87.

28. G. Nagy, Zs. Fehêr, K. Tôth, and E. Pungor. Anal. Chim. Acta, 91 (1977) 97.

29. G. Nagy, Zs. Fehêr, K. Tôth, and E. Pungor. Anal. Chim. Acta, 100 (1978) 18

30. G. Nagy, Z. Lengyel, Zs. Fehêr, K. Tôth and E. Pungor. Anal. Chim. Acta, 101 (1978) 261.

31. J. Ruzicka, E.H. Hansen and H. Mosbaek. Anal. Chim. Acta, 92 (1977) 219.

32. G. Nagy, Zs. Fehêr, K. Tôth and E. Pungor. Hungarian Scientific Instruments 46 (1979) 5.

33. G. Nagy, Zs. Fehêr, K. Tôth, E. Pungor and A. Ivaska. Talanta, 26 (1979) 11

34. Zs. Niegreisz. Diploma work. Technical University, Budapest, 1980.

35. G. Nagy, M. Gratzl, Zs. Fehêr, K. Tôth and E.Pungor, in preparation.

36. Zs. Fehêr, G. Nagy, K. Tôth, E. Pungor and A. Tôth. Analyst, 104 (1979) 56(

ENRICHMENT TECHNIQUES FOR INORGANIC MICRO- AND TRACE ANALYSIS

ATSUSHI MIZUIKE and MASATAKA HIRAIDE
Faculty of Engineering, Nagoya University, Chikusa-ku,
Nagoya 464, Japan

Abstract. Although modern instrumental techniques of determination are
highly sensitive and selective, enrichment techniques are generally required before the determination in micro- and trace analysis to extend
the lower limit of determination, to improve the precision and accuracy
of analytical results, and to widen the scope of the determination
techniques.
This paper reviews the present status of enrichment techniques, special
emphasis being given to multi element techniques and to techniques for
differentiating various chemical states of a trace element. Also,
microscale operations and mechanization and automation of enrichment
techniques are briefly discussed.

1. Introduction

Most modern instrumental techniques of determination are highly sensitive
and selective, but their direct applications are frequently difficult or
impossible in inorganic micro- and trace analyses of high-purity materials,
natural waters, biological and geological samples, etc., for example,
a) when the concentrations of the desired trace elements (the elements to
   be determined) in the sample are below the lower limits of detection,
b) when substances which interfere with the determination exist in the
   sample,
c) when the matrix is highly toxic or radioactive, or too expensive to be
   consumed,
d) when the desired trace elements are not homogeneously distributed in
   the sample,

e) when standard reference materials suitable for the calibration are not available, and

f) when the chemical and physical properties of the sample are not suitable for the direct determination.

Enrichment or preconcentration of the desired trace elements prior to the determination can overcome these difficulties. Thus use of proper enrichment techniques extends the lower limit of determination, improves the precision and accuracy of analytical results, and widens the scope of the determination techniques.

The aim of this paper is to review the present status of enrichment techniques in inorganic micro- and trace analysis. Several reviews have been published in this field [1-8]. In this paper, special emphasis is given to multielement techniques and techniques for differentiating various chemical states of a trace element. Also, microscale operations and mechanization and automation of enrichment techniques are briefly discussed.

2. General aspects of enrichment techniques

Enrichment of trace elements is effected by the use of various separation techniques based on physical, physico-chemical, and chemical principles. They are tabulated in Table 1.

Table 1. Enrichment techniques used in micro- and trace analysis

Sample state	Separated substances	Techniques
1. Solid	Particles *	Manual selection under the microscope.Sieving. Magnetic separation. Heavy liquid separation. Flotation.
	Constituents	Selective dissolution. Electrolytic dissolution. Sublimation. Extraction of gases in metals at high temperatures. Dry ashing of organic samples. Zone melting. Fire assay.
2. Solution	Particles *	Filtration. Centrifugation. Flotation.
	Solutes	Precipitation. Coprecipitation. Electrodeposition. Adsorption. Molecular sieving. Ion exchange. Liquid-liquid extraction.Volatilization. Flotation. Normal freezing. Zone freezing. Electrophoresis. Dialysis. Reverse osmosis. Ultrafiltration. Ultracentrifugation.
3. Gas	Particles *	Filtration. Impaction. Sedimentation. Centrifugation. Thermal precipitation. Electrostatic precipitation.
	Constituents	Absorption. Adsorption. Condensation. Permeation.
* Particle diameter > ca. 0,5 µm.		

Important considerations in selecting or evaluating enrichment techniques
are:
- recovery of the desired trace element,
- enrichment coefficient,
- contamination hazard, and
- operational simplicity.

2.1. Recovery of the desired trace element

Loss of the desired trace element may occur during the enrichment and related steps by adsorption of the trace element on the walls of containers,
evaporation of the trace element during decomposition of the sample, incomplete decomposition, imperfect separation, etc.   Generally speaking,
the smaller the quantity of the trace element, the more the danger of loss.
Thus, the recovery or yield of the trace element in the enrichment step is
often less than 100 %. The required recovery depends on the purpose of the
analysis as well as the determination technique used. Recoveries larger
than 95 % are acceptable in most trace analyses. Much lower and even
variable recoveries are permissible in isotope dilution analysis and
radiochemical separations using isotopic carriers. The recovery and loss
of trace elements are best investigated by the radioactive tracer technique.

2.2. Enrichment coefficient of the desired trace element

This coefficient is defined as the ratio of the recovery of the desired
trace element to the recovery of matrix. The enrichment coefficient
required depends on the concentration level of the desired trace element
in the sample and also on the determination technique used. Enrichment
coefficients greater than 100,000 are easily obtained in some enrichment techniques. However, enrichment coefficients of 100/10,000 are sufficient in
most trace analyses because of high sensitivity and selectivity of modern
instrumental determination techniques. Enrichment coefficients can be increased by using proper multistage separations without appreciable loss
of the desired trace element.

The word "enrichment" does not necessarily mean the increase in concentrations of the desired trace elements in the sample, when conversion of
the original matrix into another one, e.g., solid into aqueous solution,
occurs in the enrichment step.

2.3. Contamination hazard

During the enrichment and related steps, the sample may become contaminated by various contaminants originating from the laboratory atmosphere, the reagents and the apparatus used, and the analyst performing the analysis [9-11]. The accurate estimation of the degree of contamination for correcting analytical results is very difficult, because most kinds of contamination are complex and not reproducible. Enrichment is often rendered useless by contamination. It is essential, therefore, to reduce the contamination level to an acceptable value.

2.4. Operational simplicity

Most analysts think enrichment procedures take much skill, labor, and time. Therefore, enrichment techniques should be as simple and rapid as possible. Smooth connections with the preceding and following analytical steps such as decomposition and determination are also quite important. Some enrichment techniques are unified with another analytical step; e.g., separation of particles in gas and liquid samples (with sampling), ashing of organic samples and fire assay of ores (with decomposition), and carrier distillation in atomic emission spectrometry and stripping voltammetry (with determination).

3. Simultaneous multielement enrichment

To investigate the role, synergetic action, and correlation of trace elements in high-purity materials, environmental and biological samples, etc., analytical information is generally required on the extreme variety of trace elements that may be present in the sample. Simultaneous multi-element determination techniques, which include atomic emission spectrometry, mass spectrometry, X-ray emission and fluorescence spectrometry, and neutron activation-gamma ray spectrometry, are suitable for this purpose, because time, labor, and sample size required for the analysis are minimized. When direct determinations by these techniques are lacking in sensitivity, susceptible to interferences from coexisting substances, etc., simultaneous multielement enrichment techniques are effectively applied. Multielement enrichment techniques are also usefully combined with sequential determination of each trace element in a separate aliquot by rapid single-element (one-element-at-a-time) determination techniques such as atomic absorption spectrometry.

Useful techniques for simultaneous multielement enrichment follow.

3.1. Liquid-liquid extraction

This technique occupies an important position among multielement enrichment techniques because of its simplicity, rapidity, and wide applicability. Numerous extraction systems have been studied extensively from fundamental and practical standpoints. Either trace or matrix elements are selectively extracted with an organic solvent from an aqueous solution.

For the group extraction of trace elements in natural waters, high-purity materials, etc., organic chelating agents, 8-hydroxyquinoline, dithizone, and dithiocarbamates, are most widely used. Combined use of two or more organic chelating agents and successive extractions at several pH values increase the number of extractable trace elements.

On the contrary, matrices are removed by extraction into ethers or other organic solvents from hydrochloric, hydrobromic, or hydriodic acid solutions for the enrichment of a large number of trace impurities in iron, gallium, gold, indium, mercury, and bismuth. Application of metal chelates to the matrix extraction is limited because of their moderate solubility in the organic phase.

The applications of liquid-liquid extraction to multielement enrichment are summarized in tables 2 and 3 [1,3,8,12].

Table 2. Group extraction of trace elements

Extraction systems	Trace elements	Matrices
8-Hydroxyquinoline	Al, Bi, Cd, Co, Cu, Fe, Mn, Ni, Pb, Zn	P, V
Cupferron [13]	Fe, Mo, Sn, Ti, V	Water
N-Benzoyl-N-phenylhydro-xylamine	Hf, Nb, Ti, V, Zr	U
Dithizone	Ag, Cd, Co, Cu, Hg, Ni, Pb, Zn	Sea water and ashed biological samples [14], Si, U, W
DDTC*	Ag, Bi, Cd, Co, Cu, Fe, Mn, Ni, Pb, Zn	Water, Nb, Si, Ta, Ti, V, W, Alkaline earth metals
APDC**	Cd, Co, Cr, Cu, Fe, Mo, Ni, Pb, V, Zn	Water [15], Urine
Sodium tetramethylene-dithiocarbamate	Bi, Cd, Cu, Fe, Ga, Ni, Pb, Sb, V, Zn	Al

(Table 2, continued)

Extraction systems	Trace elements	Matrices
Hexamethyleneammonium hexamethylenedithio- carbamate [16]	Ag, Bi, Cd, Co, Cu, Ni, Pb, Tl, Zn	Water
Diethyldithiophosphoric acid	As, Bi, Cd, Pb, Sb	Zn
8-Hydroxyquinoline + dithizone	Ag, Al, Bi, Cd, Co, Cu, Fe, Ni, Pb, Zn	Se
8-Hydroxyquinoline + DDTC	Bi, Cd, Co, Cu, In, Mn, Ni, Pb, Tl, Zn	Na, K
Dithizone + APDC	Bi, Cd, Co, Cr, Cu, Fe, Mn, Ni, Pb, Zn	Al, Ti, Zr
APDC + diethylammonium diethyldithiocarbamate [17]	Cd, Co, Cu, Fe, Ni, Pb, Zn	Sea water
Fatty acids	Bi, Cu, Fe, Pb, Sn, Zn	Co, Ni
Chloride	Fe, Ga, Tl	In

* Sodium diethyldithiocarbamate.
** Ammonium pyrrolidinedithiocarbamate. More correctly ammonium tetramethylene-dithiocarbamate.

Table 3. Extraction of matrices

Extraction systems	Matrices	Trace elements
Chloride	As, Au, Fe, Ga, Mo, Sb, Tl	Al, Bi, Cd, Co, Cr, Cu, Mn, Ni, Pb, Zn
Bromide	As, Au, Fe, Ga, In, Sb, Tl	Al, Bi, Cd, Co, Cr, Cu, Mn, Ni, Pb, Zn
Iodide	Bi, Cd, Hg	Al, Co, Cu, Fe, Mg, Mn, Ni, Pb, Ti, Zn
Fluoride	Nb, Ta	Ag, Ba, Cd, Cr, Fe, Ga, In, Sr, Ti, W
Trioctylamine oxide	Re	Al, As, Ba, Be, Co, Cr, Mg, Mn, Ni, Pb
Quaternary ammonium salt	Mo, W	Al, Ba, Cd, Cu, Fe, Ga, In, Mg, Mn, Ni
TBP*	Bi, Nb, Th, U, V, Y	Al, Ca, Co, Cr, Cu, Mg, Mn, Ni, Pb, Zn
O-Isopropyl-N-methylthio-carbamate [18]	Ag	Bi, Cd, Co, Cr, Cu, Fe, Mn, Ni, Pb, Zn
Cupferron [13]	Nb, Ti, Zr	Al, Ca, Co, Cr, In, Mg, Mn, Ni, U, Zn

* Tri-n-butyl phosphate.

3.2. Selective dissolution

Either the matrix or a group of trace elements in a solid (or liquid metal) sample is selectively dissolved in an appropriate solvent. This technique is also called solid-liquid extraction when the sample is solid.

Jackwerth et al. [19-27] proposed partial dissolution of the matrix for enrichment of trace elements, electrochemically nobler than the matrix, in pure metals including aluminium, manganese, zinc, gallium, cadmium, lead, and mercury. The metal sample is first coated with a thin layer of mercury except for liquid metals, and then most of the matrix is dissolved in mineral acids such as hydrochloric and nitric acids. The trace elements remain almost quantitatively with a small amount of the matrix as residue. An example is the enrichment of traces of Ag, Au, Bi, Cd, Co, Cu, Fe, Ga, In, Ni, Pb, Pd, and Tl in high-purity manganese [27].

Solid-liquid extraction of trace elements in the sample is also useful [28-32]. A solid sample is first dissolved in mineral acids or water, and the resulting solution is evaporated to dryness to redistribute the trace elements to be extracted on the surfaces or in the interstitial space of the agglomerates of the pure matrix crystals (metal chlorides, basic nitrates, nitrates, sulfates, oxides, etc.). Then the trace elements, together with a small amount of the matrix, are dissolved in water, mineral acids, organic solvents, or their mixtures in an ultrasonic field. This technique has been applied to the enrichment of trace elements such as Co, Cu, Fe, and Zn in sodium, potassium, barium, nickel, cadmium, lead, bismuth, and tellurium matrices.

Trace recoveries of greater than 95 % and enrichment coefficients of 100/1,000 are generally achieved by both techniques mentioned above.

3.3. Precipitation

Precipitation of matrix elements, leaving trace elements in the original sample solution, can be successfully applied to multielement enrichment, provided that loss of the trace elements by coprecipitation and contamination due to a large amount of precipitant are minimized by choosing the proper precipitant and precipitating conditions. Slow evaporation of the solution and precipitation from homogeneous solution are effective in some cases to form large well-shaped crystals of the pure matrix. Recently, Jackwerth [33] compared three methods for trace enrichment from

high-purity lead by precipitation of the matrix as either the nitrate,
chloride, or sulfate salts. In general, the most suitable method was the
precipitation of lead nitrate in concentrated nitric acid from the view-
points of trace recoveries, simplicity, and rapidity. Trace recoveries of
more than 95 % were obtained with enrichment coefficients of about 1,000
Table 4 shows some examples of the matrix precipitation for trace enrichment [8,12].

Table 4. Precipitation of matrices

Precipitate form	Matrices	Trace elements
Chloride	Ag, Na*, Pb [33]	Al, Au, Cd, Co, Cu, Fe, Ga, In, Mn, Ni
Iodide	Bi, Tl	Al, Co, Cu, Fe, In, Mn, Ni, Pb, Sn, Zn
Nitrate	Pb [33]	Ag, Al Bi, Cd, Co, Cu, Fe, Mn, Ni, Pd
Sulfate	Pb [33]	Al, Cd, Co, Cu, Ga, In, Mn, Ni, Pd, Zn
Hydrous metal oxide	Te	Ag, Au, Cd, Co, Cu, Fe, In, Ni, Pb, Zn
Oxybromide	Bi	Al, Cd, Co, Cu, Fe, In, Mg, Mn, Ni, Zn
Metal **	Hg, Te	As, Cd, Cu, Fe, Ga, In, Mn, Ni, Sb, Tl
* with gaseous hydrogen chloride.		
** by reduction with formic acid or SO_2 aq.		

Precipitation of a group of trace elements, leaving matrix elements in the
original sample solution, is generally carried out after addition of
milligram quantities of a carrier element. The precipitant reacts with
the carrier element to form gathering precipitates, which quantitatively
coprecipitate the desired trace elements by such mechanisms as mixed-crystal
formation, adsorption, and occlusion. Two or more precipitants can be
simultaneously used to collect the maximum number of trace elements. Masking
of the precipitation of the matrix is sometimes necessary.

Addition of a carrier element is not always required. A water-insoluble
organic precipitant in a water-miscible organic solvent is added to an
aqueous sample solution to precipitate the precipitant itself, which co-
precipitates the trace elements. In other cases, trace elements are
quantitatively coprecipitated with small part of the matrix element, the

latter acting as a carrier element.

After the collection of trace elements, gathering precipitates are generally separated from sample solutions by filtration and centrifugation. The separation of flocculent precipitates by both techniques are, however, time-consuming or troublesome, especially for large volume sample solutions. We have proposed the coprecipitation-flotation technique [34,35] to overcome this difficulty. For example, microgram quantities of Cd, Co, Cr(III), Cu(II), Mn(II), Ni, and Pb in 1200-ml water or sea water samples are quantitatively coprecipitated with indium hydroxide at pH 9.5. After adding ethanolic solutions of sodium oleate and dodecyl sulfate, the precipitates are completely floated with numerous tiny nitrogen bubbles (diameter below 0.5 mm) to the solution surface within a few minutes, and then collected in a small sampling tube (Fig.1).

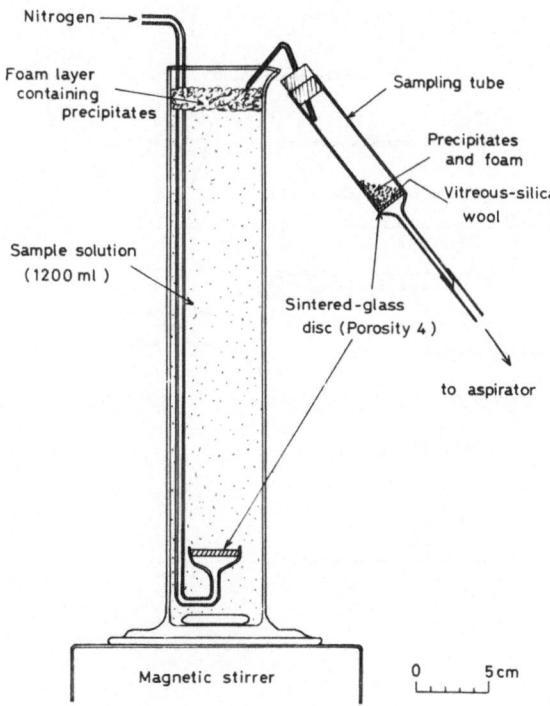

Fig. 1: Flotation cell and sampling tube

The precipitates are dissolved and analyzed by inductively coupled plasma-atomic emission spectrometry. The concentrations of the heavy metals are increased 240-fold, and those of alkali and alkaline earth metals in sea water are reduced to 1/50 - 1/20 for Na and K, and 1/2 for Mg, Ca, and Sr. Proper selection of surfactants and the pH of solution is essential to achieve successful flotation, as shown in Fig. 2. Similar techniques are useful with many other floatable metal hydroxide and organic gathering precipitates [36].

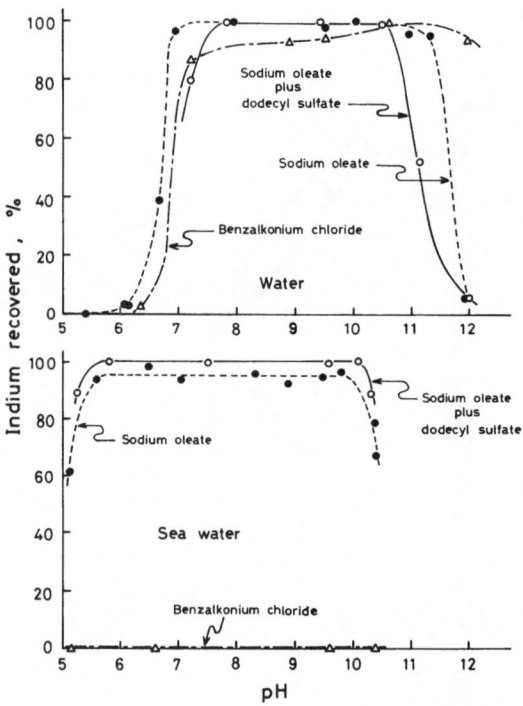

Fig. 2: Flotation of indium hydroxide precipitates

Table 5 lists examples of group precipitation of trace elements [1,3,8,12].

Table 5. Group precipitation of trace elements

Gathering precipitates	Trace elements	Matrices
Metal hydroxides (or hydrous metal oxide)		
$Al(OH)_3$	Cd, Co, Cr, Cu, Fe, Mn, Ni, Pb, Zn	Water and sea water [34]
	As, Bi, Cr, Ge, In, Pb, Sn, Te, Ti, V	Cd, Zn
$Fe(OH)_3$	Bi, Cr, Pb, Se, Te, V	Sea water, Ag
$In(OH)_3$ [35]	Cd, Co, Cr, Cu, Mn, Ni, Pb,	Water, Sea water
$Sn(OH)_4$	Co, Cu, Fe, Zn	Mg
$La(OH)_3$	As, Bi, Fe, Pb, Sb, Se, Sn, Te	Cu
$MnO_2.aq$	As, Bi, Ga, In, Ni, Pb, Sb, Sn, Te, Ti	Cd
Metal sulfides		
CuS	Au, Bi, Cd, Hg, Mo, Pb, Pd, Sb, Sn	In, Ta
In_2S_3*	Ag, Au, Cd, Cu, Pb, Zn	In
Other inorganic precipitates		
YF_3	Rare earths	Be, Ti, U, Zr
Pb* ** [37]	Ag, Au, Bi, Cu, Pd	Pb
Ammonium phosphomolybdate*	Ge, Nb, Ta, Ti, W	Mo
Organic precipitates* **		
PAN**** [38]	Cr, Cu, Eu, Hg, Mn, Ni, Zn	Water
Cupferron (+Cu)	Bi, Fe, Mo, Nb, Sn, Ta, Ti, V, W, Zr	In
APDC***** (+Co,Cu,Fe,or Mo) [39 - 42]	Bi, Cd, In, Ni, Pb, Tl, Zn	Water, Sea water, Al
8-Hydroxyquinoline + thionalide (+ Al)	Cu, Fe, Mn, Ni, Pb, Sn, Zn	K, Na
8-Hydroxyquinoline + tannic acid + thionalide (+Al or In)	Be, Co, Cr, Ga, Mo, Ni, Pb, Ti, V, Zn	Water, Ashed biological samples, K

* Partial precipitation of matrix.
** by reduction with $NaBH_4$.
*** Carrier elements in parentheses.
**** 1-(2-Pyridylazo)-2-naphthol.
***** Ammonium pyrrolidinedithiocarbamate. More correctly ammonium tetra-methylenedithiocarbamate.

3.4. Electrodeposition

Mercury cathode electrolysis has wide applicability because of the high overvoltage of hydrogen on mercury [1]. A group of heavy metal impurity elements (Bi, Cd, Co, Cu, Fe, Ni, Pb, Zn) in uranium, vanadium, magnesium, and aluminium is quantitatively deposited on a mercury cathode from dilute sulfuric or perchloric acid solutions. The trace elements thus collected are separated from mercury by evaporation of mercury or by anodic stripping. On the contrary, deposition of the matrix on a mercury cathode is useful for enrichment of traces of Al, B, Ti, V, W, alkaline earths, and rare earths in iron and nickel matrices.

Platinum and carbon electrodes are also employed for the deposition of the matrix or trace elements. Carbon (high-purity graphite, pyrolytic graphite, and glassy carbon) lies between mercury and platinum in the hydrogen over-voltage. Trace elements were deposited on a graphite rod [43] or on the inner wall of a graphite tube [44,45] through which the sample solution was circulated by a pump during the electrolysis.

3.5. Ion exchange and sorption

Sorption methods based on such mechanisms as ion exchange, complex formation, and physical sorption, are also frequently used as multielement enrichment techniques. Useful sorbents include ion exchange resins, chelating resins, cellulose or silica gel or glass beads with immobilized chelating functional groups on the surfaces, activated carbon, metal oxides, and metal sulfides.

Column operation is most widely used for sorption and desorption. The sample solution is passed through a column packed with numerous small beads of a sorbent to collect either the matrix or desired trace elements or sometimes both. In the first case, the percolated solution contains the trace elements. In the second case, the trace elements are eluted from the column with appropriate solutions (eluents). In the last case, selective desorption or chromatographic separation is carried out. A second mode is the batch operation, in which the sample solution is equilibrated with a sorbent in a vessel with the aid of mechanical stirring, shaking, or ultrasonic irradiation. Sometimes the sample solution is filtered through a thin layer of a sorbent or a filter impregnated with a sorbent to selectively collect the desired trace elements. Desorption is carried out likewise with appropriate solvents.

Desorption is sometimes omitted. Thus, after sorption of the desired trace elements on a sorbent, the latter is dried and directly employed in X-ray spectrometry, neutron activation analysis, etc., or dissolved, or incinerated.

The ion exchange or sorption behavior of an element depends greatly on the composition of the solution as well as the nature of the sorbent. Under proper experimental conditions, trace recoveries of more than 95 % are easily obtained with enrichment coefficients of more than 10,000.

Typical applications are summarized in Tables 6 - 8 [1,3,8].

Table 6. Sorption of matrices

Sorbents	Solutions	Matrices	Trace elements
Anion exchange resin	HCl	U	Mn, Ni, Pb, Rare earths
	HNO_3	Pu, Th	Al, Ba, Ca, Cr, Cu, Ni, Sc, Ti, V, Y
	HNO_3-HF	Ta	Al, Be, Cr, Fe, Hf, Nb, Ni, Ti, W, Zr
Anion exchange resin + TBP*-coated trifluorochloroethylene polymer	HNO_3	Pu, Th, U	B, Be, Cd, Ga, Mo, Ni, Re, Rh, Ru, Zn
Cellulose phosphate	HCl	U	Cd, Cu, Pb, Zn
* Tri-n-butyl phosphate.			

Table 8. Sorption of matrices and trace elements followed by selective desorption

Sorbents	Solutions	Matrices	Trace elements	Eluents
Cation exchange resin*	H_2O	Ba,K,Na,Sr	Ag, Al, Cd, Co, Cu, In, Mn, Ni, Rb, Zn	HCl
	H_2O	La,Ni,Sc, Y, Alkali and alkaline earth metals	Ag, Cd, Co, Fe, Hf, In, Sn, Zn	Dioxane-HCl, Dioxane-ethanol -HCl
Anion exchange resin	HCl	Pu, U	Ba, Ca, Cr, Cu, K, Mn, Ni, Sc, Th, Y	HCl, HNO_3, H_2O
* Precipitation ion exchange.				

Table 7. Group sorption of trace elements

Sorbents	Solutions	Trace elements	Matrices
Cation exchange resin	HCl	Cu, Fe, Ni, Pb	Ir, Pd, Pt, Rh
	HF	Rare earths	Zr
	$HF-H_2O_2$	Co, Cu, Fe, Ni, Zn	Mo, W
	$H_3PO_4-HNO_3-H_2O_2$	Co, Cu, Fe, Mg, Mn, Ni, Pb, Zn	Mo
Anion exchange resin	HCl	Bi, Cd, Co, Cu, Fe, Pb, Sn, Zn, Platinum metals	Al, Cr, Mg, Mn, Ni
	HF	B, Mo, Nb, Sb, Sn, Ta, Ti, W, Zr	Fe, Ga
Arsonic acid resin [46]	pH 5	Cd, Co, Cu, Fe, Mn, Ni, Pb, Zn	Water, Sea water
Chelating resin* [47-49]	pH 5-8	Cd, Co, Cr, Cu, Fe, La, Mn, Ni, Pb, Zn	Water, Sea water
Poly(dithiocarbamate) resin [50]	pH 6	Cd, Cu, Hg, Ni, Pb, U	Urine
Hydrophilic glycol-methacrylate gel with bound 8-hydroxyquinoline [51]	pH 5-7	Co, Mn, Ni, Pb, U	Water
Cellulose with bound Hyphan** or chromotropic acid [52-54]	pH 4-8	Co, Cu, Fe, Mn, Ni, Pb, Sr, Ta, U, Zn	Water
Glass beads or silica gel with bound chelating groups*** [55-58]	pH 5-10	Co, Cu, Hg, Mn, Ni, Pb, Se, Zn	Water
Activated carbon	pH 7-8 [59]	Ag, Bi, Cd, Cu, In, Mg, Mn, Pb	Water
	pH 4-9**** [59-65]	Bi, Cd, Co, Cu, Fe, Hf, Hg, In, Ni, Pb	Water, Ag, Al, Cr, Mn, Tl, Zn, Alkali and alkaline earth metals
Titanium (IV) oxide hydrate, zirconium (IV) oxide hydrate, or aluminium oxide [66]	pH 8	Co, Fe, Pb, U, Zn	Water, Sea water
Zinc sulfide or copper sulfide [67]	pH 3-6	Ag, Bi, Cd, Hg, Pb, Se, Sn	Water

* Chelex-100.
** 1-(2-Hydroxyphenylazo)-2-naphthol.
*** Ethylenediaminetetraacetic acid, dithiocarbamates, amines, diamines, etc.
**** In the presence of dithizone, Xylenol Orange, sodium diethyl dithiocarbamate, potassium xanthate, 8-hydroxyquinoline, hexamethyleneammonium hexamethylenedithiocarbamate, or thioacetamide.

3.6. Volatilization

Various matrix elements are removed as gaseous compounds from liquid or solid samples, leaving the desired trace elements in the residue. Applications of this technique are listed in Table 9 [1,3,8].

Table 9. Volatilization of matrices

Volatile form	Medium	Matrices	Trace elements
Distillation from aqueous solution			
Chloride	$HCl-HNO_3-HClO_4$, $HCl-HNO_3$	Ge	Ag, Al, Bi, Cu, Fe, In, Mg, Mn, Ni, Pb
Oxychloride	$HCl-HClO_4$	Cr	Al, Cu, Fe, Mg, Mn, Ni, Ti, V
Bromide	HBr	Se	Cd, Cu, Fe, Ga, Pb, S, Te, Tl, Alkali and alkaline earth metals
		In	Ag, Al, Au, Bi, Cd, Co, Cu, Fe, Pb, Te
	$HCl-HNO_3-HBr$	As	Ag, Au, Cd, Co, Cr, Cu, Mn, Ni, Pb, Zn
Fluoride	$HF-HNO_3-HClO_4$, $HF-H_2SO_4$	Si	Al, Bi, Cd, Cu, Fe, In, Ni, Pb, Tl, Zn
Oxide	$HNO_3-H_2SO_4$	Se	Al, Ba, Co, Cu, Mo, Mn, Ni, Pb, Te, Zn
Dry methods			
Chloride	Hydrogen chloride	Sb	Ag, Al, Bi, Co, Cu, Fe, Mg, Mn, Ni, Pb
	Chlorine	Zr	Ag, Bi, Co, Cr, Cu, Fe, In, Mn, Ni, Zn
	Air	NH_4Cl*	Cu, Fe, Ni, Pb
	Vacuum	P**, As**. Sb**	Ca, K, Li, Na
Chloride and oxychloride	Hydrogen chloride	V	Ag, Al, Cd, Co, Cr, Cu, Fe, Mg, Mn, Pb
Oxide	Air	As	Al, Ba, Cd, Co, Cr, In, Mg, Mn, Sn, Zn
Organo-metallic compound	Vacuum	Al***	Ag, Co, Cr, Cu, Fe, Mn, Ni, Pb

* Phosphoric acid added to prevent losses of the trace elements.
** After reaction with chlorine.
*** After reaction with ethyl bromide.

Other important examples of volatilization of the matrix are evaporation
of solutions to remove water, volatile acids, and organic solvents,
lyophilization, and ashing of organic and biological samples. Since
relatively high temperatures are used in most of these techniques, special
precaution must be taken against losses of the desired trace elements due
to partial volatilization and reactions with the container walls as well
as contamination due to the containers.

Simultaneous volatilization of a large number of trace elements in non-
volatile matrices generally requires much higher temperatures, and its
applicability is rather limited (Table 10) [1,3,12].

Table 10. Group volatilization of trace elements

Matrices	Trace elements	Method
Oxides of beryllium, aluminium, zirconium, thorium, and uranium	B, Bi, Cd, Cr, Mn, Ni, Pb, Sb, Sn, Zn	Volatilization at 1500-2000°C from the sample in air or vacuum
Silicon	Al, Ca, Cu, Mg, Zn	Volatilization from the molten sample in vacuum
Silicon dioxide	Be, Co, Fe, Ga, In, Mn, Ni, Sn	Volatilization in a stream of hydrogen chloride

4. Enrichment of a trace element in a specific chemical state

It is frequently very important to differentiate several chemical states
of a trace element in such samples as natural waters and steels. Although
some techniques can be used to directly determine a trace element in a
specific chemical state, enrichment techniques are generally required
prior to determination steps.

4.1. Natural waters

In natural waters, each trace metal may exist as simple hydrated ions of
different oxidation states, inorganic and organic complex ions, nonionic
dissolved species, and colloids. Also, trace metals are frequently adsorbed on, occluded in, or included in, inorganic and organic colloidal
or suspended particles. Information on the chemical states of trace metals
in natural waters is highly required in studies of geochemistry, environ-

mental problems, biological effects of trace metals, and water treatment.

The investigation of chemical speciation in natural waters is very difficult. Considerable changes may occur during the sampling, storage, and separation steps. To obtain the total metal concentrations, organic matter is destroyed by strong oxidation and ultraviolet irradiation, and oxidation states of ions are altered by oxidation or reduction, before separation and determination steps.

Filtration is widely used to separate suspended particulate matter in natural waters [68,69]. A 0.45 μm pore size filter is now accepted as defining the suspended matter and the dissolved or soluble fraction. Difficulties sometimes arise owing to clogging of pores of the filter, adsorption of trace elements on the filter, contamination, and rupture of phytoplankton cells.

Centrifugation at 48,000G was applied to sea water [70]. From the experimental results, it seems probable that part of Cd, Cu, Pb, and Zn is associated with the colloidal and fine particulate matter.

Selective flotation may be achieved for surface active materials by simple gas bubbling without surfactants, for negatively charged suspended matter with cationic surfactants, or for positively charged colloids and ions with anionic surfactants, in natural waters. The coprecipitation-flotation technique with indium hydroxide mentioned previously enables us to collect heavy metals associated with suspended matter such as kaolin, bentonite, hydrous metal oxides, and calcium carbonate, and complexed with humic acid, as well as inorganic ions [71].

Dialysis was carried out on filtered sea water by the use of a cellulose dialysis tubing of 4.8 nm average pore size [69]. A considerable part (up to 31 %) of zinc was not dialyzed. In an in situ dialysis technique, a dialysis bag (4.8 nm pore), filled with pure water, is directly immersed into the natural water to separate truly dissolved forms of trace elements without substantial influence of adsorption on the walls of vessels and apparatus [72]. The latter technique was also employed in an investigation of the interaction between humus and trace elements in lake water [73].

Ultrafiltration with a Diaflo UM-2 membrane (cut-off molecular weight of ca. 1,000) was used to separate humus in lake water [73].

Gel filtration may be applied to the chromatographic separation of dissolved organometallic compounds based on the molecular weight differences [74,75].

Ion exchange and other sorption techniques are very useful. A substantial fraction of Cd, Cu, Pb, and Zn in sea water is not retained by a chelating resin Chelex-100, presumably adsorbed on, or occluded in, organic or inorganic colloidal and fine particulate matter [70, 76-78]. The chemical states of trace metals adsorbed on a macroreticular resin Amberlite XAD-2 are regarded as dissolved organometallic compounds. Inorganic metal ions are not adsorbed on the resin [79,80]. Ion exchange and other sorption techniques are also applied to the selective separation of dissolved trace metal ions of different oxidation states, for example, separation of Cr(III) and Cr(VI) or of Se(IV) and Se(VI). Chromium(VI) is adsorbed on an anion-exchange resin, whereas Cr(III) is not [81]. Chromium(III) and Cr(III) plus Cr(VI) are collected on preformed iron(III) hydroxide and iron(II) hydroxide precipitates, respectively [82]. Selenium(IV) is selectively adsorbed on a macroreticular resin Amberlite XAD-2 as Se(IV)-DDTC complex in the presence of Se(VI) in sea water [83].

Another useful technique is liquid-liquid extraction. Part of copper in sea water is extracted with pure chloroform, suggesting the presence of the copper-organic complex [84]. In the extraction of Cu, Mn, Pb, and Zn in sea water with DDTC-chloroform or APDC-MIBK, part of the metals is not extracted, presumably adsorbed on, or occluded in, organic or inorganic colloids or present as strongly-bound complexes in sea water [69, 76]. Liquid-liquid extraction is also employed for the separation of trace metal ions of different oxidation states, for example, separation of Cr(III) and Cr(VI), Se(IV) and Se(VI), or As(III) and As(V). Chromium(VI) is extracted with DDTC-MIBK or APDC-MIBK (or -chloroform), leaving Cr(III) in aqueous solution [85-87]. Aliquat-336 (high-molecular-weight ammonium salt)-toluene is used for the selective extraction of Cr(III) and Cr(VI) from neutral (pH 6-8, in the presence of thiocyanate) and weakly acidic (pH 2) solutions, respectively [88]. Selective extraction is also achieved for Se(IV) with DDTC-carbon tetrachloride [89], or 1,2-diamino-3,5-dibromo-benzene-toluene [90] in the presence of Se(VI), and for As(III) with APDC-MIBK (or -nitrobenzene) in the presence of As(V) [91].

Coprecipitation is used to separate different oxidation states of chromium and selenium in natural waters. Chromium(III) is coprecipitated with iron(III) hydroxide, leaving Cr(VI) in solution [92-94], whereas Cr(VI) is coprecipitated with barium sulfate after masking of Cr(III), Fe(III), and Al with salicylic acid [95]. Selenium(IV) is coprecipitated with iron(III) hydroxide, and Se(VI) remaining in solution is collected on elemental tellurium [96].

Volatilization is used to differentiate arsenic or selenium species in natural waters. Volatile arsines such as mono-, di-, and trimethylarsines can be separated from other arsenic species by simple gas stripping [97]. Selective volatilization of inorganic As(III) in the presence of As(V) is achieved by the hydride evolution method with sodium borohydride at pH 4-6 [97-100]. Inorganic As(III) plus As(V), methylarsonic acid, and dimethylarsinic acid are separated from each other by reducing simultaneously at pH 0-1 to arsine, methylarsine and dimethylarsine, respectively, collecting the arsines in a cold trap, and volatilizing in the sequence of their boiling points by slow warming of the trap [97, 99, 100]. Volatile dimethyl selenide and dimethyl diselenide can be separated from other selenium species by gas stripping. The differentiation of inorganic Se(IV) and Se(VI) is carried out by selectively volatilizing Se(IV) as hydride from 4 M hydrochloric acid solutions with sodium borohydride [101].

4.2. Steels

Various metal oxides, carbides, nitrides, and sulfides exist as inclusions of the micrometer size in steels and produce marked effects on the properties of steels [102]. For the enrichment of these inclusions before their characterization, the iron matrix is selectively dissolved either by controlled potential electrolysis or with mineral acids, methanol solutions of iodine or bromine, etc.

The resulting residues containing the inclusions are delivered to further separation by selective dissolution, magnetic separation, and sieving, prior to the final determination step.

5. Microscale enrichment techniques

Although some modern instrumental determination techniques have extremely low absolute detection limits, the maximum size of a solid or liquid sample is limited to the low microliter level. These techniques include atomic emission spectrometry, spark source mass spectrometry, atomic absorption and fluorescence spectrometry by atomization with furnaces or filaments or by flow-injection techniques, and electron and ion microprobe techniques. Enrichment techniques operated on a microscale are frequently very useful in the effective application of the above determination techniques in micro- and trace analysis, especially in trace analyses of microsamples. Microscale operations have several advantages, i.e., economy in sample, high-purity reagents and time, and minimization of experimental wastes. However, microscale operations sometimes take more skill than conventional ones.

Some recent examples in our laboratory follow. Copper at the ng/g level in a 10-mg sample of high-purity tantalum metal powder was separated on a 2 mmϕ x 35 mm column of cation exchange resin in a diluted hydrofluoric-nitric acid solution and determined by tungsten filament volatilization-microwave plasma emission spectrometry using 2-µl aliquots [103]. The solution volumes were always at the microliter level. Silver at the sub- or low ppm level in a few milligrams of high-purity lead metal was separated by selective adsorption of silver on suspended dithizone particles (ca. 5 µm particle size) from acidic sample solutions and determined by graphite-furnace atomic absorption spectrometry.

Analytical results with satisfactory precision were obtained within about 20 min by using a unified decomposition/separation vessel (Fig.3) with the aid of ultrasonics [104].

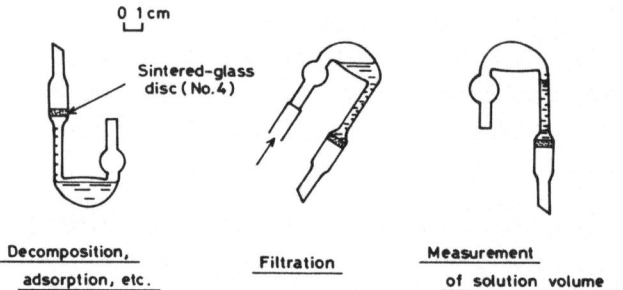

Fig. 3: Decomposition/separation vessel

In the quantitative analysis of individual steel rust particles of 1 µg or less, a drop of the sample solution was applied on a 2-µm thick film of indium oxide on a glass plate (Fig.4), and heavy metals were concentrated with a solvent (a 45:5:1 mixture of methanol, ethanol, and 6 M hydrochloric acid) into a narrow area of about 1 mm^2 quantitatively and homogeneously for measurements with an ion microprobe mass analyzer [105].

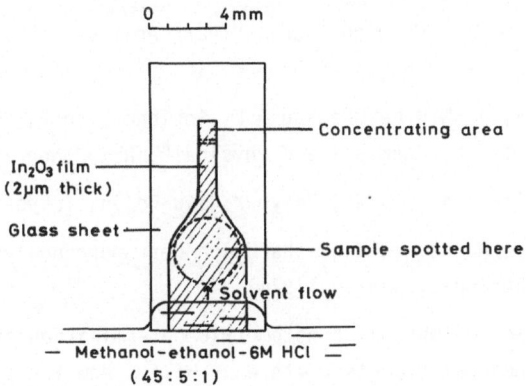

Fig. 4: Thin layer plate

6. Mechanization and automation of enrichment techniques

When enrichment of the desired trace elements is required in the routine
analysis of large numbers of essentially similar samples, mechanization
and automation may be desirable for this operation to minimize attention
time, labor, and skill of the analyst. Although several enrichment
techniques including liquid-liquid extraction, solid-liquid extraction,
distillation, ion exchange, and even precipitation and filtration, can
be incorporated into an automatic scheme, complexity of the mechanism
and lack of flexibility sometimes cancel the advantages of mechanization
and automation mentioned above. Enrichment techniques suitable for
mechanization and automation should be developed in the future.

References

1. A. Mizuike, in G.H. Morrison, ed., "Trace Analysis: Physical Methods", Wiley, New York (1965) p. 103.

2. J. Minczewski, in W.W.Meinke and B.F. Scribner, eds., "Trace Characterization: Chemical and Physical", NBS Monograph 100 (1967),p.385

3. A. Mizuike, "Trace Analysis", Tokyo Kagaku Dojin, Tokyo (1968).

4. O.G. Koch and G.A. Koch-Dedic, "Handbuch der Spurenanalyse", 2. Aufl., Springer, Berlin (1974).

5. E.B. Sandell and H. Onishi, "Photometric Determination of Traces of Metals: General Aspects", 4th ed., Wiley, New York (1978).

6. Yu.A. Zolotov, Pure & Appl. Chem., 50, 129 (1978).

7. G. Tölg, Pure & Appl. Chem., 50, 1075 (1978).

8. E. Jackwerth, A. Mizuike, Yu.A. Zolotov, H. Berndt, R. Höhn, and N.M.Kuzmin, Pure & Appl. Chem., 51, 1195 (1979).

9. M. Zief and R. Speights, eds., "Ultrapurity: Methods and Techniques", Marcel Dekker, New York (1972).

10. M. Zief and J.W. Mitchell, "Contamination Control in Trace Element Analysis", Wiley, New York (1976).

11. A. Mizuike and M. Pinta, Pure & Appl. Chem., 50, 1519 (1978).

12. M.P. Semov, in Kh. I. Zil'bershtein, ed., "Spectrochemical Analysis of Pure Substances", English translation by J.H. Dixon, Adam Hilger Ltd., Bristol (1977) p. 231.

13. Yu.A. Zolotov, "Extraction of Chelate Compounds", Japanese translation by M. Tanaka et al., Baifukan, Tokyo (1972) p. 229.

14. H. Ármannsson, Anal. Chim. Acta, 110, 21 (1979).

15. K.M. Bone and W.D. Hibbert, Anal. Chim. Acta, 107, 219 (1979).

16. A. Dornemann and H. Kleist, Z. Anal. Chem., 291, 349 (1978).

17. L.G. Danielsson, B. Magnusson, and S. Westerlund, Anal. Chim. Acta, 98, 47 (1978).

18. E.A. Startseva, N.M. Popova, I.G. Yudelevich, N.G. Vanifatova, and Yu.A. Zolotov, Z.Anal.Chem., 300, 28 (1980).

19. E. Jackwerth and A. Kulok, Z. Anal. Chem., 257, 28 (1971).

20. E. Jackwerth, E. Döring, J. Lohmar, and G. Schwark, Z. Anal.Chem., 260, 177 (1972).

21. E. Jackwerth, R. Höhn, and K. Koos, Z. Anal.Chem., 264, 1 (1973).

22. R. Höhn, E. Jackwerth, and K. Koos, Spectrochim. Acta, 29B, 225 (1974).

23. E. Jackwerth and J. Messerschmidt, Z. Anal. Chem., 274, 205 (1975).

24. R. Höhn and E. Jackwerth, Z. Anal. Chem., 282, 21 (1976).

25. R. Höhn and E. Jackwerth, Erzmetall, 29, 279 (1976).

26. E. Jackwerth and J. Messerschmidt, Anal. Chim. Acta, 87, 341 (1976).

27. E. Jackwerth, J. Messerschmidt, and R. Höhn, Anal. Chim.Acta, 94, 225
          (1977).

28. A. Mizuike, H. Kawaguchi, and T. Kono, Mikrochim. Acta (Wien), 1970,
          1095.

29. A. Mizuike and K. Fukuda, Mikrochim. Acta (Wien), 1972, 257.

30. A. Mizuike, K. Fukuda, and Y. Ochiai, Talanta, 19, 527 (1972).

31. K. Fukuda, Y. Ochiai, and A. Mizuike, Japan Analyst, 23, 75 (1974).

32. A. Mizuike and K. Fukuda, Mikrochim. Acta (Wien), 1975 I, 281.

33. E. Jackwerth, Pure & Appl. Chem., 51, 1149 (1979).

34. M. Hiraide, Y. Yoshida, and A. Mizuike, Anal. Chim. Acta, 81, 185 (1976).

35. M. Hiraide, T. Ito, M. Baba, H. Kawaguchi, and A. Mizuike, Anal.Chem.,
          52, 804 (1980).

36. M. Hiraide and A. Mizuike, Japan Analyst, 26, 47, 655 (1977); 29, 84
          (1980).

37. E. Jackwerth, R. Höhn, and K. Musaick, Z. Anal. Chem., 299, 362 (1979).

38. M.G. Vanderstappen and R.E. Van Grieken, Talanta 25, 653 (1978).

39. E.A. Boyle and J.M. Edmond, Anal. Chim. Acta, 91, 189 (1977).

40. V. Hudnik, S. Gomišček, and B. Gorenc, Anal. Chim. Acta, 98, 39 (1978).

41. A.J. Pik, A.J.Cameron, J.M. Eckert, E.R.Sholkovitz, and K.L.Williams,
          Anal. Chim. Acta, 110, 61 (1979).

42. H. Berndt and J. Messerschmidt, Z. Anal. Chem., 299, 28 (1979).

43. Y. Thomassen, B.V. Larsen, F.J. Langmyhr, and W. Lund, Anal. Chim.Acta,
          83, 103 (1976).

44. G. Volland, P. Tschöpel, and G. Tölg, Anal. Chim. Acta, 90, 15 (1977).

45. G.E. Batley and J.P. Matousek, Anal. Chem., 49, 2031 (1977).

46. J.S. Fritz and E.M. Moyers, Talanta, 23, 590 (1976).

47. H.M. Kingston, I.L. Barnes, T.J. Brady, T.C. Rains, and M.A.Champ, Anal. Chem., 50, 2064 (1978).

48. C. Lee, N.B. Kim, I.C. Lee, and K.S. Chung, Talanta, 24, 241 (1977).

49. R.E. Sturgeon, S.S. Berman, A. Desaulniers, and D.S. Russell, Talanta, 27, 85 (1980).

50. R.M. Barnes and J.S. Genna, Anal. Chem., 51, 1065 (1979).

51. Z. Slovák and S. Slováková, Z. Anal. Chem., 292, 213 (1978).

52. P. Burba and K.H. Lieser, Z. Anal. Chem., 286, 191 (1977).

53. P. Burba and K.H. Lieser, Z. Anal. Chem., 297, 374 (1979).

54. K.H. Lieser, H.M. Röber, and P. Burba, Z. Anal. Chem., 284, 361 (1977).

55. M.M. Guedes da Mota, M.A. Jonker, and B. Griepink, Z. Anal. Chem., 296, 345 (1979).

56. D.E. Leyden and G.H. Luttrell, Anal. Chem., 47, 1612 (1975).

57. D.E. Leyden, G.H. Luttrell, W.K. Nonidez, and D.B. Werho, Anal. Chem., 48, 67 (1976).

58. D.E. Leyden, G.H. Luttrell, A.E. Sloan, and N.J. DeAngelis, Anal. Chim. Acta, 84, 97 (1976).

59. E. Jackwerth, J. Lohmar, and G. Wittler, Z. Anal. Chem., 266, 1 (1973).

60. E. Jackwerth, Z. Anal. Chem., 271, 120 (1974).

61. E. Jackwerth and H. Berndt, Anal. Chim. Acta, 74, 299 (1975).

62. H. Berndt, E. Jackwerth, and M. Kimura, Anal. Chim. Acta, 93, 45 (1977).

63. M. Kimura, Talanta, 24, 194 (1977).

64. B.M. Vanderborght and R.E. Van Grieken, Anal. Chem., 49, 311 (1977).

65. H. Berndt and E. Jackwerth, Z. Anal. Chem., 290, 369 (1978).

66. K.H. Lieser, S. Quandt, and B. Gleitsmann, Z. Anal. Chem., 298, 378 (1979).

67. A. Disam, P. Tschöpel, and G. Tölg, Z. Anal. Chem., 295, 97 (1979).

68. G.E. Batley and D. Gardner, Water Res., 11, 745 (1977).

69. J.F. Slowey and D.W. Hood, Geochim. Cosmochim. Acta, 35, 121 (1971).

70. M.I. Abdullah, O.A. El-Rayis, and J.P. Riley, Anal. Chim. Acta, 84, 363 (1976).

71. A. Mizuike, M. Hiraide, and J. Mizutani, unpublished work.

72. P. Beneš and E. Steinnes, Water Res., 8, 947 (1974).

73. P. Beneš, E.T. Gjessing, and E. Steinnes, Water Res., 10, 711 (1976).

74. M.E. Bender, W.R. Matson, and R.A.Jordan, Environ.Sci.Technol., 4, 520 (1970).

75. J.L. Means, D.A. Crerar, and J.L. Amster, Limnol. Oceanogr., 22, 957 (1977).

76. T.M. Florence and G.E. Batley, Talanta, 23, 179 (1976).

77. G.E. Batley and T.M. Florence, Anal. Lett., 9, 379 (1976).

78. T.M. Florence, Water Res., 11, 681 (1977).

79. Y. Sugimura, Y. Suzuki, and Y. Miyake, Deep-Sea Res., 25, 309 (1978).

80. Y. Sugimura, Y. Suzuki, and Y. Miyake, J. Oceanogr. Soc. Jpn., 34, 93 (1978).

81. J.F. Pankow and G.E.Janauer, Anal.Chim.Acta, 69, 97 (1974).

82. R.E. Cranston and J.W.Murray, Anal.Chim.Acta, 99, 275 (1978).

83. Y. Sugimura and Y. Suzuki, J. Oceanogr. Soc. Jpn., 33, 23 (1977).

84. J.F. Slowey, L.M. Jeffrey, and D.W. Hood, Nature, 214, 377 (1967).

85. K. Hiiro, T. Owa, M. Takaoka, T. Tanaka, and A. Kawahara, Japan Analyst, 25, 122 (1976).

86. S. Osaki, T. Osaki, S. Shibata, and Y. Takashima, Japan Analyst, 25, 358 (1976).

87. H. Bergmann and K. Hardt, Z. Anal. Chem., 297, 381 (1979).

88. G.J. de Jong and U.A.Th. Brinkman, Anal. Chim. Acta, 98, 243 (1978).

89. T. Kamada, T. Shiraishi, and Y. Yamamoto, Talanta, 25, 15 (1978).

90. Y. Shimoishi and K. Tôei, Anal. Chim. Acta, 100, 65 (1978).

91. T. Kamada, Talanta, 23, 835 (1976).

92. L. Chuecas and J.P. Riley, Anal. Chim. Acta, 35, 240 (1966).

93. R. Fukai, Nature, 213, 901 (1967).

94. R. Fukai and D. Vas, J. Oceanogr. Soc. Jpn., 23, 298 (1967).

95. H. Yamazaki, Anal. Chim. Acta, 113, 131 (1980).

96. O. Yoshii, K. Hiraki, Y. Nishikawa, and T. Shigematsu, Japan Analyst, 26, 91 (1977).

97. M.O. Andreae, Anal. Chem., 49, 820 (1977).

98. J. Aggett and A.C. Aspell, Analyst, 101, 341 (1976).

99. R.S. Braman, D.L. Johnson, C.C. Foreback, J.M. Ammons, and J.L.Bricker, Anal. Chem., 49, 621 (1977).

100. A.U. Shaikh and D.E.Tallman, Anal. Chim. Acta, 98, 251 (1978).

101. G.A. Cutter, Anal. Chim. Acta, 98, 59 (1978).

102. I. Taguchi, Bunseki, 1979, 370.

103. E. Kitazume, T. Sakamoto, H. Kawaguchi, and A. Mizuike, Japan Analyst, 27, 566 (1978).

104. A. Mizuike, M. Hiraide, and S. Kawakubo, Mikrochim. Acta (Wien). 1979 II, 487.

105. S. Kawakubo, Y. Yamaguchi, and A. Mizuike, Japan Analyst, to be published.